T0291832

CAMBRIDGE LIBRARY COLLECTION

Books of enduring scholarly value

Life Sciences

Until the nineteenth century, the various subjects now known as the life sciences were regarded either as arcane studies which had little impact on ordinary daily life, or as a genteel hobby for the leisured classes. The increasing academic rigour and systematisation brought to the study of botany, zoology and other disciplines, and their adoption in university curricula, are reflected in the books reissued in this series.

Water Plants

Agnes Arber (1879–1960) was a prominent British botanist specialising in plant morphology and comparative anatomy. In 1946, she became the first female botanist to be elected a Fellow of the Royal Society. First published in 1920, this volume provides a detailed anatomical study of aquatic flowering plants, with a discussion of their evolutionary history. Arber describes the general anatomical and reproductive organs, life histories and physiological adaptations of aquatic plants in detail, with interpretations informed from her previous experimental work. The final section of this volume discusses the evolutionary history of aquatic plants in the light of affinities to terrestrial flowering plants. Arber's account of aquatic plants was first general description of these plants published, and provides a classic example of the comparative anatomy studies which were central to botanical investigation during the early twentieth century. An extensive bibliography and over 170 illustrations are included in this volume.

Cambridge University Press has long been a pioneer in the reissuing of out-of-print titles from its own backlist, producing digital reprints of books that are still sought after by scholars and students but could not be reprinted economically using traditional technology. The Cambridge Library Collection extends this activity to a wider range of books which are still of importance to researchers and professionals, either for the source material they contain, or as landmarks in the history of their academic discipline.

Drawing from the world-renowned collections in the Cambridge University Library, and guided by the advice of experts in each subject area, Cambridge University Press is using state-of-the-art scanning machines in its own Printing House to capture the content of each book selected for inclusion. The files are processed to give a consistently clear, crisp image, and the books finished to the high quality standard for which the Press is recognised around the world. The latest print-on-demand technology ensures that the books will remain available indefinitely, and that orders for single or multiple copies can quickly be supplied.

The Cambridge Library Collection will bring back to life books of enduring scholarly value (including out-of-copyright works originally issued by other publishers) across a wide range of disciplines in the humanities and social sciences and in science and technology.

Water Plants

A Study of Aquatic Angiosperms

Agnes Arber

CAMBRIDGE UNIVERSITY PRESS

Cambridge, New York, Melbourne, Madrid, Cape Town, Singapore,
São Paolo, Delhi, Dubai, Tokyo, Mexico City

Published in the United States of America by Cambridge University Press, New York

www.cambridge.org
Information on this title: www.cambridge.org/9781108017329

This edition first published 1920
This digitally printed version 2010

ISBN 978-1-108-01732-9 Paperback

WATER PLANTS

CAMBRIDGE UNIVERSITY PRESS
C. F. CLAY, MANAGER
LONDON : FETTER LANE, E.C. 4

LONDON : H. K. LEWIS AND CO., LTD.
LONDON : WILLIAM WESLEY AND SON
NEW YORK : THE MACMILLAN CO.
BOMBAY
CALCUTTA } MACMILLAN AND CO., LTD.
MADRAS
TORONTO : THE MACMILLAN CO. OF CANADA, LTD.
TOKYO : MARUZEN-KABUSHIKI-KAISHA

Nymphaea lutea, L. The Yellow Waterlily, showing rhizome and submerged leaves from a woodcut in Otto von Brunfels' *Herbarum vivae eicones*, 1530 (reduced).

WATER PLANTS

A STUDY OF AQUATIC ANGIOSPERMS

BY

AGNES ARBER, D.Sc., F.L.S.

FELLOW OF NEWNHAM COLLEGE, CAMBRIDGE,
AND KEDDEY FLETCHER-WARR STUDENT OF THE
UNIVERSITY OF LONDON

WITH A FRONTISPIECE AND
ONE HUNDRED AND SEVENTY-ONE TEXT-FIGURES

CAMBRIDGE
AT THE UNIVERSITY PRESS
1920

TO THE MEMORY OF

E. A. N. A.

PREFACE

IT was affirmed a few years ago, by one of the most eminent of living biologists, that it "is no time to discuss the origin of the Mollusca or of Dicotyledons, while we are not even sure how it came to pass that *Primula obconica* has in twenty-five years produced its abundant new forms almost under our eyes." To this statement I venture to demur. I yield to none in my admiration for the results achieved by the analytical methods introduced by Mendel, and I do not doubt the possibility that the direct experimental study of variations and their inheritance may eventually play a large part in bringing the tangled problems of evolution into the full daylight for which we all hope. But this is no reason for condemning those countless uncharted routes which may lead, even if circuitously, to the same goal. Any step towards the solution of the essentially *historical* problems of Botany—for example those concerned with the origin and development of such morphological groups as the Dicotyledons, or of such biological groups as the Aquatic Angiosperms—must necessarily contribute some mite to our conceptions of the course of evolution. These less direct methods of approaching the central problem of biology may perhaps, at the best, bring only a faint illumination to bear upon it, but in the deep obscurity involving all evolutionary thought at the present time, we cannot afford to despise the feeblest rush-light; even the glimmering of a glow-worm may at least enable us to read the compass, and learn in which direction to expect the dawn.

I approached the study of Water Plants with the hope that the consideration of this limited group might impart some degree of precision to my own misty ideas of evolutionary processes. Botanists seem to be universally agreed that the

Aquatic Angiosperms are derived from terrestrial ancestors, and have adopted the water habit at various times subsequent to their first appearance as Flowering Plants. The hydrophytes thus present the great advantage to the student, that they form a group for whose history there is a generally accepted foundation. Throughout the present study I have constantly borne phylogenetic questions in mind, and the first three Parts of this book may be regarded as a clearing of the ground for the more theoretic considerations concerning the evolutionary history of water plants to which the Fourth Part is mainly devoted. In that section of the book, and sporadically in the earlier chapters, I have set down such speculations as have been borne in upon me in the course of a study of water plants with which I have been occupied more or less continuously for the last ten years.

The literature relating to Aquatic Angiosperms has now grown to such formidable proportions that I have felt the necessity of trying to provide some clue to the labyrinth. With this end in view I have given a bibliography of the principal sources, which includes a brief indication of the nature and scope of each work, with page numbers showing where it is cited in the text. For the convenience of those seeking information about any particular plant, I have indexed the families and genera named in the titles enumerated, and in the notes regarding the contents of each memoir. I found it impracticable to compile a subject index to the bibliography, but the references under the individual chapters to some extent serve this purpose.

It is a pleasure to express my grateful appreciation of the kindness of those botanists who have helped me in various ways during the preparation of this book. I am particularly indebted to Professor A. C. Seward, F.R.S. for valuable suggestions and advice; to Dr H. B. Guppy, F.R.S. for reading the pages in Part IV which treat of Distribution; to the Hon. Mrs Huia Onslow (Miss M. Wheldale) for some helpful criticism of the chapters dealing with physiological questions; to Mr F. W. Lawfield, M.A. for aid in fenland botany; and—

last but not least—to Miss Gulielma Lister, who, many years ago, showed me the winter-buds of the Frogbit in a pool in Epping Forest, and awoke in me the desire to know more of the ways of water plants.

I have to thank the Councils of the Linnean Society, and the Cambridge Philosophical Society, and the Editors of *The Annals of Botany*, *The Journal of Botany*, and *The American Naturalist*, for permission to incorporate in this book parts of the text and illustrations of certain of my papers which have appeared in their publications.

Of the figures in the present book, about one-third are original; these are indicated by the initials A. A. The sources of the others are acknowledged in the legends, but I must take this opportunity of expressing my obligation to the numerous authors from whose memoirs they are derived. I am indebted to the Clarendon Press for the use of the block for Fig. 127. The photographic reproduction of a number of the illustrations has been carried out by Mr W. Tams, while some have been re-drawn by Miss Evelyn McLean. I have to thank my sister, Miss Janet Robertson, for the design reproduced on the cover, which is based upon a wood-cut of the Yellow Waterlily in Lobel's "Kruydtboeck," of 1581. I am much indebted to my father for reading and criticising my manuscript and proofs.

To my husband, E. A. Newell Arber, I owed the original impulse to attempt the present study, which arose out of his suggestion that life in Cambridge offered unique opportunities for the observation of river and fenland plants. To his memory I dedicate this book.

AGNES ARBER.

BALFOUR LABORATORY,
CAMBRIDGE.
March 1, 1920.

CONTENTS

PART I

WATER PLANTS AS A BIOLOGICAL GROUP, WITH A CONSIDERATION OF CERTAIN TYPICAL LIFE-HISTORIES

PART II

THE VEGETATIVE AND REPRODUCTIVE ORGANS OF WATER PLANTS, CONSIDERED GENERALLY

LIST OF ILLUSTRATIONS

PART I

WATER PLANTS AS A BIOLOGICAL GROUP,

WITH A CONSIDERATION OF CERTAIN TYPICAL LIFE-HISTORIES

"If...an inquiry into the Nature of *Vegetation* may be of good Import; It will be requisite to see, first of all, What may offer it self to be enquired of; or to understand, what our *Scope* is: That so doing, we may take our aim the better in making, and having made, in applying our Observations thereunto."

Nehemiah Grew, *The Anatomy of Plants*, 1682.

CHAPTER I

WATER PLANTS AS A BIOLOGICAL GROUP

(1) INTRODUCTION

WE are living at the present day in what may be described botanically as the Epoch of Angiosperms, or Flowering Plants. The members of this group now represent the dominant type of vegetation and are distributed over nearly all the land surfaces of the globe. The vast majority are typically terrestrial, carrying on their existence with their flowers and leafy shoots in the air, but with their roots embedded in soil of varying degrees of moisture, from which they derive their water supply. This water supply is one of the prime necessities of their life, and in their relation thereto, the plasticity of their organisation is notably exhibited. At one end of the scale there are plants which can withstand long periods of drought and are capable of flourishing under desert conditions in which the water supply is minimal. At the other extreme we meet with hydrophytes—plants which have exchanged terrestrial for aquatic life. Those which have embraced this change most thoroughly, live with their leafy shoots completely submerged, and have, in some cases, ceased to take root in the substratum, so that all their vegetative life is passed floating freely in the water—which is to them what atmosphere and soil are to terrestrial plants. The ultimate term in the acceptance of aquatic conditions is reached in certain hydrophytes with submerged flowers, in which even the pollination is aquatic—water replacing air as the medium through which the pollen grain is transferred to the stigma. These fundamental changes in habit are necessarily associated with marked divergences from the structure and life-history of land plants. The result has been that the aquatic flowering plants have come to form a distinct assemblage, varying widely among themselves, but characterised, broadly speaking, by a number of features associated with their peculiar mode of life. It is the biological group thus formed which we propose to study in the present book.

There is good reason to assume that the Angiosperms were originally a terrestrial group and hence that the aquatic Flowering Plants existing at the present day can trace back their pedigree to terrestrial ancestors. If this be the case, we may interpret the various gradations existing within the hydrophytic group as illustrating a series of stages leading from ordinary terrestrial life to the completest adoption of an aquatic career. At one end of the series we have plants which are normally terrestrial, but which are able to endure occasional submergence, while at the other end we have those wholly aquatic species whose organisation is so closely related to water life that they have lost all capacity for a terrestrial existence. Between these extremes there is an assemblage of forms, bewildering in number and variety. In order to clear one's ideas, it is necessary to make some attempt to classify hydrophytes according to the degree to which they have become committed to water life. It must be realised, however, that, though such a scheme is convenient and helpful in 'pigeon-holeing' the known facts about aquatics, little stress ought to be laid upon it, except as illustrating the striking variety of form and structure met with among these plants. A classification of aquatics on biological lines is highly artificial, and, since it sometimes places in juxtaposition plants which are quite remote in natural affinity, it has only an indirect bearing on questions of phylogeny.

The classification of aquatics which forms the second part of the present chapter, is based upon a scheme put forward by Schenck[1] more than thirty years ago, which in its main outlines has never been superseded. But the wider knowledge of the group, which has been acquired since that date, has resulted, as is so often the case, in the blurring of the sharp lines of demarcation between the individual bionomic classes recognised at an

[1] Schenck, H. (1885).

earlier stage. The present writer has freely modified Schenck's scheme, and has carried the sub-division to a further point. The various types met with amongst aquatics are arranged in a linear series for the sake of simplicity; but this plan is obviously open to the same criticisms as all other linear systems, whether biological or phylogenetic. The following classification is outlined with the utmost brevity, and aims merely at supplying a key to the biological forms encountered. The life-histories of typical plants illustrating the characters of the more important subdivisions will be considered in some detail in Chapters II–X; but the order in which the life-histories are grouped in these chapters has been determined mainly by reasons of natural affinity, and thus bears no close relation to the following scheme.

(2) BIOLOGICAL CLASSIFICATION OF HYDROPHYTES

I. *Plants rooted in the soil.*

A. Plants which are essentially terrestrial, but which are capable of living as submerged water plants, though without marked adaptation of the leaves to aquatic life.

E.g., *Achillea ptarmica*, L. (Sneezewort).
Cuscuta alba, J. and C. Presl (Dodder).
Glechoma hederacea, L. (Ground Ivy).

B. Plants which are sometimes terrestrial, but sometimes produce submerged leaves differing markedly from the air type. The air leaves are associated with the flowering stage.

E.g., Certain Umbelliferae, such as *Sium latifolium*, L. (Water Parsnip).

C. Plants which produce three types of leaf, (*a*) submerged, (*b*) floating and (*c*) aerial, according to the conditions—internal or environmental.

(i) Plants in which the *aerial* type of leaf is generally associated with the flowering stage.

E.g., Many Alismaceae, such as *Sagittaria sagittifolia*, L. (Arrowhead).

(ii) Plants in which the *floating* type of leaf is generally associated with the flowering stage.

E.g., *Nymphaea lutea*, L. (Yellow Waterlily).
Castalia alba, Greene (White Waterlily).
Various Batrachian Ranunculi (Water Buttercups).
Callitriche verna, L. (Water Starwort).
Potamogeton natans, L. (Pondweed).

D. Plants which may, in certain cases, occur as land forms, but are normally submerged and are characterised by a creeping axis bearing long, branching, leafy shoots with no floating leaves, or by a plexus of leafy, rooting shoots without a creeping rhizome.

(i) Leafy aerial shoots produced at the flowering period.

E.g., *Myriophyllum verticillatum*, L. (Water Milfoil).
Hippuris vulgaris, L. (Mare's-tail).

(ii) Inflorescence raised out of the water, but no aerial foliage leaves except in the land forms.

E.g., *Myriophyllum* (except *M. verticillatum*) (Water Milfoil).
Hottonia palustris, L. (Water Violet).
Many Potamogetons (Pondweeds).

(iii) Inflorescence submerged, but essential organs raised to the surface.

E.g., *Elodea canadensis*, Michx. (Water Thyme).

(iv) Inflorescence entirely submerged and pollination hydrophilous.

E.g., *Naias*.
Zannichellia (Horned Pondweed).
Zostera (Grass-wrack).
Callitriche autumnalis, L. (Water Starwort).
Halophila.

E. Plants which in some cases may occur as land forms, but which are very commonly submerged, and are characterised by an abbreviated axis from which linear leaves arise.

(i) Inflorescence raised above the water or borne on a land plant.

E.g., *Lobelia Dortmanna*, L. (Water Lobelia).
Littorella lacustris, L.
Sagittaria teres, Wats.

(ii) Inflorescence sometimes raised above water or sometimes submerged.

E.g., *Subularia aquatica*, L. (Awlwort).

F. Plants which are entirely submerged as regards the vegetative organs and which have a thallus (morphologically either of root or shoot nature) attached to the substratum. The flowers are aerial.

Tristichaceae and Podostemaceae.

II. *Plants which are not rooted in the soil, but live unattached in the water.*

(A transition between I and II is found in *Stratiotes aloides*, L. (Water Soldier), which is rooted during part of the year but floats freely during another part. There are also a number of rooted plants, such as *Hottonia palustris* and *Elodea canadensis*, which are capable of living unattached for considerable periods.)

A. Plants with floating leaves or leaf-like shoots. Flowers raised into the air.

(i) Roots not penetrating the soil.

E.g., *Hydrocharis Morsus-ranae*, L. (Frogbit).
Spirodela polyrrhiza, Schleid., } (Duck-
Lemna minor, L. and *L. gibba*, L. } weeds).

(ii) Rootless.

Wolffia (Rootless Duckweed).

B. Plants entirely or partially submerged.

(i) Rooted, but roots not penetrating the soil. Floating shoots, formed at flowering time, which raise the flowers into the air.

Lemna trisulca, L. (Ivy-leaved Duckweed).

(ii) Rootless.

(*a*) Inflorescence raised out of the water.
Aldrovandia.
Utricularia (Bladderwort).

(*b*) Flowers submerged; hydrophilous pollination.
Ceratophyllum (Hornwort).

CHAPTER II

THE LIFE-HISTORY OF THE ALISMACEAE

THE Alismaceae[1] are perhaps the most typically amphibious of all water plants and they vary in appearance according to their environment in a thoroughly protean fashion. The Arrowhead, *Sagittaria sagittifolia*, L., may be chosen for description as a characteristic member of the family. Seen in ditches and backwaters in the late summer, its fine sagittate leaves and bold inflorescences[2] (Fig. 1, p. 10) make it one of the most striking of our water plants. It is apparently insect pollinated, but the records on the subject seem to be confined to the statement that, in the Low Countries, certain species of Fly have been observed to visit the flowers[3]. The present writer has once noticed a Water-snail crawling over a female flower and engaged in eating the perianth; it is conceivable that these animals may play an occasional part in pollination. The large fruits, whose hassock-shaped receptacles are completely clothed with compressed, winged achenes, give the plant a highly individual character (Fig. 2, p. 10).

In complete contrast to the flowering form, is the guise which the Arrowhead assumes in deep and rapidly-flowing water. As long ago as 1596[4] a tuber, bearing strap-shaped leaves, was described by Gaspard Bauhin under the name of "Gramen bulbosum," while in 1620[5] he published a figure of it,

[1] For a systematic review of the Alismaceae see Buchenau, F. (1903[1]), and, for a general study of their life-history, Glück, H. (1905); Glück's work has been largely drawn upon in the present chapter.

[2] On the detailed structure of the reproductive organs see Schaffner, J. H. (1897).

[3] MacLeod, J. (1893).

[4] Bauhin, G. (1596).

[5] *Ibid.* (1620).

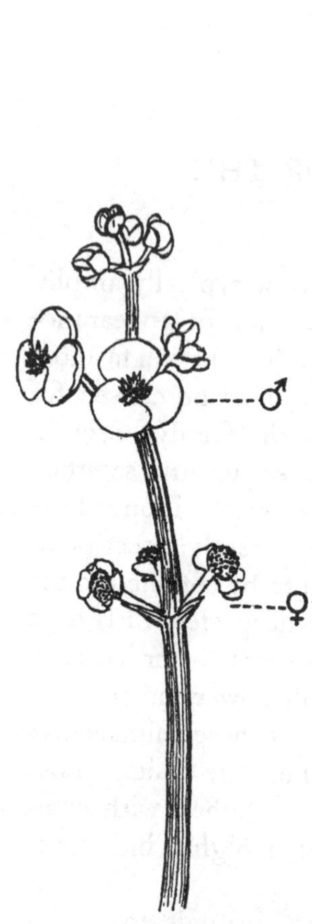

FIG. 1. *Sagittaria sagittifolia*, L. Top of inflorescence, August 17, 1917. ♂ = whorl of male flowers; ♀ = whorl of female flowers with withered perianths. ($\frac{2}{3}$ nat. size.) [A. A.]

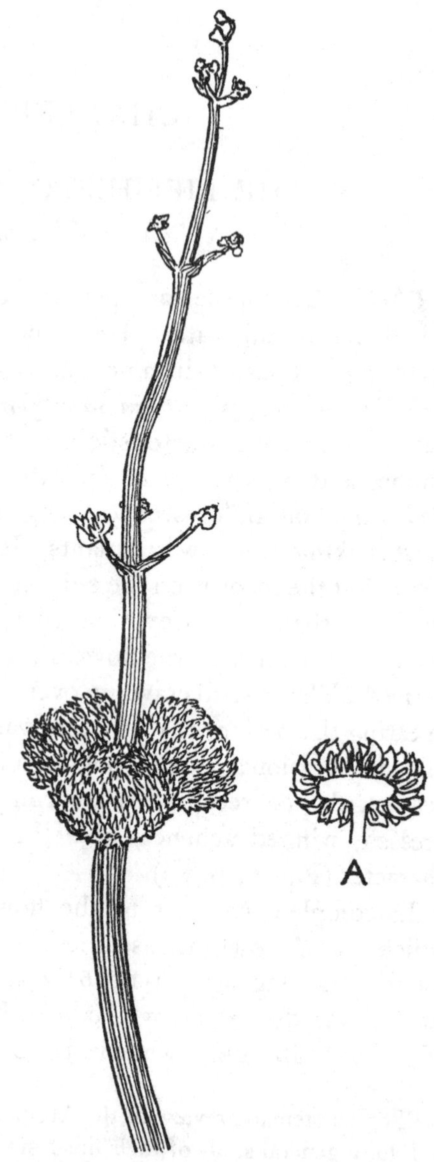

FIG. 2. *Sagittaria sagittifolia*, L. Top of infructescence, September 8, 1917. ($\frac{2}{3}$ nat. size.) *A*, Longitudinal section of fruit. [A. A.]

which is here reproduced (Fig. 3). A century later, Loeselius[1] recognised these strap-shaped leaves as belonging to the Arrowhead; under the name of "*Sagittaria aquatica foliis variis*,"

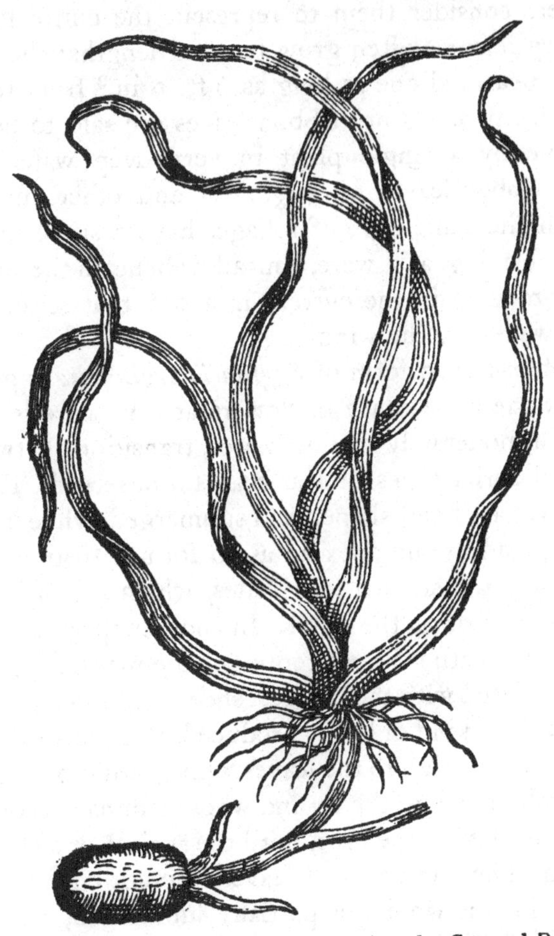

FIG. 3. *Sagittaria sagittifolia*, L. An illustration given by Gaspard Bauhin in the *Prodromos Theatri Botanici*, 1620, under the name of "Gramen bulbosum aquaticum," but which in reality represents a germinated tuber of the Arrowhead, bearing ribbon-leaves.

he figured a plant bearing both ribbon-leaves and leaves of sagittate shape. The ribbon-leaved, deep-water form has been distinguished as f. *vallisneriifolia*. An opportunity of examining

[1] Loeselius, J. (1703).

the plant in its submerged state sometimes occurs when weeds are being cleared out of a river. The semi-transparent leaves—which have been regarded by some authors as purely petiolar[1], while others consider them to represent the entire leaf in a rudimentary form[2]—often grow to great lengths; the present writer has measured one as long as 6 ft. 9 in.[3] from the river Cam. As many as twenty ribbon-leaves are said to be sometimes borne by a single plant in very deep water[4]. The streaming ribbon-leaves of *Sagittaria* and other submerged plants with the same type of foliage, have a singular beauty when seen forming, as it were, a meadow beneath the surface of the water, moving in the current in a way that recalls a field of wheat swayed by the wind.

The ribbon-leaved form of *Sagittaria sagittifolia* is generally sterile, but the appearance of flowers at this stage is not unknown[5]. In moderately shallow water, transitions between the aquatic and aerial types of leaf may be observed. The first-formed leaves are band-shaped and submerged, while later ones begin to spread at the apex so as to form a distinct lamina. Some of these transitional leaf-blades, which are of lanceolate to ovate form, float on the water. In another species, *Sagittaria natans*[6], these floating leaves represent the mature type of leaf and are associated with the inflorescence, but, in the Arrowhead itself, yet a third kind of leaf is produced. The abbreviated axis gives off, in succession to the leaves with floating blades, others whose petioles rise into the air and whose laminae become more and more sagittate at the base, until the typical arrowhead form is achieved. The band-shaped leaves, though characteristic of the plant which is wholly or partially submerged, are not confined to it. The first leaves produced by a germinating seed or tuber are ribbon-like, whether the plantlet develops in air or water. At the end of May, the present writer has found young

[1] Candolle, A. P. de (1827). [2] Goebel, K. (1880).
[3] A length of two metres (6 ft. 6 in.) has been recorded by Costantin, J. (1886). [4] *Ibid.* (1886).
[5] Kirschleger, F. (1856). [6] Wächter, W. (1897[1]).

plants growing from tubers, among the drift at the edge of a river, with a varying number of ribbon-like leaves, succeeded in some cases by one or two of slightly spathulate form (Fig. 4). Fig. 5, p. 14 represents a young plant found in July which shows a series of leaf stages between the early band-like form and the

FIG. 4. *Sagittaria sagittifolia*, L. Plant with soft submerged leaves growing from a tuber, *t*; from river drift at the edge of the Cam near Waterbeach, May 31, 1911. ($\frac{2}{3}$ nat. size.) [A. A.]

mature 'arrowhead' type. The significance of this heterophylly and its relation to the environment will be discussed in Chapter XI.

Sagittaria, like the other Alismaceae, is characterised by the presence of mucilage-secreting trichomes, in the form of scales, in the axils of the leaves. In a paper published a few years ago,

Fig. 5. *Sagittaria sagittifolia*, L. Young plant, July 16, 1910, showing transitions from ribbon-shaped to arrowhead type of leaf. (Reduced.) [A. A.]

two American writers[1], in describing the seedling of *Sagittaria variabilis*, allude to the occurrence of a cellular plate just within the cotyledonary sheath. They refer to this as "a vestigial structure" and interpret it as probably representing a second cotyledon. It appears, however, to the present writer that it is much more reasonable, judging from the figure and description given, to suppose that this scale is merely one of the "squamulae intravaginales," whose existence in the seedlings of *Sagittaria* was placed on record by Fauth[2]. These structures, which are so common among water plants, belong to the category of hairs; they contain no vascular tissue and cannot be homologised with a foliar organ such as the cotyledon.

Plants of the Arrowhead, carefully dug up in the late summer, are found to show preparations for the winter's rest and for next season's growth[3]. From among the bases of the crowded leaves arising on the short main axis, a number of white stolons protrude (*s*, Fig. 6, p. 16), distinguished from the roots by their greater thickness. They each bear one or more scale-leaves and terminate in a bud (*t*). The present writer measured a stolon on July 16, 1910, which had reached a length of 25 cms.[4]. Later on, the two internodes below the terminal bud swell up and form a tuber which may be 5 cms. long. As many as ten stolons may arise from the base of a single plant, so that, where *Sagittaria* grows freely, a very large quantity of tubers are produced. One author[5] records that he collected two to three litres of tubers on digging up soil whose superficial area was one square metre. By a downward curve of the stolons, these reproductive bodies are carried some depth into the mud, where they pass the winter. The mature tubers are coloured blue by anthocyanin, which

[1] Coulter, J. M. and Land, W. J. G. (1914).

[2] Fauth, A. (1903).

[3] Nolte, E. F. (1825), Walter, F. (1842) and Münter, J. (1845).

[4] The stolons seem to develop earlier in terrestrial plants than in plants growing in water. The present writer has found that vigorous plants growing in water may show only quite short stolons in the middle of August.

[5] Klinge, J. (1881).

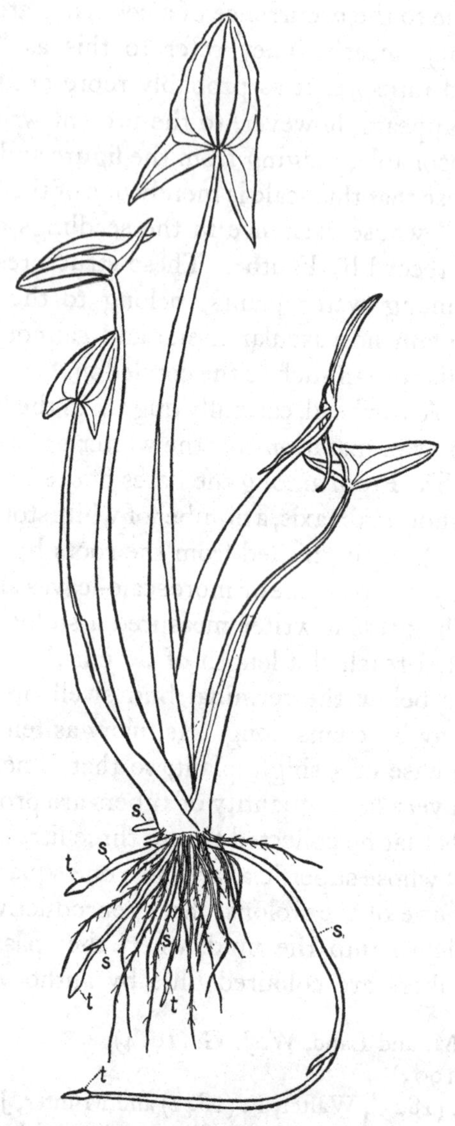

FIG. 6. *Sagittaria sagittifolia*, L. Plant dug up July 16, 1910, with five stolons (*s*) growing from its base among roots, and terminating in young tubers (*t*). (½ nat. size.) [A. A.]

occurs in the epidermis. The blue tint seems very constant; it is recorded by European writers and is shown in the coloured illustrations to that splendid Japanese flora, "Honzo Zufu[1]."

The store of reserve material, laid up in the tuber for the succeeding year's growth, makes the Arrowhead a potential food plant. In Germany the tubers are sometimes employed to feed pigs under the name of "Bruch-Eicheln[2]." They are used in Japan[3], while in China the plant is actually grown for the sake of its tubers, which, in cultivation, reach the size of a clenched fist[4]. The tubers of the related *Sagittaria variabilis*, sometimes called "Swan's Potatoes[5]," are said to be eaten by the American Indians under the name of "Wapatoo[6]."

By winter time, the decay of the stolons sets the tubers free from the parent plant, which does not itself survive until the next season. In the spring, the apical region of the tuber grows out into an elongated axis bearing scale leaves, and carrying the terminal bud up to the surface of the mud, where it produces a new plant. Food is absorbed from the parent tuber for some time; it is possible to find a plant still attached to the tuber from which it arose (Fig. 7, p. 18) and already itself producing the stolons (st_2) which will develop into the tubers of the next generation. At this stage the parent tuber (*t*) has given up its stores of food material and is in a dry, spongy, exhausted state. The conditions which influence tuber formation will be discussed in Chapter XVII, when the wintering of water plants comes under consideration.

The Arrowhead is reproduced by seed as well as vegetatively. The tubers suffice for colonisation of a limited area, but the seeds serve to distribute the species over greater distances. The mericarps, which each enclose a single seed, are flattened and air-containing; they are suitable for dispersal either by wind or water. Their specific gravity is still further lowered by the presence of an oil in the secretory ducts of the pericarp. The

[1] Anon. (1828).
[2] Walter, F. (1842).
[3] Anon. (1895).
[4] Osbeck, P. (1771).
[5] Paillieux, A. and Bois, D. (1888).
[6] Buchenau, F. (1882).

surface of the mericarps is non-wettable and they often float for long periods, sometimes until frost produces waterlogging of the fruit wall. After the decay of the latter, the embryo is still protected by the cuticularised testa[1].

The petioles of *Sagittaria sagittifolia* contain lacunae crossed at intervals by diaphragms (*D* in Fig. 8). A peculiarity, which has been recorded in connexion with the life-history, is that not

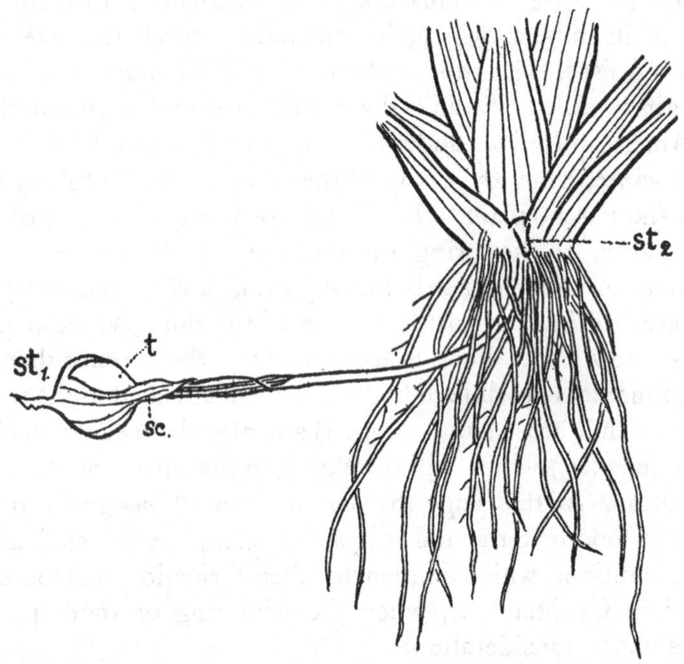

FIG. 7. *Sagittaria sagittifolia*, L. Base of plant dug up July 16, 1910, showing remains of old stolon (st_1) from plant of previous year bearing tuber (*t*) with scale leaves (*sc*); the plant of the current year has also produced a stolon (st_2) which will give rise to a tuber later in the autumn. At this stage the old tuber is dry and spongy in texture, having contributed all its reserves to the plant which has sprung from it. (⅔ nat size.) [A. A.]

only roots but also stolons may sometimes break through the diaphragms of the leaf-sheath of living leaves and penetrate as far as 10 cms., running in the petiole parallel to its long axis[2]. It would be interesting to know whether any significance is to

[1] Fauth, A. (1903). [2] Klinge, J. (1881).

be attached to this observation, which, as its author points out, suggests a case of auto-parasitism.

After the Arrowhead, probably the best known British member of the Alismaceae is the Water Plantain, *Alisma Plantago*, L.[1] According to modern views[2], this Linnean species includes two plants which are each worthy of specific rank—*Alisma Plantago*, (L.) Michalet, and *A. graminifolium*, Ehrh. The former is more suited to land life, while the latter is typically a water plant. *A. Plantago*, (L.) Mich. generally lives in shallow water, where air leaves form the chief assimilatory organs. These are preceded, however, by band-shaped primary leaves (Fig. 101 *B*, p. 153) and then generally some swimming leaves (Fig. 102, p. 153), so that the Water Plantain, like the Arrowhead, produces three distinct types of leaf. In dark situations the swimming leaves may be replaced by submerged leaves differing from the ordinary submerged band-leaves in possessing a distinct lamina. This species only flowers successfully in relatively shallow water in which air leaves can be produced; in moderately deep water, in which submerged and swimming leaves occur, a reduced inflorescence is occasionally formed, but, in very deep water, where all the leaves are band-shaped and submerged, flowers are always absent. *Alisma graminifolium*, on the other hand, has its optimum growth in

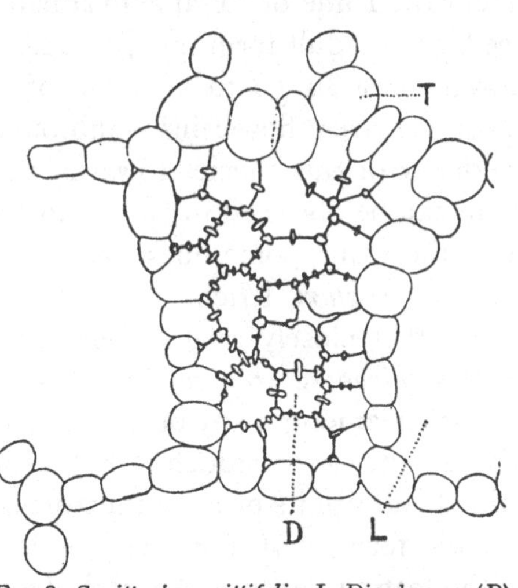

FIG. 8. *Sagittaria sagittifolia*, L. Diaphragm (*D*) of petiole in transverse section. *T* = wall of lacuna; *L* = lacuna. [Blanc, M. le (1912).]

[1] On the details of fertilisation, etc., in this species see Schaffner, J. H. (1896).

[2] Glück, H. (1905).

deeper water than *A. Plantago* and is capable of flowering at a stage when it bears band-shaped leaves alone. It was figured in this condition by Loeselius[1], more than two hundred years ago, under the name of "Plantago aquatica." It never, either in the seedling or adult form, produces floating leaves. It grows and flowers best in 50 to 70 cms. of water; at a greater depth (2 to 4 metres) flowering is inhibited. In spite of the marked tendency of this species towards a strictly aquatic life, a land form can be obtained in cultivation; this proves to be identical with the plant, sometimes found wild, which has been called *Alisma arcuatum*, Mich.

Another closely related genus is represented by the pretty little *Echinodorus ranunculoides*, whose different forms can be observed, among many other water plants, at Wicken Fen near Cambridge—an untouched fragment of fenland, which has retained many of its primitive features. Fig. 9 *C* shows the young aquatic form, with both narrow submerged leaves and leaves with floating blades. An entirely submerged form has been described, which may flower under water at a depth of three feet[2]. Fig. 9 *B* shows the luxuriance which the mature plant may attain, when it grows in water, but raises its leaves and flowers into the air, while Fig. 9 *A* indicates the general dwarfing of the land form. Fig. 147, p. 224, shows the transitions which sometimes occur in this species between inflorescences and entirely vegetative rosettes. The related genus *Elisma*, with its single species, *E. natans*, (L.) Buchenau, is chiefly interesting on account of a similar intimate relationship between the inflorescence and the vegetative shoot. The bracts of the inflorescence are in whorls of three; flowers typically arise in the axils of two of the bracts, while a leafy shoot is developed in the axil of the third. The inflorescences are thus partly reproductive and partly vegetative; there are also certain purely vegetative off-shoots, which may be interpreted, in a morphological sense, as inflorescences which have become wholly sterile.

[1] Loeselius, J. (1703). [2] West, G. (1910).

FIG. 9. *Echinodorus ranunculoides*, (L.) Engelm. *A*, plant from a dried up fen, August 5, 1911; *B*, plant growing in water in a ditch, with aerial leaves only, and very long petioles and flower stalks, June 27, 1914; *C*, plant with submerged and floating leaves only, from a shallow pool, June 27, 1914. (Reduced.) [A. A.]

Another case, in which the replacement of the inflorescence by vegetative structures has been carried much further, is that of *Caldesia parnassifolia*, (Bassi) Parl., a plant which is somewhat widely distributed in Southern Europe, but does not reach Britain. When it grows in water 30 to 60 cms. deep, the 'inflorescences' often bear, instead of flowers, vegetative buds about 2 cms. long, which are able to reproduce the plant (Figs. 148 and 149, p. 225). Sometimes these 'turions' as they are called, and also flowers, may occur in the same whorl. Glück, to whose work on the Alismaceae we owe so much, regards these buds as flower rudiments, which, in consequence of submerged life, have developed in a degenerate vegetative form. This species seems to be losing its power of sexual reproduction, for, even when it bears flowers, they commonly fail to set fertile seed. It affords a good instance of a tendency, common among water plants, to substitute vegetative for sexual reproduction; this characteristic will be discussed more fully in Chapter XVII.

The range of leaf-form met with amongst the Alismaceae—not only in passing from species to species, but also in the same individual under different conditions—prompts one to ask which of these divergent types are fundamental and which are derived. Glück's study of the family has led him to the conclusion that the ribbon form of leaf is primitive, and, on this assumption, he suggests the following scheme, as representing successive phyletic stages which may have occurred in the evolution of the leaves; he admits, however, that the series may conceivably be read in the reverse order. This seriation merely illustrates possible progressive steps and, obviously, does not represent the actual phylogeny of the genera, since examples of Stage I, the most primitive leaf type, and Stage VI, the most highly evolved, are to be found within the limits of the one genus *Sagittaria*.

Stage I. Band leaves alone developed, e.g. *Sagittaria teres*, Watson.

Stage IIa. Band leaves extremely important and associated with the flower, but leaves with *lanceolate blades* also occurring, e.g. *Alisma graminifolium*, Ehrh.

Stage II b. Band leaves of considerable importance, but the flowering stage generally associated with aerial leaves with *lanceolate blades*, e.g. *Echinodorus ranunculoides*, (L.) Engelm.

Stage III. Band leaves still important and sometimes associated with the flower, but floating leaves also produced, with a *broadly elliptical* lamina, sharply marked off from the petiole, e.g. *Elisma natans*, (L.) Buchenau.

Stage IV a. Band-shaped leaves produced, as well as floating leaves and air leaves with a *slightly cordate* base, e.g. *Damasonium stellatum*, (Rich.) Pers.

Stage IV b. Similar to *Stage IV a*, but the band leaves of less importance, e.g. *Alisma Plantago*, (L.) Mich.

Stage V. Similar to *Stage IV*, but the base of the lamina *definitely cordate*, giving a Nymphaeaceae-like leaf. Band leaves extremely reduced, e.g. *Caldesia parnassifolia*, (Bassi) Parl.

Stage VI. Air leaves of *sagittate* form. In the transition from the band leaves to the mature leaves analogies can be found for all the preceding types, e.g. *Sagittaria sagittifolia*, L.

CHAPTER III

THE LIFE-HISTORY OF THE NYMPHAEACEAE AND OF LIMNANTHEMUM

THE Nymphaeaceae, like the Alismaceae dealt with in the last chapter, are a typically aquatic family, but, in the Nymphaeaceae, the water habit has become even more firmly established than in the Alismaceae, land forms being relatively rare. The dominant type of leaf has a floating blade, whereas, although this form of leaf occurs among the Alismaceae, it occupies as a rule a minor place. The rhizome again, which is seldom a conspicuous organ in the Alismaceae, assumes considerable importance in the case of some Nymphaeaceae, although the family includes also a number of annuals.

Our British Waterlilies perenniate by means of rhizomes; these are rich in starch and in the case of some foreign species are used for food[1]. That of the Yellow Waterlily is epigeal[2], with the result that small specimens are occasionally torn from their moorings and found among river drift. The hypogeal rhizomes of the White Waterlily, on the other hand, can seldom be obtained unless they are actually dragged up with a boathook out of the mud. The rhizome of *Nymphaea lutea*, L.[3] is a very striking object (Fig. 10 *A*). It is slightly flattened and of a greenish colour on the upper surface, but pallid and yellowish below. It is decorated with the scars of the leaves (*l.s.*) of previous years—punctuated by the vascular strands which supplied them—and also with the scars of the peduncles (*p.s.*), which can be distinguished by their rounded form. With each leaf-base, three roots are usually associated; at r_1 these roots can

[1] Paillieux, A. and Bois, D. (1888).
[2] Royer, C. (1881–1883).
[3] = *Nuphar luteum*, Sibth. et Sm.

be seen as rudiments and at r_3 as scars, while numerous groups of three mature roots are also shown (e.g. r_2). Fig. 10 *B* represents such a group in further detail. The root system is very elaborate, since the adventitious roots bear branches (Fig. 10*B*) which themselves branch again (Fig. 10 *C*). At the apex arises the rosette of leaves and flower stalks belonging to the current year, and lateral buds may also be produced (Fig. 10 *A*, *l.b.*). The rhizome may be as thick as a man's arm.

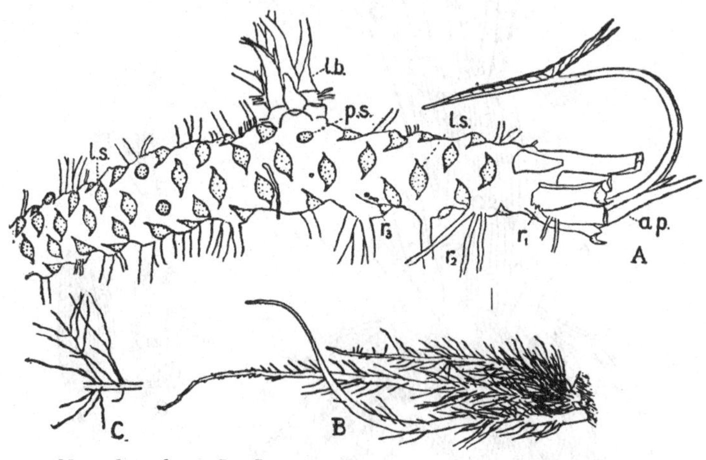

FIG. 10. *Nymphaea lutea*, L. June 30, 1913. *A*, Part of old rhizome, bases, only, of leaves indicated and all root-branches omitted; *p.s.* = scar of peduncle; *l.s.* = leaf-scar; r_1, group of three rudimentary roots arising from a leaf-base; r_2, corresponding group of three roots fully developed; r_3, scars of three dead roots; *ap*, apical region of rhizome; *l.b.*, lateral branch bearing leaves of submerged type only. *B* and *C*, roots in greater detail, placed horizontally to save space. *B*, three young roots from a single leaf-base to show laterals; *C*, part of an old root to show branching of laterals. (¼ nat. size.) [A. A.]

Castalia alba[1], Greene, has a shorter rhizome with the leaves crowded in the apical region (Fig. 11 *A*, p. 26). When the older leaves and flower stalks have been removed to expose the apical bud, the most remarkable feature revealed is the occurrence of large membranous stipules, one of which accompanies each young leaf adaxially; each appears to represent a fused pair (*st.* in Fig. 11 *B*, *C*, *D*). A larger number of roots is associated with each leaf than in the case of the Yellow Waterlily. These

[1] = *Nymphaea alba*, L.

roots may be seen in Fig. 11 *A*, and their rudiments (*r.*) in Fig. 11 *B* and *D*. In *Nymphaea lutea* stipules are absent but the

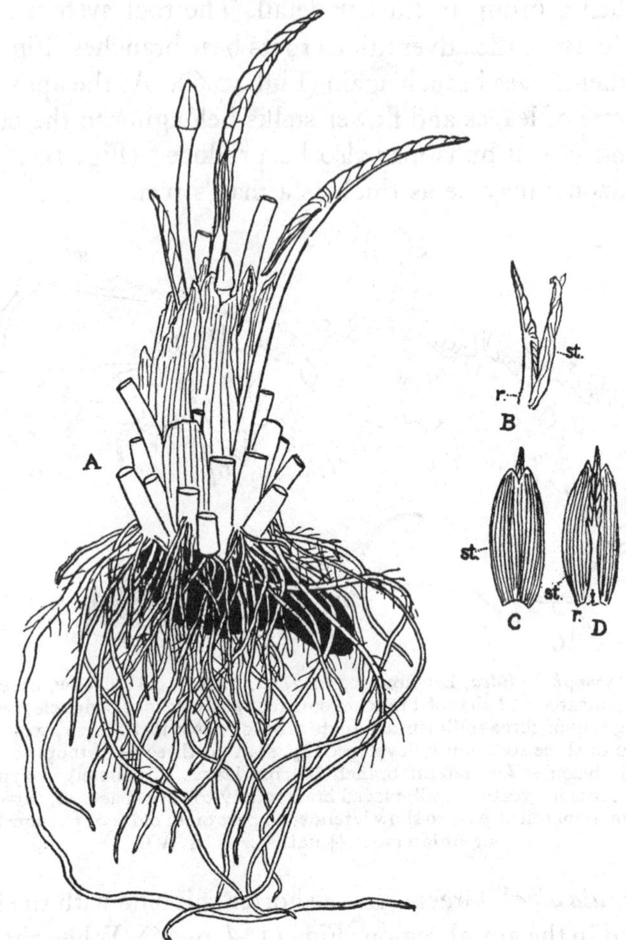

FIG. 11. *Castalia alba*, Greene. Apical part of rhizome pulled up from bottom of water, June 30, 1913. In *A*, the leaves and flowers have mostly been cut away to show the young flower buds, the young leaves and the stipules which protect them. In *B*, *C*, *D*, three views are given of a young leaf and its stipule (or pair of stipules united on the adaxial side) *st.* In *B* and *D* the rudiments of the roots, *r.*, are seen at the leaf-base. (⅓ nat. size.) [A. A.]

petioles are winged, and the stipules seem to be replaced by a silky fringe of hairs[1].

[1] Irmisch, T. (1853).

Sometimes, if a young rhizome of *Nymphaea lutea* be brought up from the bottom of the water, it will be found to bear leaves differing widely from the familiar floating type[1]. They are wholly submerged, relatively short-stalked, translucent, sinuous, and of a delicate, flaccid texture recalling the fronds of *Ulva* (Fig. 12). In a wood-cut in the famous *Herbarum vivae eicones*

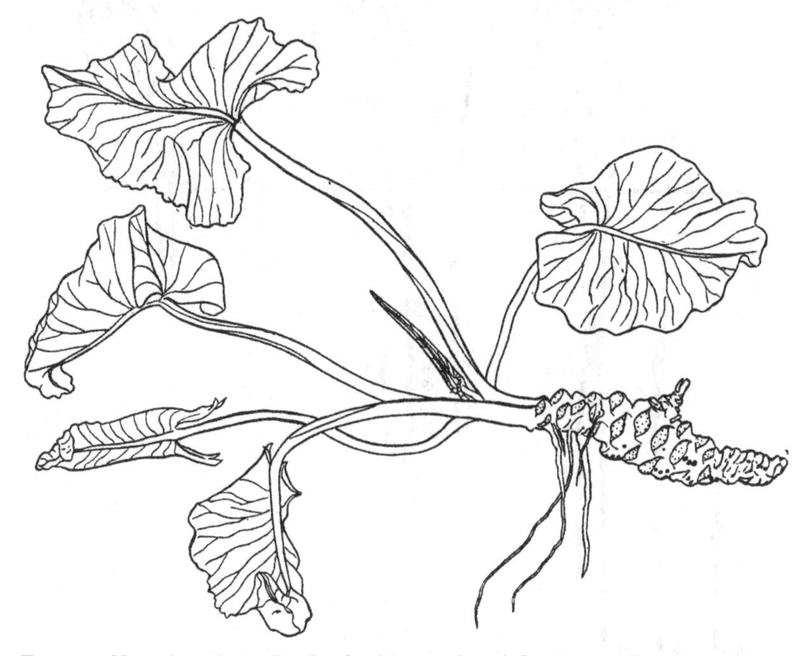

FIG. 12. *Nymphaea lutea*, L. Leafy rhizome found floating on Cam near Waterbeach, May 17, 1911. Leaves all of submerged type, flaccid, translucent and somewhat sinuous at the margin. Rhizome shows leaf-scars, and root-scars in rows of two or three on leaf-bases on under side. ($\frac{1}{3}$ nat. size.) [A. A.]

of Otto von Brunfels (1530)—reproduced in the Frontispiece of the present book—some of the outer leaves with short petioles undoubtedly belong to this type, though no description of the submerged leaves of the Waterlilies occurs in botanical literature until a hundred years later[2]. They were re-discovered—

[1] Royer, C. (1881–1883), Arcangeli, G. (1890), Brand, F. (1894).
[2] Bauhin, G. (1623). See also Desmoulins, C. (1849).

like so many matters well known to the ancients—in the nineteenth century[1].

These submerged leaves, which are stomateless, are characteristically produced in the winter and spring[2], and are usually succeeded, in the course of the season, by floating leaves; in

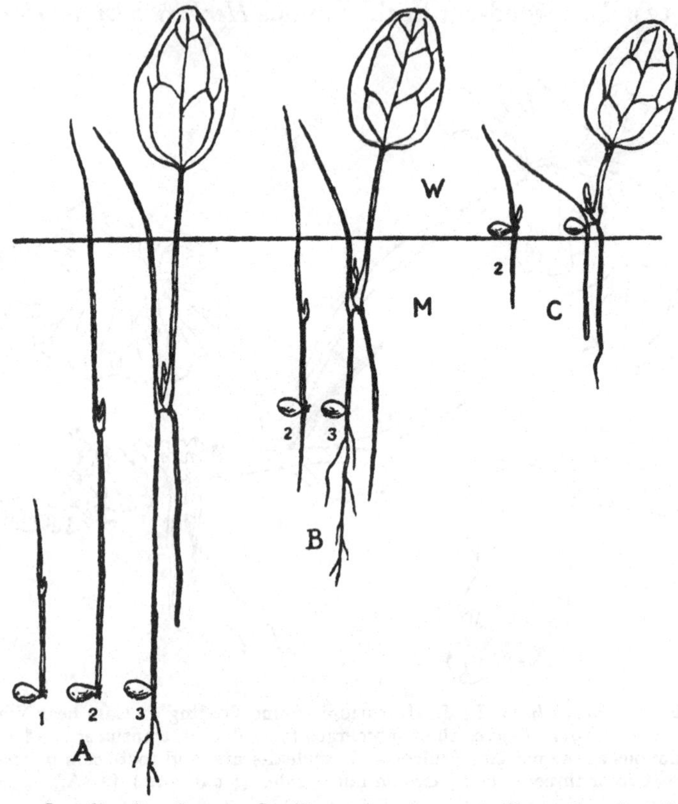

FIG. 13. *Castalia alba*, Greene. Seedlings of various ages illustrating the effect of sowing at different levels on or in the mud (*M*) at the bottom of the water (*W*); accommodation takes place in length of first internode, acicular first leaf, and petiole of second leaf with ovate lamina. [Massart, J. (1910).]

deeper, rapidly flowing water, however, foliage of the submerged type may be exclusively produced for years, even when the plant is so old as to have a massive rhizome[3]. If the water does

[1] The submerged leaves of *Nuphar minima*, Smith, were described by Spenner, F. C. L. (1827).

[2] Costantin, J. (1886). [3] Goebel, K. (1891–1893).

not freeze, the submerged leaves may vegetate throughout the winter. In 1911, the present writer observed a number of plants of the Yellow Waterlily flowering without having produced any but submerged leaves. Possibly this was associated with the peculiarly brilliant sunshine of that summer, which may have supplied the submerged leaves with unwontedly intense light for assimilatory purposes.

Castalia alba produces submerged leaves less freely than *Nymphaea lutea* and they are said to be incapable of surviving the winter; the first leaves of the seedling are of this type (Fig. 13).

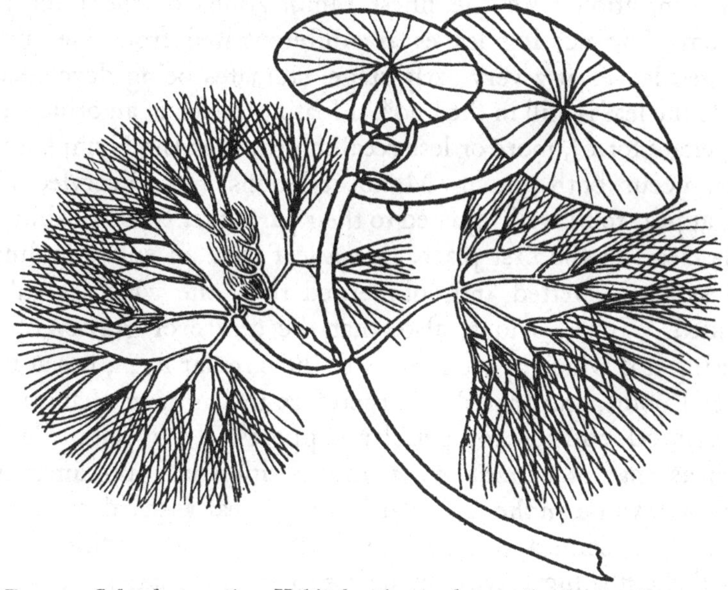

FIG. 14. *Cabomba aquatica*. Habit drawing to show entire floating leaves and dissected submerged leaves. [Goebel, K. (1891–1893).]

The American Water-shield, *Cabomba*, which is placed in a different tribe from *Castalia* and *Nymphaea*, and might, indeed, almost be assigned to a different family, produces submerged leaves of a very distinct type[1] (Fig. 14). They are finely dissected and comparable with the submerged leaves of various Batrachian Ranunculi.

[1] Goebel, K. (1891–1893) and Raciborski, M. (1894[2]).

The floating leaves of the British Waterlilies are typical of swimming leaves in general. The lamina is coriaceous and non-wettable. No leaf which attains to any size can float successfully unless it be of a strong, leathery texture, since the motion of the water exposes it to tearing, and in heavy rain it is liable to be much more severely battered than an air leaf, which can yield freely in a medium so elastic as the atmosphere[1]. The normal stomates are borne upon the upper surface of the floating leaves, where they are in contact with the air, but water stomates have been observed on the lower surface in two American species of *Nymphaea*[2]. These water pores occur in direct communication with the finest ramifications of the tracheal system. The floating leaves are differentiated from the submerged leaves at a very early stage, stomates being developed while the leaf is still in the bud[3]. Floating leaves of an orbicular or peltate form[4], more or less recalling those of the Nymphaeaceae, occur both among Monocotyledons and Dicotyledons and appear to be well adjusted to their particular type of habitat. It is clear, in the first place, that a leaf with an entire outline is less easily wetted and submerged than one which is subdivided. It is obvious, also, that the centre of gravity of a floating leaf which approximates to the circular form, lies at its central point, and that this is therefore the most mechanically economical position for petiolar support[5]. In a peltate leaf, such as that of *Victoria regia*, this position is approximately achieved, while, in the orbicular Waterlily leaf with a deep sinus at the base, some approach is made to the same condition.

All the floating leaves belonging to any associated group of plants, unlike a corresponding series of air leaves, have, without exception, to expand their laminae in one horizontal plane. The competition among the leaves for space is shown by the way in which every available square inch of water surface is

[1] Schenck, H. (1885). [2] Schrenk, J. (1888). [3] Costantin, J. (1886).

[4] For a mathematical demonstration of the physical advantages accruing to a floating leaf from a circular form, see Hiern, W. P. (1872).

[5] Jahn, E. (1897).

covered in any spot where Waterlilies abound. In the case of *Nymphaea* and *Castalia*, a place in the sun is secured through the pliability of the petioles, which vary in length according to the depth of the water, but do not rigidly determine the position of the lamina. The variation in length of the peduncles goes hand in hand with that of the petioles. The present writer has measured a peduncle of *Castalia alba* over six feet in length, and such length is by no means uncommon; Fig. 15 shows the proportion borne by peduncle to flower in this case, the peduncle being represented coiled in order to include its entire length in the sketch. It is rather curious that in the gigantic *Victoria regia* this great elongation of the petioles and peduncles does not occur; the plant flourishes in the shallows and has been recorded in the Amazon region in water only two feet deep[1].

FIG. 15. *Castalia alba*, Greene. Peduncle and flower-bud to show great proportionate length of peduncle. The peduncle, which was more than 6 feet long, is represented coiled in order to include its whole length in the diagram. (Reduced.) May 30, 1911. [A. A.]

Another result of the length and flexibility of the leaf-stalk in the Waterlilies is that the lamina can respond freely to the movement of the water and is hence less liable to be submerged. This response is also shown in many other plants which are rooted at the bottom of the water and bear floating leaves. *Potamogeton natans*[2] is a good example. Here the axis from which the leaves arise, instead of being a solid rhizome lying in or on the mud, as in the Waterlilies,

[1] Spruce, R. (1908). [2] Jahn, E. (1897).

takes the form of a slender stem occupying a slanting position in the water. The petioles arise obliquely from the flexible axis, to which they have a very pliable attachment. If the stem be pushed to and fro in the water, the leaves follow all its movements while yet retaining their position on the surface. In the case of such a hydrophyte as *Hydrocharis*, on the other hand, in which not only the leaves but the rosette as a whole floats freely, the entire plant responds to every movement of the water. In spite, however, of a form and structure suited, up to a certain point, to their environment, floating leaves still remain liable to serious risks of wetting and submergence; this is proved by the fact that plants bearing such leaves are quite unable to colonise windy and exposed surfaces where the water is liable to be rough[1].

In the summer, in addition to the floating leaves of *Castalia alba*, others may be seen which rise well above the water surface and are typical air leaves in appearance. The White Waterlily is even able, on occasion, to develop a terrestrial form which can vegetate for an entire summer without submergence[2]. The leaves of this land form are described as generally being short-stalked, with their lower surfaces almost on the ground. Eighteen centimetres is the greatest diameter recorded: the margins are inrolled towards the upper side.

We showed that in the Alismaceae it is possible to arrange the species in a series beginning with those in which the leaves are extremely simple and concluding with those in which they are highly differentiated, such as *Sagittaria sagittifolia*. We also pointed out that in the Arrowhead the successive juvenile leaves epitomised the series—recalling the various mature forms of leaf characteristic of the less highly differentiated species. In both respects the Nymphaeoideae run strictly parallel with the Alismaceae. *Victoria regia* may be regarded as occupying the same position among the Nymphaeoideae as *Sagittaria sagittifolia*

[1] See pp. 288, 289.

[2] Bachmann, H. (1896). A land form of *Nuphar pumilum* (*Nymphaea pumila*, Hoffm.) was obtained experimentally by Mer, É (1882[1]).

among the Alismaceae. The leaf-succession in the Giant Waterlily of the Amazons was long ago recorded[1], but the full appreciation of its significance we owe to Gwynne-Vaughan[2], who contributed greatly to our knowledge of the Nymphaeaceae. He pointed out that the successive leaves of the *Victoria regia* seedling show a progressive change from the acicular primordial leaf to the peltate form of the mature leaf. The following account of the series is derived from his work:

The *first leaf* is acicular and without a blade (l_1 in Fig. 16).

The *second leaf* is elongate lanceolate, sometimes with two hastate lobes, and resembles the adult leaves of *Barclaya* (l_2 in Fig. 16).

The *third leaf* is elongate-hastate to deltoid-hastate, recalling the sagittate leaves of *Castalia pygmaea*, Salisb., etc. At the base of the lamina, just above the insertion of the petiole, there is a little pocket or pouch on the adaxial side, which appears to be formed by the fusion of the auricles at their bases.

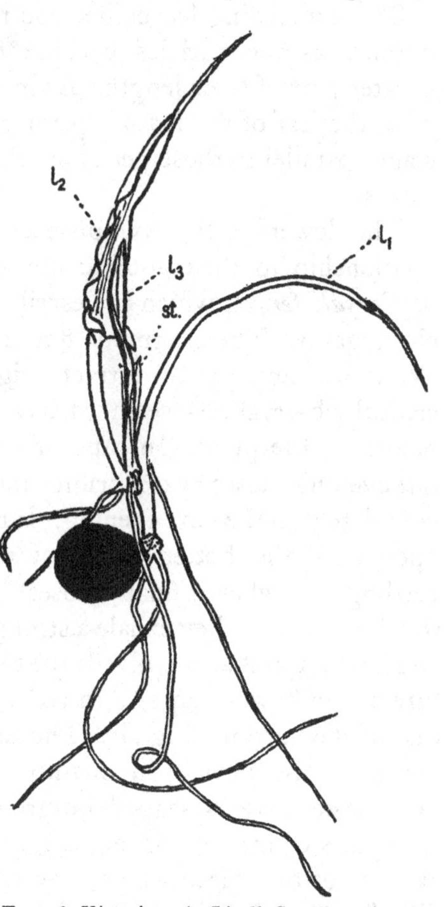

FIG. 16. *Victoria regia*, Lindl. Seedling, showing acicular first leaf l_1, and hastate second and third leaves l_2 and l_3. (According to Gwynne-Vaughan, D. T. (1897), the second leaf is more commonly lanceolate.) *st.*, stipules of second leaf which protect the third. (Nat. size.) [A. A.]

The *fourth leaf* is the first swimming leaf, and is distinctly

[1] Trécul, A. (1854). [2] Gwynne-Vaughan, D. T. (1897).

peltate, by the further fusion of the two auricles. It recalls the adult leaf of many Castalias, e.g. *Castalia Lotus*, Tratt. It is the first leaf to bear spines.

The succeeding leaves become more and more orbicular in outline, as the auricles become fused along a successively greater part of their length. As in the case of *Sagittaria sagittifolia*, the leaf of the mature plant passes, in its youth, through stages parallel to those permanently retained by the embryonic leaves.

The flowers of the Nymphaeaceae do not show any obvious relationship to their aquatic life, except perhaps in the case of *Euryale ferox*[1], which is described as exhibiting submerged cleistogamy. The enormous flowers of *Victoria regia*, the Giant Waterlily, apparently attract night-flying insects, but no critical observations seem to have been made in the native haunts of the plant. In captivity, each flower partially opens one evening, closes next morning and opens completely on the next evening. It remains open until the hotter hours of the succeeding day, when it finally closes[2]. When the flowers open they exhale a strong scent, and much heat is also evolved; the temperature of the flower may rise to 10° C. above that of the surrounding air. The heat and perfume are developed mainly in the carmine-red, sigma-shaped outgrowths at the apices of the carpels, apparently at the expense of the starch which they contain[3]. The flower sinks after pollination, and the fruit ripens in the water about six weeks after flowering[4].

FIG. 17. *Nymphaea lutea*, L. Fruit showing persistent calyx. August 11, 1910. (½ nat. size.) [A. A.]

The fruits and seedlings[5] of our British Waterlilies are of considerable interest, although the young plants do not display

[1] Goebel, K. (1891–1893).
[2] Seidel, C. F. (1869).
[3] Knoch, E. (1899).
[4] Seidel, C. F. (1869).
[5] For very early and good figures of the primordial leaves of the White and the Yellow Waterlily see Tittmann, J. A. (1821).

such an extensive series of leaf-forms as the seedling of *Victoria regia*. The green bottle-shaped fruits which succeed the yellow flowers of *Nymphaea lutea* (Fig. 17) are usually found floating just at the surface of the water. Water-fowl are occasionally seen pecking at them[1]. In order to follow the dehiscence and germination, the present writer brought some fruits collected on October 1, 1914, into the laboratory, and kept them in an aquarium. In the course of the first few days the pericarp began to disintegrate. The green fruit-wall burst irregularly at the base and the torn segments gradually curled right up round the stigmatic disc, disclosing the seed-containing loculi. These, which were snow-white, owing to the presence of air in their walls, soon became detached from the fruit, and for a time floated on the water, either singly or in groups; but, in a couple of days or so, they had become water-logged and had sunk to the bottom of the bell-jar[2]. It has been shown[3] that these detached loculi are clothed with thin-walled cells which secrete much mucilage outwards. The cells have at first

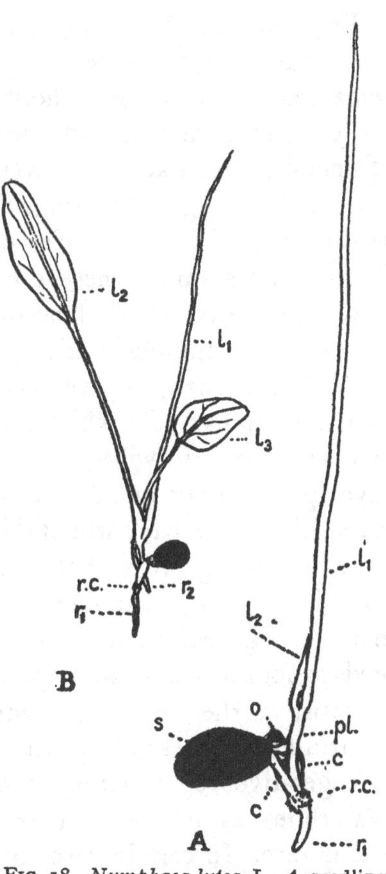

FIG. 18. *Nymphaea lutea*, L. *A*, seedling April 23, 1915 (× 2). *B*, seedling May 28, 1915 (nat. size); r_1 = primary root; *r.c.* = collar of root hairs; *c*, *c* = cotyledons; *pl.* = plumule; l_1, l_2, l_3, first, second and third leaves of plumule; *s* = seed; *o* = operculum. [A. A.]

[1] Guppy, H. B. (1893).

[2] On dehiscence of *Castalia alba* see p. 302.

[3] Raciborski, M. (1894[2]).

a rich starch content, which diminishes as the mucilage is developed—suggesting that the mucilage is formed at the expense of the starch.

By November 24, 1914, the loculi under observation had mostly decayed completely, and the seeds were set free. They remained dormant throughout the winter, but by April 23, 1915, a large number had germinated and there was a forest of slender, grass-like, first leaves arising from the bottom of the bell-jar. Seedlings at this stage are shown in Fig. 18 *A*, p. 35; the seed-coat opens by means of an operculum (*o*) to emit the radicle. During the summer these seedlings developed a number of submerged leaves with lanceolate blades (Fig. 18 *B*), which increased in number until, on September 18, some of the plants had as many as seven such leaves. In spite of the unnatural conditions under which they were living, many of the little plants survived the winter and, by the spring of 1916, they had developed distinct but miniature rhizomes marked with leaf-scars. The leaves were still of the submerged type only. But the most interesting event of this second spring was the germination of a very large number of seeds which had remained dormant for eighteen months. This delay in the sprouting of the seeds is not unusual in water plants (see p. 243). Unfortunately the frost of the very severe winter 1916–1917 destroyed the aquarium, and these observations came to an abrupt end.

Vegetative reproduction, though not so universal among Waterlilies as in some other groups of aquatics, is by no means rare. In certain cases tubers are formed as part of the ordinary course of development of the species, while in *Castalia Lotus*[1] the flowers may, under the abnormal conditions due to cultivation, be replaced by tubers which can reproduce the plant (Fig. 19). Like the seedlings, these young plants developed from a germinating tuber have a simple type of first leaf (l_1).

The anatomy of the Nymphaeaceae has been investigated by

[1] Barber, C. A. (1889).

Gwynne-Vaughan[1]. The rhizomes contain an indescribable confusion of bundles, which he suggests may have been derived from a simpler structure previously existent in a stem with longer internodes; the adoption of a rhizomic habit, associated with telescoping of the internodes, might well lead to this extreme complexity. The most interesting anatomical feature of the family, however, is the occurrence of polystely. In the rhizome of *Victoria regia* "all the root-bearing bundles belonging to the same leaf-base are grouped together so as to form a structure having the appearance of a definite and distinct stele," in which about twenty bundles form a ring. However the most typical polystely occurs, not in the rhizomes themselves, but in the elongated tuber-bearing stolons, which certain species of *Castalia* produce as lateral branches. In the stolons of *Castalia flava*, for instance, the bundles are arranged in four or five widely separated groups or steles, each enclosed in an endodermis and surrounding a protoxylem canal. In *Cabomba*, on the other hand, it is the rhizome in which polystely occurs, though in the simplest possible form; two steles occur throughout, each consisting of a single pair of bundles. The significance of polystely in aquatic plants will be considered in Chapter XIII.

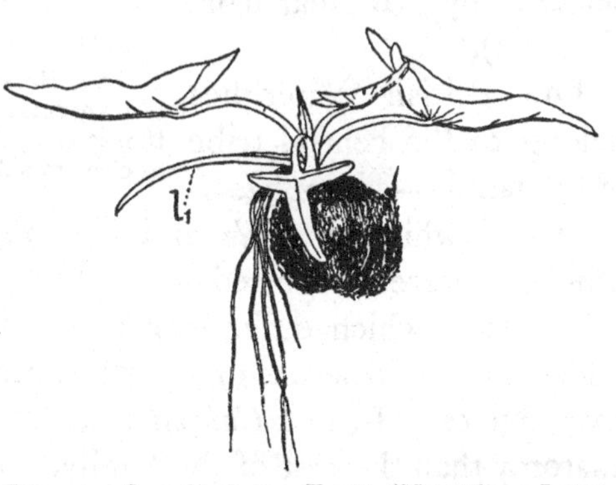

FIG. 19. *Castalia Lotus*, Tratt. (*Nymphaea Lotus*, L.) var. *monstrosa*. Germination in spring of a tuber which has developed in place of a flower; l_1, simple first leaf. (Reduced.) [Barber, C. A. (1889).]

The Nymphaeaceae have a remarkably well-developed aerating system in their leaf- and flower-stalks. The long peduncles of Waterlily flowers are said to have been sold in the bazaars at Cairo as tobacco pipes: the base of the flower, which was

[1] Gwynne-Vaughan, D. T. (1897); see also Trécul, A. (1845) and (1854), Wigand, A. (1871), Blenk, P. (1884), Strasburger, E. (1884), etc.

destroyed, formed a hollow for the lighted tobacco, and the smoke passed through the air-spaces in the stalk[1].

The mucilage which coats the young organs in most of the Nymphaeaceae will be considered later[2]. It is secreted by glandular hairs (Fig. 20).

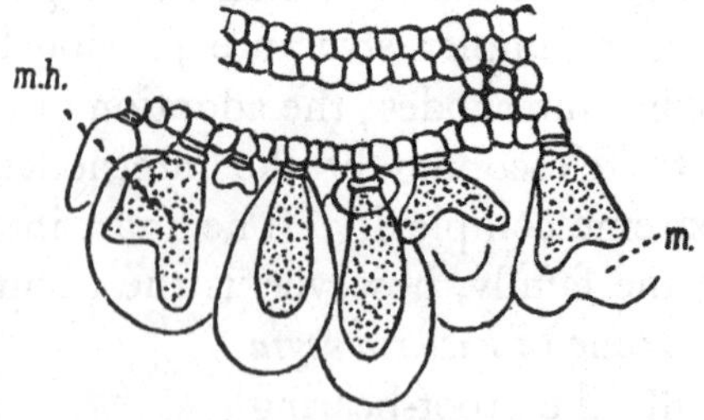

FIG. 20. *Brasenia Schreberi*, J. F. Gmel. Part of transverse section of young leaf to show the secretory hairs, *m.h.*, surrounded by a layer of clear mucilage, *m.* [Goebel, K. (1891–93).]

Our British Waterlilies belong to the central tribe of the family—Nymphaeoideae—of which *Euryale* and *Victoria* also form part. Two other tribes are recognised—the Cabomboideae and the Nelumbonoideae—which differ markedly from the Nymphaeoideae. The Cabomboideae are in many respects relatively simple; they have free carpels, and *Cabomba* also has a less complex type of anatomy than the rest of the family. *Brasenia Schreberi*, which belongs to this tribe, is notable for the enormous development of surface mucilage (Fig. 20)[2].

The Nelumbonoideae include the Sacred Lotus, *Nelumbo Nelumbo* and one other living species belonging to the same genus. In Cretaceous and Tertiary times the genus had, however, a cosmopolitan range (Fig. 21)[3]. This tribe, and the Waterlilies proper, differ so much that they have been described as having nothing in common except the number of cotyledons, the polypetalous flowers, the numerous stamens, and the medium in which they live[4]. The acyclic arrangement of the petals and stamens might also be mentioned as constituting a similarity to some of the Nymphaeoideae. The exalbuminous seeds[5] and the carpels sunk in the curious obconical receptacle, are indeed difficult to reconcile with the characters of the other Waterlilies. Gwynne-Vaughan[6] pointed out that *Nelumbo* shows an

[1] Raffeneau-Delile, A. (1841).
[2] See pp. 271, 272.
[3] Berry, E. W. (1917).
[4] Trécul, A. (1854).
[5] Wettstein, R. von (1888).
[6] Gwynne-Vaughan, D. T. (1897).

almost complete absence, both in leaf and stem, of these features that may be regarded as primitive for the family. *Nelumbo* may possibly be interpreted as the most highly differentiated of the Waterlilies, and part of its peculiarities may perhaps be due to the fact that it is rather a marsh plant than a true aquatic. Possibly it is a genus descended from aquatic ancestors, which has reverted in some degree towards a terrestrial life[1].

Another genus which, though extremely distant from the Waterlilies in its systematic position, yet in its life-history resembles them in some degree, may be mentioned at this

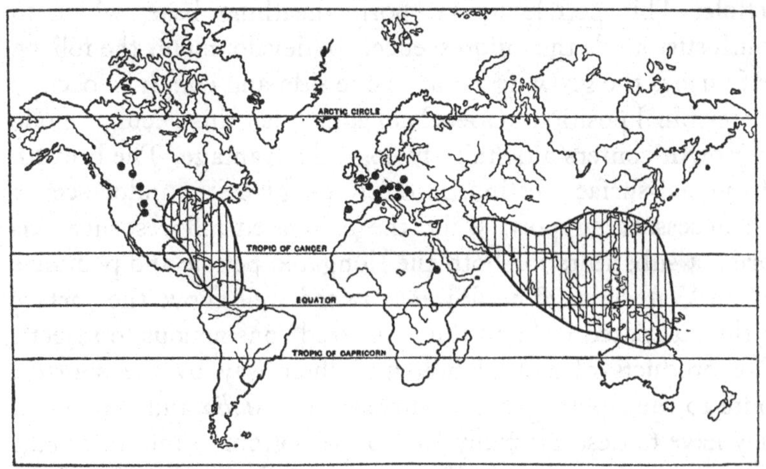

FIG. 21. Sketch map showing the existing and geologic distribution of *Nelumbo*. The obliquely lined areas represent the range of the two existing species, while the Cretaceous and Tertiary records which occur outside these areas are marked by solid black circles. [Berry, E. W. (1917).]

point. This is *Limnanthemum* (*Villarsia*), a member of the Gentianaceae, which is represented in Britain by the beautiful *L. nymphoides* with its fringed yellow flowers. Like *Castalia* and *Nymphaea* it has a rhizome at the bottom of the water while its leaves float at the surface (Fig. 22, p. 41). The length of the internodes of the rhizome varies with the time of year[2] (Fig. 23, p. 41). In the autumn, the leaves are closely packed and the adventitious roots hold the axis with its abbreviated internodes close to the ground. In the spring, elongated internodes are

[1] Dollo, L. (1912). [2] Wagner, R. (1895).

produced and the axis ends in a cymose inflorescence with a terminal flower. The shoot morphology is somewhat puzzling, and remained obscure until it was elucidated by Goebel[1] who studied *L. indicum* and other species from this point of view. In plants of *Limnanthemum*, examined at the flowering season, it is found that a long stalk given off from the rhizome appears to bear both a lamina and flowers, or, in other words, that the flowers seem to arise laterally from a leaf-stalk. In reality this long stalk is however the axis of the inflorescence, and only the short segment of leaf-stalk above the inflorescence is actually the petiole. This petiole has a short, sheathing base, which in youth surrounds the inflorescence. In development, the foliage leaf pushes the growing point to one side and comes to occupy the terminal position. Goebel considers that this peculiar mode of growth confers a definite biological advantage. The breadth of the leaf-surface resting on the water gives the inflorescence the necessary support, while the elongated inflorescence axis forms a substitute for both the elongated petiole and peduncle of the Waterlilies. The flower is raised well above the surface of the associated leaf and thus rendered conspicuous to insects. The products of assimilation find their way by the shortest route to the ripening fruit, whereas in *Castalia* and *Nymphaea* they have to descend many feet to the bottom of the water and then rise again a similar distance to the flower, because there is no connexion between lamina and flower, except *viâ* the rhizome. But, as Goebel suggests, such an arrangement as that met with in *Limnanthemum* would have less value in the case of the Waterlilies, because the Nymphaeaceae store so much food in their rhizomes that the ripening fruit is not dependent upon the products of contemporaneous assimilation. It would be utterly unsafe, however, to suppose that the morphological differences between the Waterlilies and *Limnanthemum* are to be explained on such simple adaptational lines, though it is obvious, from the success which both families achieve, that their respective types of construction must be well suited to aquatic life.

[1] Goebel, K. (1891) and (1891–1893).

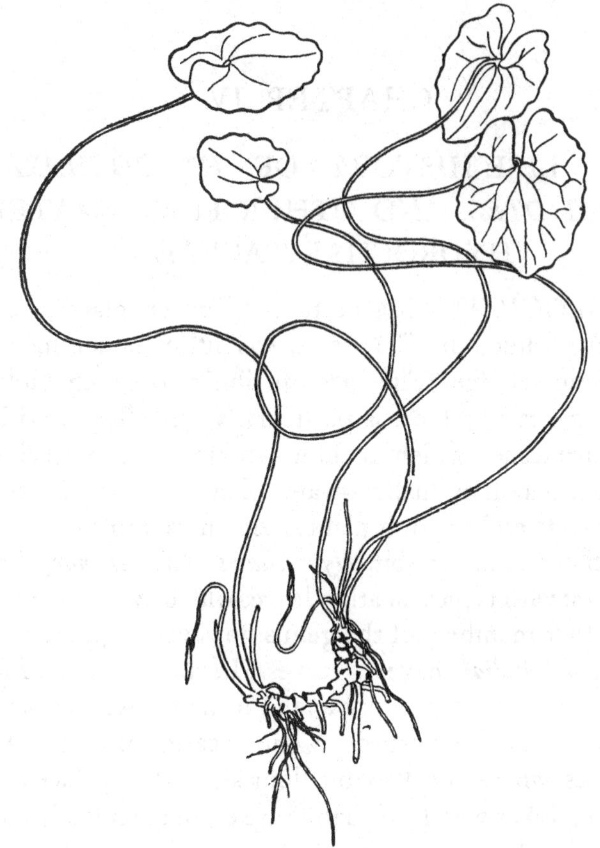

FIG. 22. *Limnanthemum nymphoides*, Hoffmgg. and Link, showing rhizome and leaf-scars. River Ouse. May 30, 1911. (Reduced.) [A. A.]

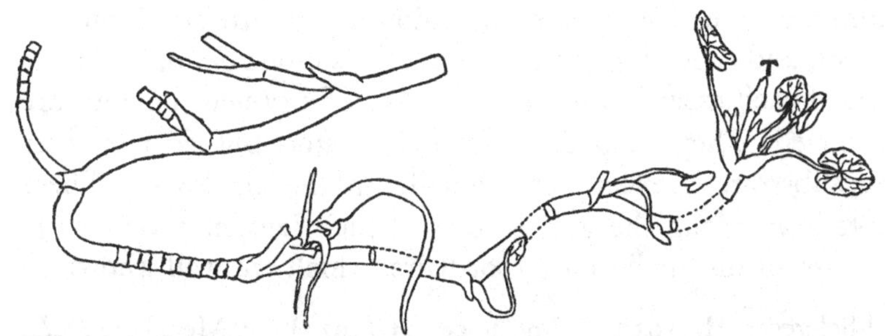

FIG. 23. *Limnanthemum nymphoides*, Hoffmgg. and Link. Rhizome with long and short internodes; *T*, terminal flower. (Reduced.) [Wagner, R. (1895).]

CHAPTER IV

THE LIFE-HISTORY OF *HYDROCHARIS*, *STRATIOTES*, AND OTHER FRESH-WATER HYDROCHARITACEAE

A BIOLOGICAL classification of water plants, such as that outlined in Chapter I, has little in common with any phyletic scheme. The incompatibility between biological and phylogenetic systems is particularly well illustrated in the Hydrocharitaceae, which include—besides some marine genera—both marsh or shallow-water plants with air leaves, submerged plants and floating plants. As an example of the latter we may choose the Frogbit, *Hydrocharis Morsus-ranae*, the only British plant with typical floating leaves which swims freely in the water. Other members of the genus however, e.g., *H. asiatica*[1] and *H. parnassifolia*[2], have air leaves. In the case of *H. Morsus-ranae* it is possible to produce a land form artificially[3], and this form has also been recorded on one occasion in nature[4].

In places where the Frogbit flourishes, the surfaces of the ditches and dykes which it inhabits are often completely covered by its leaves, which resemble a miniature edition of those of the White Waterlily. These leaves are produced in rosettes from a tiny, abbreviated stem, which gives rise during the summer to numerous lateral stolons, each ending in a rosette similar to the parent, and repeating the production of stolons *da capo*. At the base of each rosette, a number of roots of a greenish colour are produced. They hang down into the water, but do not enter the substratum except occasionally in the shallows[5]. These roots bear, along the greater part of their length, a very large number of unusually long root-hairs, which are well known as

[1] Solereder, H. (1913). [2] Solereder, H. (1914). [3] Mer, É. (1882[1]).
[4] Glück, H. (1906). [5] Goebel, K. (1891–1893).

favourable material for observing the rotation of protoplasm. The roots, with their thick mat of root-hairs, get much tangled together, and the countless stolons growing in every direction are similarly enlaced, with the result that *Hydrocharis* forms a thick carpet which can scarcely be submerged even by rough movements of the water. Detritus collects between the root-

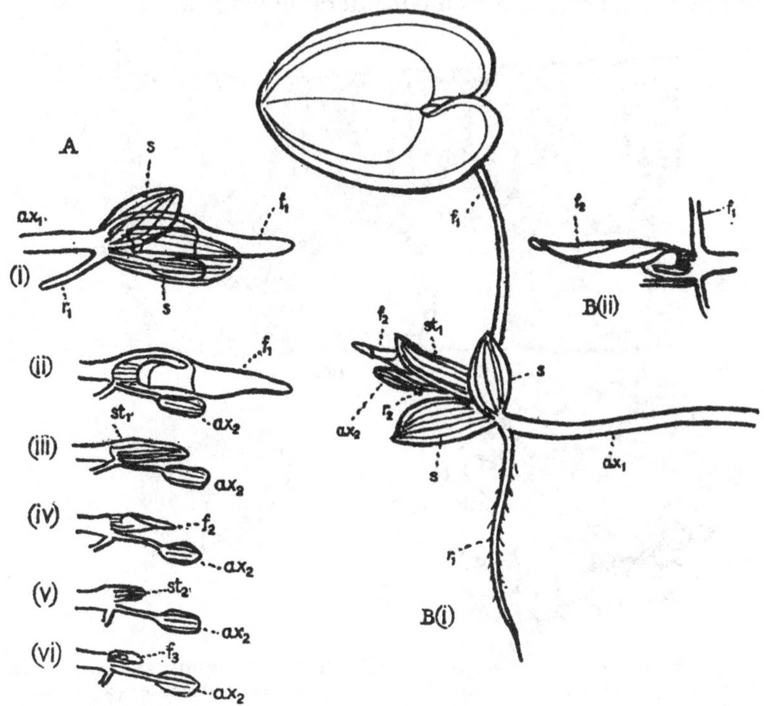

FIG. 24. *Hydrocharis Morsus-ranae*, L. *A*, dissection of a summer bud, just opening; (i)–(vi) show the result of removing successive members. *B* (i), a bud of which one leaf has unfolded; *B* (ii) shows the result of removing the outer scale leaves and the stipules of the first foliage leaf; ax_1 and ax_2, stolons terminating in first and second bud; f_1, f_2, f_3, successive foliage leaves; st_1, st_2, stipules belonging to f_1 and f_2; *s* and *s*, outer scale leaves; r_1 and r_2, roots belonging to first and second bud. [A. A.]

hairs and may serve as a source of food. This colonial mode of growth offers serious resistance to the intrusion of other water plants.

The bud, in which each stolon terminates, is enclosed in two delicate, membranous scales (*s* and *s* Fig. 24 *A* (i)). These are interpreted as paired axillary stipules, whose leaf-blade is generally rudimentary[1]. They are succeeded by a foliage leaf, with its blade tightly inrolled (f_1), whose stipules (st_1) enclose the next foliage leaf (f_2). The young stolon of the next generation (ax_2) is also present in the bud. Fig. 24 *B* (i) shows a bud at a later stage in which the first foliage leaf is fully expanded and the root has grown to a considerable length.

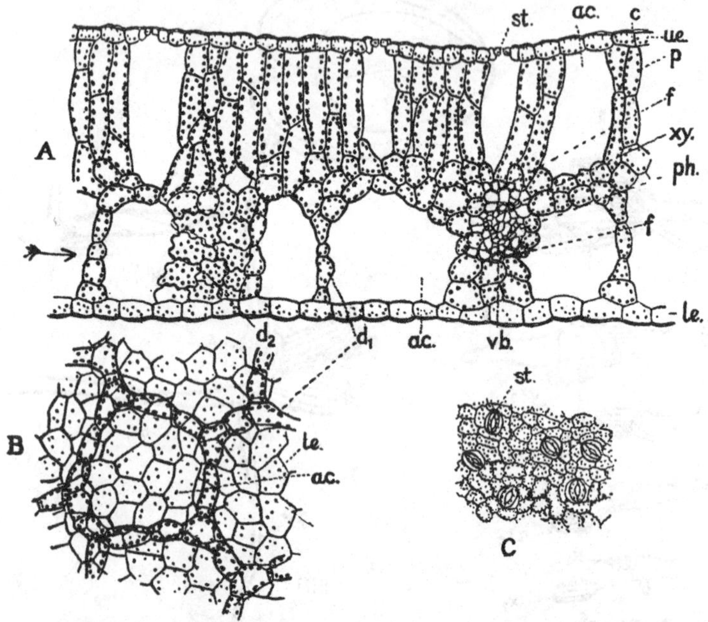

FIG. 25. *Hydrocharis Morsus-ranae*, L. *A*, T.S. leaf; *B*, tangential section through leaf at level of arrow in *A*; *C*, upper epidermis with open stomates (× 78 *circa*); *st.* = stomate; *ac.* = air cavity; d_1 = diaphragm in section; d_2 = diaphragm in surface view; *f* = fibres; *vb.* = vascular bundle; *xy.* = xylem; *ph.* = phloem; *ue.* = upper epidermis; *le.* = lower epidermis; *c* = thin layer of cuticle on upper surface; *p* = palisade parenchyma. [A. A.]

The structure of the lamina of *Hydrocharis* may be described in some detail as an example of the anatomy of a floating leaf (Fig. 25). The upper surface is clothed with an epidermis whose cells contain a few chlorophyll grains. The outer wall is sculptured internally, and bears a delicate layer of cuticle externally. The stomates, which are confined to the upper

[1] Glück, H. (1901).

surface of the leaf, have slightly prominent, external, cuticular ridges (Fig. 26); it is probable that here, as in *Trianea* and in certain other plants with floating leaves, the closure of the stomates is brought about by the approximation of these ridges, rather than by the bulging of the ventral walls[1]. Haberlandt has suggested that this form of stomate is adapted to diminish the risk of capillary occlusion of the aperture by water.

The palisade parenchyma, which lies beneath the upper epidermis, is, in normal leaves of *Hydrocharis*, extremely well differentiated (Fig. 25). On one occasion, however, in the latter part of May, the present writer found a number of plants which were entirely submerged, the winter buds having

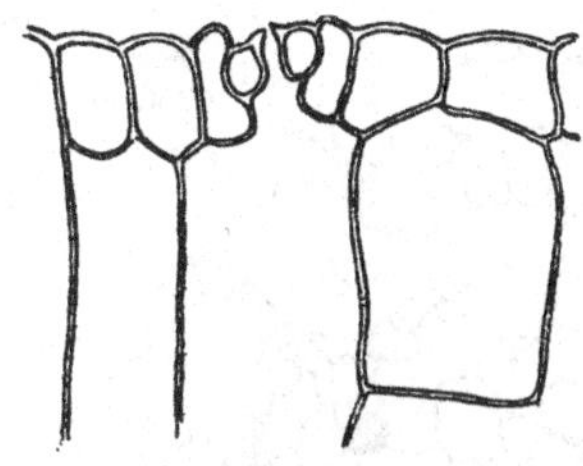

FIG. 26. *Hydrocharis Morsus-ranae*, L. T.S. upper epidermis passing through a stomate. (×318.) [A. A.]

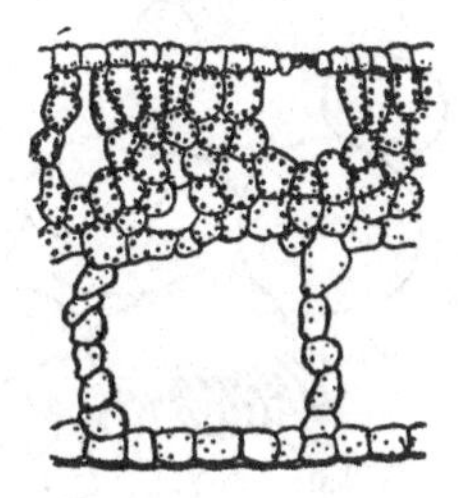

FIG. 27. *Hydrocharis Morsus-ranae*, L. T.S. leaf of young plant growing entirely submerged at the bottom of a ditch, May 17, 1911. (×78 *circa*.) [A. A.]

apparently been caught in an algal tangle at the bottom of a ditch, so that they were unable to reach the surface, but unfolded beneath the water. The green colour of these leaves was unusually pale, and a section of one of them revealed the fact that the palisade region was poorly differentiated, the cells being scarcely elongated (Fig. 27); it was, in fact, a typical 'shade leaf.' The spongy mesophyll was developed normally. In *Hydrocharis* this tissue is not distributed in the irregular fashion with which we are familiar in land plants, but it takes the form of plates of cells disposed in a polygonal mesh-work over the lower epidermis, which itself contains a small amount of chlorophyll (Fig. 25 *B*). Attention has been drawn by

[1] Haberlandt, G. (1914).

Solereder[1] to an anatomical peculiarity of the laminae, the occurrence, namely, of small inversely orientated bundles in the mesophyll (Fig. 28)[2].

Hydrocharis is generally described as dioecious, but further observations are needed on this point. A botanist who examined the species in Sweden records that, if the male and female 'plants' are removed from the water without breaking the intermediate stolons, they are found in reality to be shoots

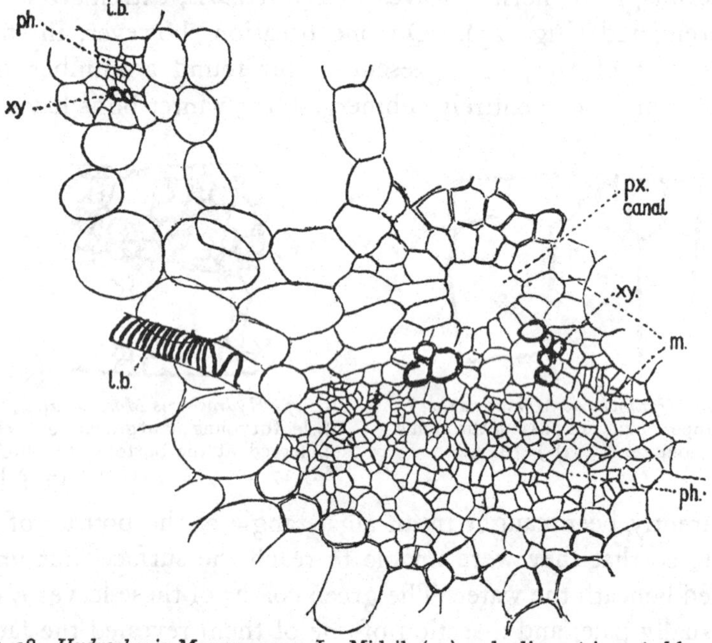

FIG. 28. *Hydrocharis Morsus-ranae*, L. Midrib (*m.*) and adjacent inverted bundle (*i.b.*) from transverse section of leaf. *xy.* = xylem; *ph.* = phloem; *l.b.* = lateral branch of midrib; *px.* = protoxylem. (× 198 *circa.*) [A. A.]

belonging to the same complicated vegetative system, and not separate individuals[3].

Though the flowers of the Frogbit are not uncommon, seed is hardly ever set in this country. The ripened seed vessels are to

[1] Solereder, H. (1913).
[2] Cf. *Stratiotes*, p. 52. See also pp. 337–345.
[3] Lindberg, S. O. (1873).

be found, however, in Continental stations; dehiscence is said to be brought about through the pressure of a slimy mucilaginous mass produced from the testas[1]. As in so many water plants, vegetative reproduction is the chief method of continuance of the species; it occurs by means of winter buds or 'turions,' which in the late summer begin to replace the ordinary buds (Fig. 29). The turions differ from the leaf-buds, which

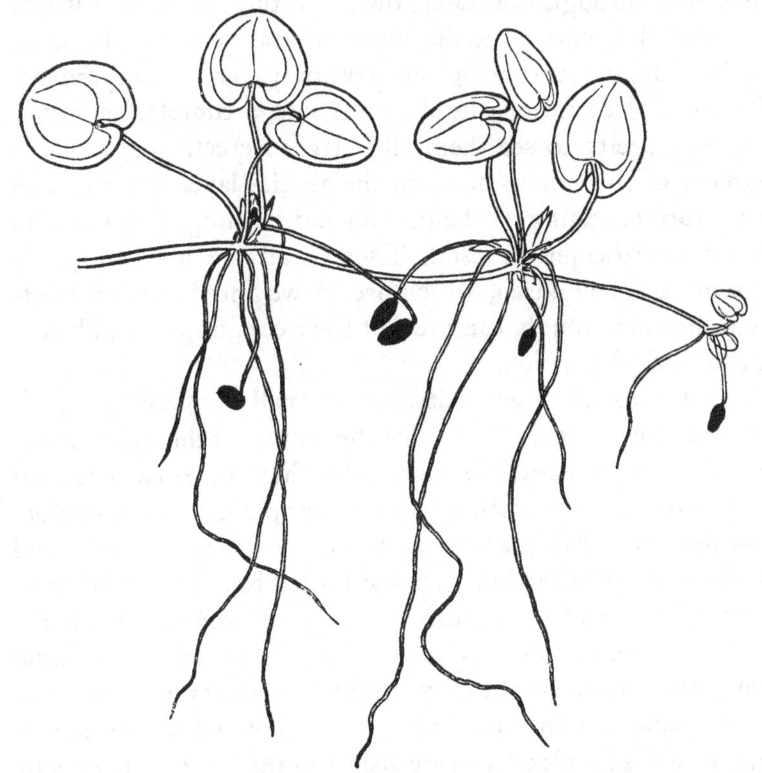

FIG. 29. *Hydrocharis Morsus-ranae*, L. Part of plant, October 1, 1910, showing turions, marked solid black. (Reduced.) [A. A.]

precede them throughout the spring and early summer, in the fact that the two first scale leaves do not unfold, but firmly enwrap the succeeding leaves, while the roots, instead of being developed at once, remain within the axis as rudiments. The

[1] Goebel, K. (1891–1893).

cells of the short, thick stem are packed with large, compound starch grains. The stolons bearing winter buds are readily distinguishable, since they incline downwards in the water, whereas those bearing the summer buds are horizontal or turn slightly upwards. By the early autumn (e.g. October 1), the turions are ripe and a slight touch detaches them at the absciss layer, which traverses the stolon close to the base of the bud. They sink through the water, owing to the starch with which they are laden, and, since the centre of gravity lies in the solid, basal region, the morphological apex always remains uppermost. If a handful of turions be dropped into a tumbler of water, it is very pretty to see them all balanced erect, only the tiny segment of the stolon, between the absciss layer and the base of the turion, resting on the bottom and forming, as it were, an almost microscopic pedestal. They recall the little tumbling toys made for children, which are so weighted that no treatment, however rough, can prevent their coming to equilibrium in the vertical position.

The turions, which are protected externally by a layer of mucilage, pass the winter in the mud at the bottom of the water. It has been demonstrated experimentally that they can remain dormant for at least two years without losing their power of germination. The dormancy has been shown to be due to lack of light[1] and can be induced if the buds are not buried but are merely darkened. The increased sunshine of spring or early summer is the signal for renewed development. The present writer has found that these turions will readily survive the winter at the bottom of an ordinary rain-water tub. It was noticed in one season that, whereas no plantlets were visible in the tub on May 10, by May 15 about seven had risen to the top and were unfolding. This occurred after a long period of warm weather. The development of the little plants coincided remarkably in point of time; on May 16 they were practically all at the same stage (Fig. 30). In each case the three outer scales had turned back

[1] Terras, J. A. (1900). See also p. 280.

so that their tips were below the base of the bud and four or five foliage leaves had unfolded. The two first of these leaves had tiny laminae; no roots were yet developed.

The rare land form of *Hydrocharis Morsus-ranae* produces turions earlier in the year than the water form; they are generally subterranean[1].

Stratiotes aloides, another British member of the Hydrocharitaceae, resembles *Hydrocharis* very closely in its flower, but is quite unique in vegetative structure. One of its names, "Water Aloe," vividly suggests the character of its appearance.

FIG. 30. *Hydrocharis Morsus-ranae*, L. Young plant developed from a turion, showing the stage reached on May 16, 1911. (Nat. size.) [A. A.]

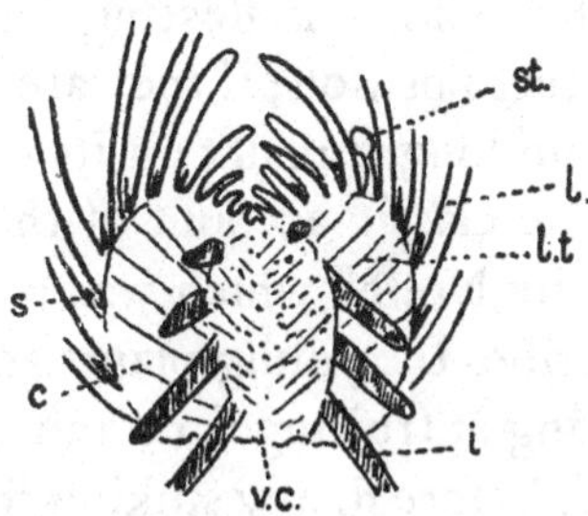

FIG. 31. *Stratiotes aloides*, L. Semi-diagrammatic sketch of stem, as it appears in August, bisected longitudinally (*v.c.*=vascular region of stem; *c*=stem cortex; *l.t.*=leaf-trace; *l*=leaf; *st.*=young stolon; *s*=squamula intravaginalis; *i*=adventitious root). (Slightly enlarged.) [Arber, A. (1914).]

From an abbreviated, almost tuberous stem (Fig. 31) arise a very large number of long, linear leaves, serrated at the edge so sharply as to demand a caution in handling which justifies the plant's generic name and also its commonest English title—"Water Soldier." The leaves may be nearly two feet long. Though the plants of *Stratiotes* live submerged for the greater part of the year, the present writer has noticed, in cultivating them among other aquatics, that their aloe-like form has the effect of keeping the water surface above them clear of swimming plants.

[1] Glück, H. (1906).

From the lower part of the stem of the Water Soldier, numerous green, unbranched roots hang down into the water. These roots may attain great lengths. On August 11, 1910, the present writer measured three roots, each over 40 inches long, growing from the base of one big plant, while on June 30, 1913, seven roots belonging to a single plant, were found to attain an average of nearly 33 inches in length. The rate of growth of these roots is singularly rapid; an elongation of over 2 inches in 24 hours was several times recorded in the case of plants growing under somewhat uncongenial conditions in a London garden[1]. There is no doubt that, at stages when the Water Soldier is floating freely, these long roots balance it in an erect position; if they are destroyed it is found that the plant can no longer maintain its equilibrium.

The classic account of the life-history of *Stratiotes aloides* is that by Nolte[2] which was published nearly a century ago. He describes the young plants as rising to the surface in the spring, sinking at fruiting time and rising again, after the seed has been shed, before finally sinking for the winter. The process appears, however, to be much less regular than would be gathered from Nolte's description[3] and no later observer seems to have witnessed the rising of the Water Soldier *twice* during the year. The plants certainly show a gradual rise in the spring and summer, while they sink again in the autumn, but the movements probably vary with the depth and composition of the water, and they may be influenced by the achievement or failure of fertilisation. The actual mechanism of the rising and sinking process has now been ascertained[4]. *Stratiotes* is apt to frequent water rich in lime[5] and the sinking in autumn is brought about by the deposition of calcium carbonate upon the surface of the leaves, until a point is reached at which the specific gravity of the plant becomes higher than that of the surrounding

[1] Arber, A. (1914). [2] Nolte, E. F. (1825).

[3] Geldart, A. M. (1906) and Kirchner, O. von, Loew, E. and Schröter, C. (1908, etc.).

[4] Montesantos, N. (1913). [5] Davie, R. C. (1913).

water[1]. It has been shown experimentally[2] that, if the chalky deposit be carefully removed from the surface of a plant which is stationed at the bottom of the water, it immediately rises to the top. In nature, the rising of the plant in spring is brought about by the relative lightness of the young leaves, due to the absence of a surface layer of calcium carbonate. As these young leaves become more and more numerous in proportion to the old leaves with their heavy deposit, the specific gravity of the plant becomes less and less, until at last it is lighter than water and floats up to the surface.

The incrustation of the leaves of *Stratiotes* is by no means unique; it has long been known that aquatic plants living in 'hard' water are liable to become covered with a chalky coat. The generally recognised explanation is that, since calcium carbonate is scarcely soluble except in water containing carbonic acid, the abstraction of carbon dioxide, by the green organs of aquatics, leaves the chalk as a deposit on their surfaces. This theory is due to Pringsheim[3], who demonstrated the truth of his view by a series of very delicate experiments, in which he actually observed microscopically the deposition of crystals of calcium carbonate upon the surface of moss leaves, algal filaments, etc., immersed in water containing carbon dioxide and calcium carbonate in solution.

Owing to the curious mode of life of *Stratiotes*, its youngest leaves are usually entirely submerged, but when mature, they may be submerged for part of the year but raised above the surface for another part. It was formerly supposed that the distribution of the stomates on the leaves could be directly traced to the action of the environment. For instance, it has been stated[4] that, in a single leaf which was partly submerged

[1] In justice to Nolte, it ought to be mentioned that he anticipated the discovery that the rising and sinking of the plant was due to differences in specific gravity between the old and young leaves, but he made the mistake of supposing that the greater weight of the old leaves was due to waterlogging.

[2] Montesantos, N. (1913).

[3] Pringsheim, N. (1888).

[4] Costantin, J. (1885[3]) and (1886).

and partly aerial, the exposed region bore stomates, while the submerged part had none. Recent work has shown that this is altogether too simple an account of the position. It has been demonstrated[1], for instance, that leaves which are entirely, or almost entirely submerged, may nevertheless have stomates throughout their entire length. On the other hand, in the case of a plant which was growing at the bottom of the water, and of which the outer leaves were partly aerial, it was found that these outer leaves bore no stomates whatever, but a transition to stomate-bearing leaves was observed among the younger leaves; the youngest, which were also the deepest in the water, bore the most numerous stomates. The interpretation suggested by the writer to whom we owe these observations, is that the leaf with stomates is the higher form, which can only be developed in favourable surroundings, while the stomate-free leaves are primary leaves, occurring typically under conditions of poor nutrition. We shall return to this subject later on, in considering heterophylly in general[2].

Besides the epidermis, the other tissues of the leaf show certain interesting features. The vascular skeleton consists of five, or more, strong longitudinal veins united by transverse connexions. Spirally thickened tracheids occur in the bundles even in the submerged leaves. In the transverse section of the rather thick lamina, besides the main row of normally orientated bundles, there are two rows of small bundles, one row lying near the under side and normally orientated, and the other towards the upper surface and inversely orientated[3]. The occurrence of these inverted bundles in the leaves of the Hydrocharitaceae is significant in connexion with the 'phyllode theory' of the Monocotyledonous leaf[4].

In the axil of each leaf of the Water Soldier are found the mucilage-secreting scales (*squamulae intravaginales*) characteristic of the Helobieae[5].

[1] Montesantos, N. (1913).
[2] See pp. 156–160.
[3] Solereder, H. (1913).
[4] Arber, A. (1918); see also p. 46.
[5] Nolte, E. F. (1825) and Irmisch, T. (1858[2]).

If vigorous plants of *Stratiotes* be examined in the late summer, they will be found to have produced numerous lateral stolons terminating in buds[1] (Fig. 32). These buds do not, like those of *Hydrocharis*, pass the winter in a closed condition, but open at once, and may be described as winter-buds

FIG. 32. *Stratiotes aloides*, L. Plant after flowering in August, bearing five plantlets at the ends of stolons. (Reduced.) [Modified from Nolte, E. F. (1825).]

which germinate while attached to the parent plant. There is, in fact, no interruption in the vegetative life, since the daughter shoots, as soon as they become free from the parent axis in autumn or winter, begin to form new winter-buds themselves. In North Germany, the Water Soldier was described in 1860

[1] Glück, H. (1906).

as so abundant as to be a troublesome weed, the plantlets surviving the hardest winter[1].

In the great majority of localities the continued existence of *Stratiotes* depends absolutely upon bud-formation, since the plant is dioecious, and only in a small part of its range is it found with both male and female flowers. In England only the female plant is usually met with (Fig. 33). There are some records of the occurrence of hermaphrodite flowers[2], but ripe

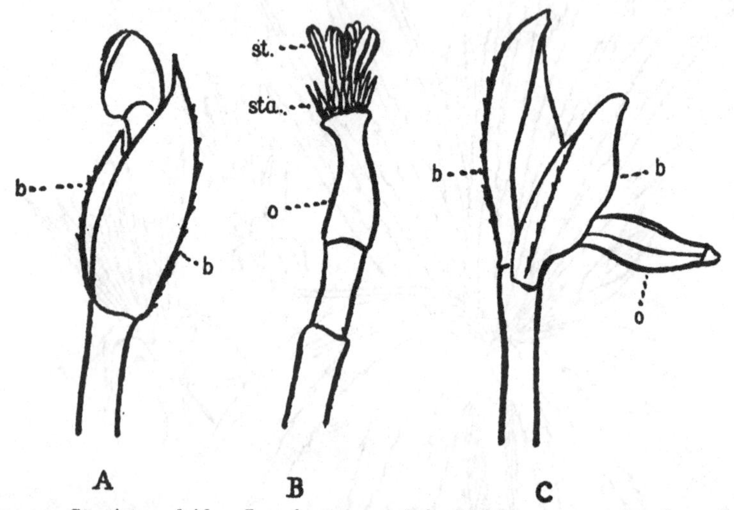

FIG. 33. *Stratiotes aloides*, L. *A*, unopened female flower emerging from two bracts (*b*). *B*, female flower with bracts and perianth removed to show ovary (*o*), stigmas (*st.*) of which there are six, each bifurcated to base, and staminodes (*sta.*). *C*, unfertilised fruit (*o*) emerging obliquely from the bracts. [A. A.]

seed does not seem to be formed in this country at the present day, though fruits with seeds are known from Pliocene and Pleistocene deposits[3]. The geographical distribution of the sexes is rather curious. According to Nolte[4], in the northerly part of the range of the species only female plants occur, while at the southern extremity the plants are either predominantly or entirely male. In an intermediate area both sexes occur.

In addition to the Frogbit and the Water Soldier, *Hydrilla*

[1] Klinsmann, F. (1860). [2] Geldart, A. M. (1906). [3] Reid, C. (1893).
[4] Nolte, E. F. (1825); see also Caspary, R. (1875).

verticillata, another member of the Hydrocharitaceae, has recently been recorded from one station in Britain, though it is typically a plant of warm climates[1]. But a fourth genus, *Elodea*, represented by the Canadian Waterweed, a submerged plant, which was apparently introduced into this country about 1843[2], has become very much more common than any other member of the family. In nearly all the localities in Britain, only the female plant is found, though the male has been recorded as occurring near Edinburgh[3]. The reproduction of *Elodea canadensis*, which is amazingly rapid, is thus entirely vegetative; the snapping of the slender, brittle stems sets free fragments which live independently, while special winter-shoots may also be produced (Fig. 34). The small leaves, which are arranged in whorls of three, are only two cells thick and it is to their extreme delicacy that the plant probably owes its incapacity to produce a land form[4].

FIG. 34. *Elodea canadensis*, Michx. Wintering shoot. [Raunkiaer, C. (1896).]

The pollination mechanism of the genus *Elodea* is of some significance, owing to the different phases met with in different species. Most of the species have inconspicuous flowers. The male flowers either become detached and rise separately to the surface of the water, e.g. *E. canadensis*[5], or they are carried up by the growth of their thread-like stalks, e.g. *E. ioensis*[6] (Fig. 35, p. 56). The stigmas reach the surface owing to the elongation of the floral tube which in *E. canadensis* may reach a length of 30 cms.[7]. In an Argentine species, *E. callitrichoides*[8], in

[1] Bennett, A. (1914).

[2] Marshall, W. (1852) and (1857), Caspary, R. (1858[2]) and Siddall, J. D. (1885). See pp. 210–213 for a further account of the spread of this plant in the British Isles.

[3] Douglas, D. (1880).

[4] Schenck, H. (1885).

[5] Wylie, R. B. (1904).

[6] Wylie, R. B. (1912), also *E. canadensis* according to Douglas, D. (1880).

[7] Wylie, R. B. (1904).

[8] Hauman-Merck, L. (1913[2]).

which the pollination has been described in detail, the submerged male buds are found to be each occupied by a bubble of gas, probably carbon dioxide. Directly the flower reaches the

FIG. 35. *Elodea ioensis*, Wylie. 1, open staminate flower attached to plant. 2, mature staminate flower enclosed within the spathe. 3, staminate flower emerging from the spathe. 4, detached and empty staminate flower floating on the water with elongated axis trailing. [Wylie, R. B. (1912).]

surface by the elongation of its filiform axis[1], it opens suddenly and at the same moment the pollen sacs dehisce explosively. It thus comes about that abundant pollen floats on the surface of

[1] This axis is mentioned by Caspary, R. (1858[2]) with the incorrect description "tubus calicis filiformis."

the water, and surrounds the stigmas of the female flower. It has been suggested that perhaps the pollen may be attracted to the receptive surfaces by currents due to some secretion from the stigmas. It has been shown in *E. canadensis* that the spines on the outer coat of the pollen-grain hold back the surface-film from contact with the body of the spore and thus imprison enough air to keep it afloat[1].

A somewhat different method, in which water also plays a part, is found in *Vallisneria*[2], while in *Hydromystria* the pollination is sometimes effected by wind and sometimes by water[3]. In *Elodea densa*[4], the large white flowers contain nectar, and insect pollination occurs; this genus thus shows transitions between the entomophilous members of the family, such as *Hydrocharis*, and the hydrophilous and anemophilous genera.

Among vegetative characters, perhaps the most notable feature of the Hydrocharitaceae is the great variation in the form and mode of life of the leaf in the different genera. To illustrate this we may briefly enumerate the leaf characters of a few genera selected entirely from the fresh-water members of the family.

Hydrocharis. In certain species, heart-shaped floating leaves alone.

Stratiotes. Stiff, serrated, linear leaves, sometimes entirely submerged, sometimes partially aerial.

Boottia. Lower leaves short-stalked and submerged; upper leaves long-stalked and often aerial.

Ottelia. Leaves differentiated into submerged leaves, with a narrow blade, and stalked leaves with broader blades, which may be submerged, floating or aerial.

Vallisneria. Leaves entirely submerged, ribbon-like, growing in rosettes.

Hydrilla and *Elodea*. Leaves entirely submerged, short and linear, growing on elongated axes.

Three genera of the Hydrocharitaceae, *Enhalus*, *Halophila* and *Thalassia*, live in salt water; these we shall consider in Chapter X.

[1] Wylie, R. B. (1904). [2] See p. 235.

[3] Hauman, L. (1915). [4] Hauman-Merck, L. (1913[2]).

CHAPTER V

THE LIFE-HISTORY OF THE POTAMOGETONACEAE OF FRESH WATERS[1]

POTAMOGETON, the central genus of the Potamogetonaceae, includes the very numerous Pondweeds, so common in temperate waters, and is the richest in species of all our native aquatic genera. The Pondweeds are an exceedingly difficult group from the point of view of the student of systematic botany, as the numerous species can, in many cases, only be discriminated as the result of much experience. A character which increases the difficulty of identifying them is the capacity for variation in form shown by one and the same individual. The present writer took a typical shoot of *Potamogeton perfoliatus* from the Cam in July, and kept it floating in a rain-water tub. By October 1 most of the large perfoliate leaves had decayed and those on the new shoots were so much narrower and less perfoliate as to make it difficult to believe that they belonged to the same species (Fig. 36). This power of variation in leaf-form within one individual is a well-known feature of *P. perfoliatus*. It has been recorded that an isolated plant in a newly-dug clay-pit, observed during several years, changed so much in the shape, colour and texture of the leaves as to give rise to the idea that all the British forms of the species which have been described, may possibly be mere states and not variations[2].

The most obvious difference between the Potamogetons and the water plants hitherto considered, lies in the extreme complexity of the shoot systems of the Pondweeds. The rhizomes

[1] The marine Potamogetonaceae are considered in Chapter x.

[2] Fryer, A., Bennett, A. and Evans, A. H. (1898–1915). This account of the British Potamogetons is of the first importance.

form mats at the bottom of the water, retaining the soil in their meshes and thus consolidating it, while, from these rhizomes, a forest of leafy shoots rises into the water[1]. An examination of the individual axes shows the branch system to be sympodial[2]. The shoots are of two kinds; the first is horizontal, more or less buried in the soil, colourless and scale-bearing, while the second is erect, floating to some degree, and producing perfect leaves. Fig. 37, p. 60, illustrates the general scheme of branching. The creeping stem is a sympodium formed by the union, end to end, of the two first internodes of successive generations (I, II,

FIG. 36. *Potamogeton perfoliatus*, L. Detached floating shoot, October 1, 1910, showing how much the plant at this time of year may depart from the perfoliate leaf type. Several "winter shoots" have developed. (½ nat. size.) [A. A.]

III, etc.), the succeeding internodes constituting the erect stem. In one season a great many of these rhizome units may be formed. The first scale leaf of each erect shoot (*a*, *a*′, *a*″, *a*‴) bears a reserve bud on its axil, which may give rise to another segment of rhizome, again repeating the entire process, so that the whole ramification becomes extremely complicated. In Fig. 37, II′, III′, represents a reserve shoot, arising in the axil of *c*, the third scale leaf of Shoot I. By the decay of the older parts of the rhizomes fresh individuals become separated, and even the

[1] Graebner, G. in Kirchner, O. von, Loew, E. and Schröter, C. (1908, etc.). [2] Irmisch, T. (1858[3]) and Sauvageau, C. (1894).

erect shoots, if detached from the parent, can form new plants. The leafy shoots branch relatively sparsely in the large-leaved forms, but more freely in those with small leaves.

FIG. 37. General branch system of a typical *Potamogeton*. I, II, III, . . . the different shoot-generations; *a*, *b*, *c*: *a′*, *b′*, *c′*, . . . the three first scale-leaves borne by each shoot-generation; II′, III′ is a reserve shoot arising in the axil of leaf *c* belonging to shoot I. [Adapted from Sauvageau, C. (1894).]

The various species of *Potamogeton* show transitions between plants with floating leaves, capable of producing a land form, and plants with submerged leaves, living entirely beneath the water-surface, except that they raise their flowers slightly into the air. *Potamogeton natans* may be taken as a type of the Pondweeds with floating leaves; these consist of a sheathing base with stipules, a long petiole and an elliptical to lanceolate blade, leathery in texture. The early leaves on each shoot, which do not reach the water-surface, are phyllodic and represent only the petioles of the perfect leaves. Intermediate leaf-forms also occur, with small, spoon-like expansions of the apex[1]. The relation between the narrow submerged leaves and the broad floating leaves is identical with that subsisting between the two corresponding leaf-types in *Sagittaria*. The land form of *Potamogeton natans* is shown in Fig. 125, p. 196.

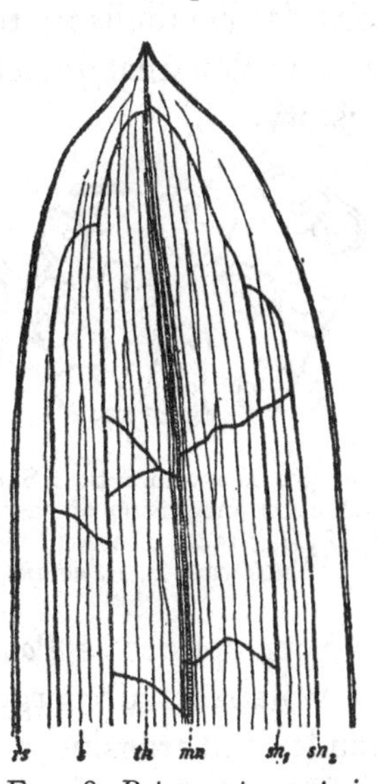

FIG. 38. *Potamogeton zosterifolius*, Schum. Upper part of leaf; *mn*, sn_1, sn_2, *tn*, vascular bundles; *s*, bast bundles; *rs*, bast bundle along margin. (×12 *circa*.) [Raunkiaer, C. (1903).]

Another species of *Potamogeton*, *P. pulcher*, Tuckerm., of N. America, produces not only broad floating leaves but broad submerged leaves, while others, such as *P. heterophyllus*, Schreb., have ovate or oblong floating leaves, but their submerged leaves are of a narrower type.

The more completely aquatic species form submerged leaves alone, with laminae of variable breadth. Examples of this group are *P. lucens*, *P. perfoliatus* and *P. crispus*. In these and related species the blade is exceedingly thin, often with only one plate of cells between the two epidermal layers, but it is supported by fibrous strands running the length of the leaf (*s* in Fig. 38). The lamina is often crisped or undulated at the margin in a

[1] Schenck, H. (1885); see also Fig. 168, p. 339.

graceful way. A similar undulation is characteristic of *Aponogeton ulvaceus*, Baker[1]. A curious feature of the leaves of various species, e.g. *P. lucens* and *P. praelongus*, is their shining oily surface[2], which is due to the presence, in the epidermal cells, of large oil drops secreted by special colourless plastids. The non-wettable, slippery surface thus produced may be, it is suggested, a protection against water animals and micro-parasites. It has also been supposed that the oil may hinder diffusion and hence prevent the soluble products of assimilation from being washed out of the leaf. But it seems to the present writer more probable that the oil is a mere by-product of the plant's metabolism; there is no valid reason for making the assumption that it performs any special function in the life-history.

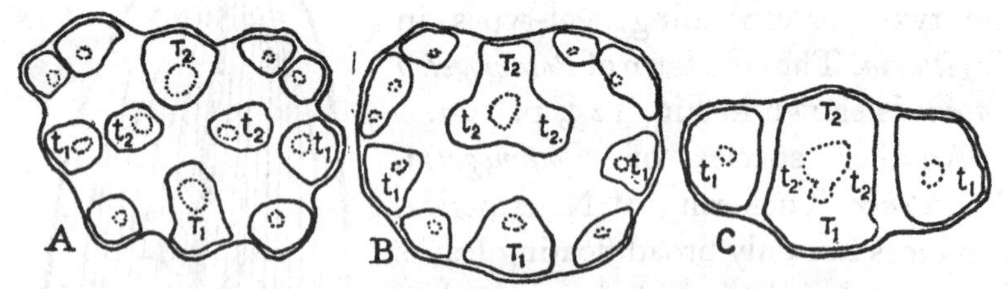

FIG. 39. Diagrammatic T.S. of stem stele of three species of *Potamogeton* to show reduction and fusion of vascular strands. t_1, T_1, t_1, traces of next higher leaf; t_2, T_2, t_2, traces of second higher leaf; remaining strands cauline. *A*, *P. pulcher*, Tuckerm. *B*, *P. natans*, L. *C*, *P. crispus*, L. [Chrysler, M. A. (1907).]

Such species as *Potamogeton trichoides* and *P. pectinatus* have very narrow submerged leaves which are linear in form and tender and translucent in texture.

The species belonging to *Potamogeton* and the allied genus *Zannichellia* can be arranged, according to the anatomy of their stems and roots, in a reduction series, beginning with the types with floating leaves, whose axes show a complicated internal structure, and ending with entirely submerged, narrow-leaved species, in which the anatomy is reduced to a state of extreme simplicity[3]. But it is uncertain whether this sequence completely

[1] Krause, K. and Engler, A. (1906). [2] Lundström, A. N. (1888).
[3] Schenck, H. (1886) and Raunkiaer, C. (1903).

represents the evolutionary history, since it is possible that certain forms with floating leaves may have had a submerged ancestry. The species whose central cylinder diverges least from a normal terrestrial type, seems to be *Potamogeton pulcher*[1] (Fig. 39 *A*). Here a section across an internode of the leafy shoot reveals, within the central cylinder, three distinct bundles (t_1, T_1 and t_1) which are the traces of the leaf immediately above, and three more (t_2, T_2 and t_2) which entered at a still higher node. In addition there are several bundles which are purely cauline. The type represented by our native *P. natans* (Fig. 39 *B*) differs from that of *P. pulcher* in the fact that the traces belonging to each leaf do not so fully retain their independence in the central cylinder. *P. perfoliatus* belongs to the type of *P. natans*. In *P. crispus* (Figs. 39 *C* and 40 *A*, p. 64) the stele is more condensed, the bundles being collected into three groups. In very slender stems of this species, the two passages in each group representing the xylem may fuse so that the distinctness of the bundles is maintained by the phloems alone. *P. lucens*[2] (Fig. 40 *B*) has a median and two lateral bundle-groups, but these are more reduced—the median group consisting of one xylem passage and two phloem regions, and the laterals, of one xylem passage, and one patch of phloem. In this species the tendency to concentric arrangement begins to make itself felt. In *P. pusillus* (Fig. 40 *C*) the lateral bundles are entirely fused with the median, as far as the xylem is concerned, but the phloems still remain distinct. In *P. pectinatus* (Fig. 40 *D*) the ultimate term in the reduction series is reached: a ring of phloem surrounds a single xylem passage. *Zannichellia* closely resembles *P. pectinatus*; ephemeral xylem vessels have been detected in the apical region of the stem[3]. In the case of the related genus *Althenia*[4], vessels are also retained in this region and in the nodes.

[1] Chrysler, M. A. (1907).

[2] On this and other species, Sauvageau, C. (1894) should be consulted. His account diverges in some points from that of Schenck.

[3] Schleiden, M. J. (1837). [4] Prillieux, E. (1864).

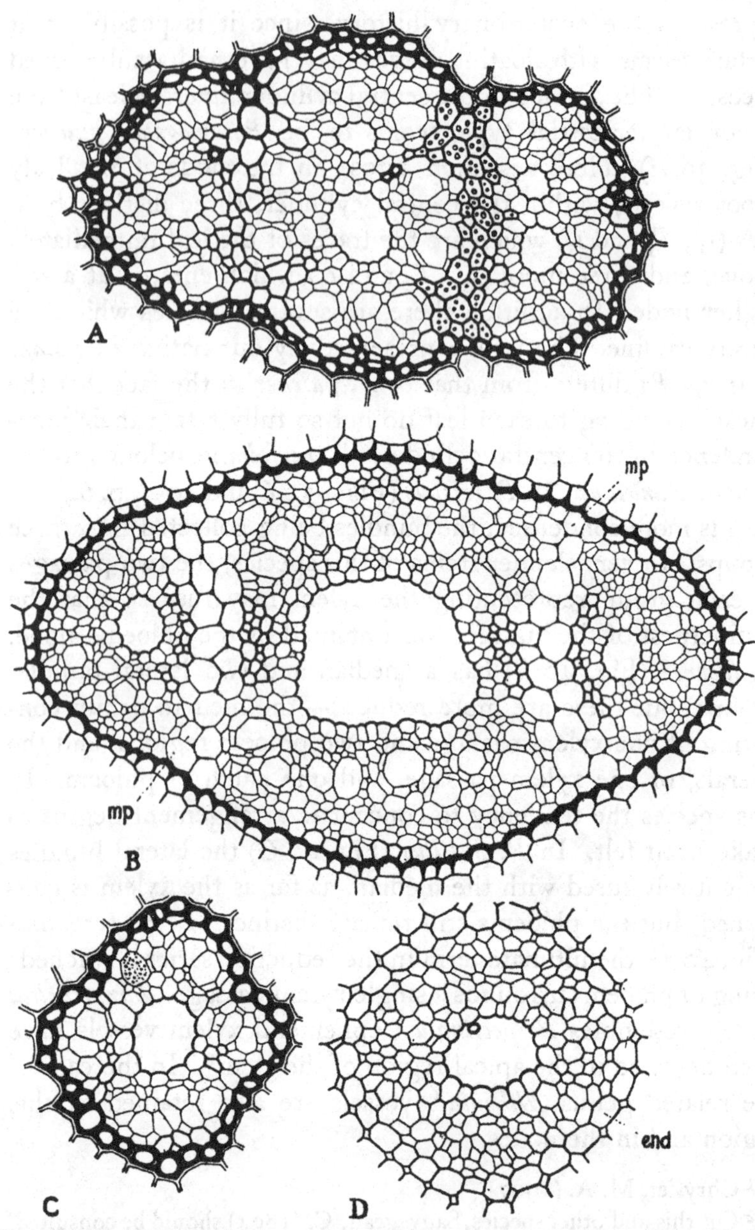

FIG. 40. Reduction series in central cylinder of stem in *Potamogeton*. *A*, *P. crispus*, L. (cf. diagrammatic Fig. 39, *C*) (× 160); *B*, *P. lucens*, L., in which fusion of the strands has gone further, so that each of the three bundle groups has one xylem only; *mp* = conjunctive tissue (× 130); *C*, *P. pusillus*, L., in which the xylems of all the individual bundles form a single central passage (× 290); *D*, *P. pectinatus*, L., completely concentric structure in which all trace of the component bundles is lost; *end* = endodermis (× 290). [Schenck, H. (1886).]

The tendency to condensation and simplification of the stem stele, which is so well illustrated among the Potamogetons[1], is, as we shall see in Chapter XIII, a characteristic of many aquatics. The stem of the Pondweeds is, however, peculiar in that the bundles are not confined to the central cylinder. In some species there is a complicated system of cortical strands, occurring at the intersection of the diaphragms separating the lacunae. These cortical bundles communicate with one another and with the axial strand by means of anastomoses at the nodes.

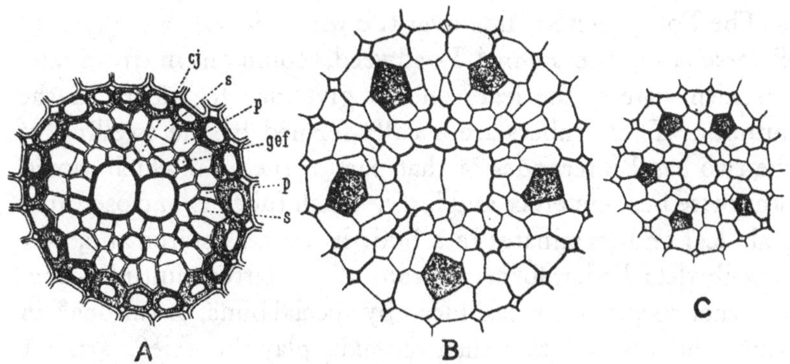

FIG. 41. Structure of central cylinder of root in *Potamogeton*. *A*, *P. natans*, L. *gef*, vessel; *s*, sieve tube with companion cell; *p*, pericycle; *cj*, conjunctive tissue (× 470). *B*, *P. densus*, L. Similar to *P. natans*, but vessels and endodermis thin-walled; sieve tubes shaded (× 470). *C*, *P. pectinatus*, L., xylem reduced to single vessel (× 470). [Schenck, H. (1886).]

We have so far been considering the anatomy of the leafy shoot alone. It should be noted that the structure of the horizontal rhizome and of the inflorescence axis are often markedly different. For instance, in the creeping stem of *P. pulcher*, the central cylinder takes the form of "a truly dicotyledonous looking ring of collateral bundles," while the flowering axis of *P. natans* also has its vascular strands arranged in a regular ring[2].

[1] Sanio, C. (1865) first recognised that the apparently simple axial strand of certain Potamogetons was really the reduced representative of a whole system of bundles.

[2] Raunkiaer, C. (1903) and Chrysler, M. A. (1907).

A similar reduction series to that met with in the central cylinder of the stem can be traced in the root[1]. In *Potamogeton natans* (Fig. 41 *A*, p. 65) the root is pentarch and the walls of all the elements, except the sieve tubes, are thickened. *P. densus* (Fig. 41 *B*) has the same type of structure, but the cell-walls remain thin. In *P. pectinatus* (Fig. 41 *C*) the five protoxylem elements are absent, and the xylem is represented merely by a single central vessel with delicate, spiral thickening[2]. The structure of the root of *Zannichellia* is similar, but the axial vessel is unthickened.

The Potamogetons tide over the winter in various ways. In *P. pectinatus*, the Fennel Pondweed, common in fresh and brackish waters, the leafy shoots give rise to tubers in the autumn. These tubers are usually formed by the swelling of the two basal internodes of that part of the axis which would otherwise become erect and leafy. Each tuber is enclosed in a scale leaf and terminates in a bud; it contains starch and, as it is easily detached, it forms a means of vegetative multiplication.

Other species are reproduced by special buds, or turions[3], in which the leaves, rather than the axis, play the chief part. A group of submerged Pondweeds with linear leaves, of which *P. pusillus* and *P. trichoides* are examples, is characterised by winter-buds enclosed in scales corresponding morphologically to axillary stipules accompanied by rudimentary laminae. In this group of species there is no rhizome, branching sympodially in the mud, the only part corresponding to such a rhizome being the elongated axis of the turion; the branched leafy shoots play the chief *rôle* in the axial development. The whole vegetative body in these species dies off in the autumn and the turions alone remain. These buds are formed in great numbers, and

[1] Schenck, H. (1886).

[2] Sauvageau, C. (1889[2]) describes the roots of *P. pectinatus* as having, in general, a less degraded type of structure than that attributed to them by Schenck, H. (1886).

[3] Glück, H. (1906) deals comprehensively with the turions of the genus.

often many thousands lie on the soil at the bottom of the water. They germinate without rising to the surface. The formation of winter-buds in this group of Pondweeds, as indeed in aquatics in general, is encouraged by unfavourable conditions[1]. For instance, if the environment is otherwise satisfactory, but the depth of the water is excessive, causing the plant to exhaust itself in the production of long axes, turion formation may occur unusually early in the year.

Potamogeton crispus[2] is related, in its wintering habits, to the group just dealt with, but its turions are singular in certain respects. The word 'bud' seems in this case to be a misnomer, as the thick, toothed leaves of the turion do not enfold one

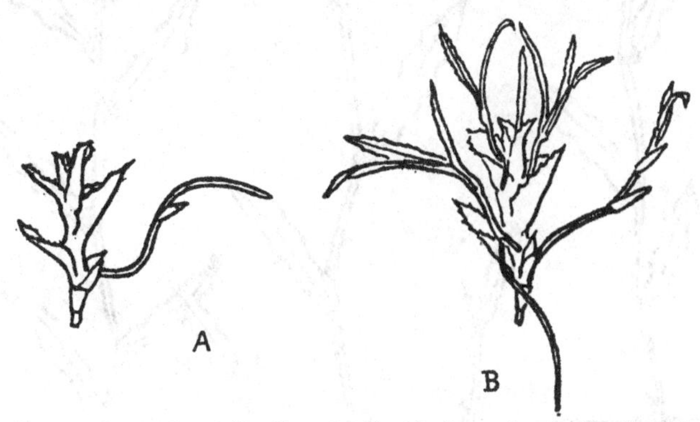

FIG. 42. *Potamogeton crispus*, L. Germinating turion. *A*, a turion from bottom of water, March 16, 1912, with one lateral branch. *B*, the same turion, April 11, 1912, when it had developed a number of lateral branches and a root. (Nat. size.) [A. A.]

another, but stand out at a wide angle from the axis. They are of unusual consistency, being hard and horny. The turions may be from 10 to 50 mm. long and bear three to seven leaves. As their discoverer, Clos, pointed out more than sixty years ago, their mode of germination is quite peculiar, since there is no elongation of the axis, and further development is due entirely to the production of axillary branches. The process of germination can be followed in Fig. 42 and Fig. 43, p. 68. Figs.

[1] See pp. 222–224.

[2] Clos, D. (1856), Treviranus, L. C. (1857), Hildebrand, F. (1861), Cöster, B. F. (1875) and Glück, H. (1906).

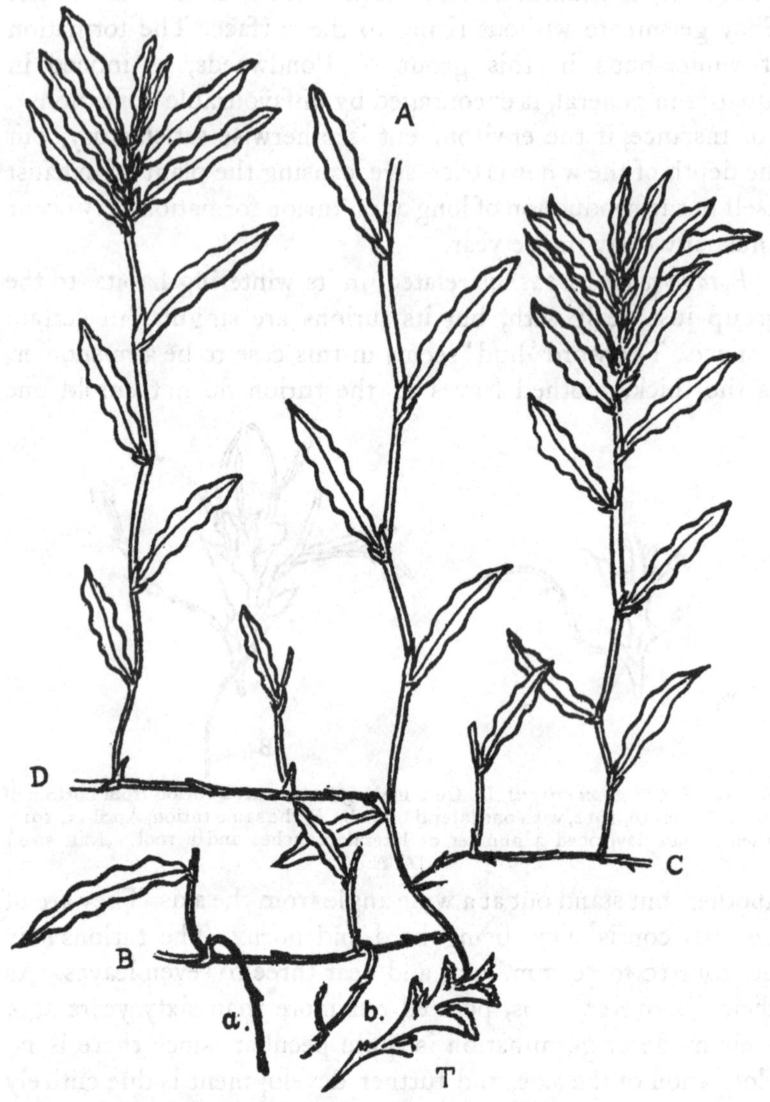

FIG. 43. *Potamogeton crispus*, L. Advanced stage in the germination of a turion (reduced). The first shoot, *A*, produced from the turion, *T*, has given rise to three lateral sympodia, *B*, *C*, *D*. The first and second shoot-generations of *B* have given rise to two reserve shoots, *a* and *b*. [Adapted from Sauvageau, C. (1894).]

42 *A* and *B* were drawn from a bud which had passed the winter at the bottom of a rain-water tub in the present writer's garden. The turions of this species seem to be primarily reproductive bodies, and to be only secondarily concerned with tiding over the winter, for large numbers germinate without a resting period. Not only the rhizomes, but certain of the leafy shoots, are capable of lasting over the cold season, if they are not actually frozen. The special winter branches differ somewhat from the summer shoots in having leaves without a crisped margin, and they have hence been sometimes mistaken for a distinct species.

A second group of Pondweeds is characterised by winter-buds whose enclosing scales consist merely of axillary stipules, the corresponding blades having wholly disappeared. Fig. 44 represents a transverse section of a turion of *Potamogeton rufescens*, which conforms to this type. In this species the winter-buds are formed chiefly on the underground rhizome, while in *P. fluitans*, Roth—a species closely related to *P. natans*—they occur in this situation only.

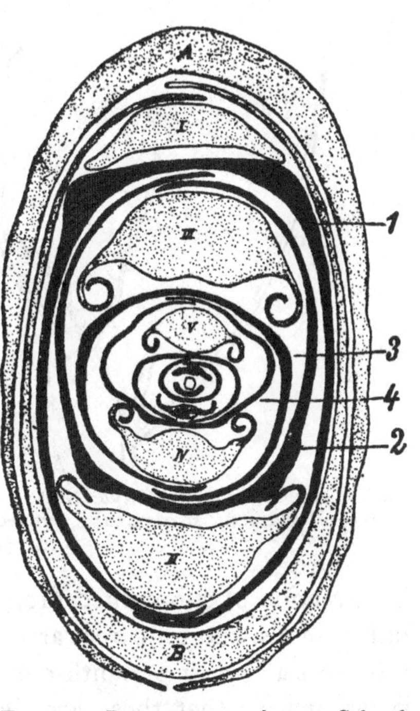

FIG. 44. *Potamogeton rufescens*, Schrad. T.S. through a turion. *A* and *B*, outer scale leaves equivalent to stipules; I–IV, foliage leaves, whose stipules are marked 1–4 and put in in solid black. *Squamulae intravaginales* are omitted. (Enlarged.) [After Glück, H. (1906), *Wasser- und Sumpfgewächse*, Bd. II, p. 160, Fig. 23.]

Potamogeton perfoliatus, L. forms winter-buds which are not deciduous but unfold *in situ* (Fig. 36, p. 59).

In flower structure[1], as well as in anatomy, a reduction series

[1] Schenck, H. (1885).

can be traced in the Potamogetonaceae. This series ranges from forms such as *Potamogeton natans*, with an erect spike of numerous flowers, through various intermediate types, to the related genus *Ruppia*, in which the pollen floats, and the stigmas are raised to the surface to receive it, and ultimately to *Zannichellia* and various marine members of the family, in which the pollination is entirely submerged. Even within the genus *Potamogeton* itself, there are a number of gradations in

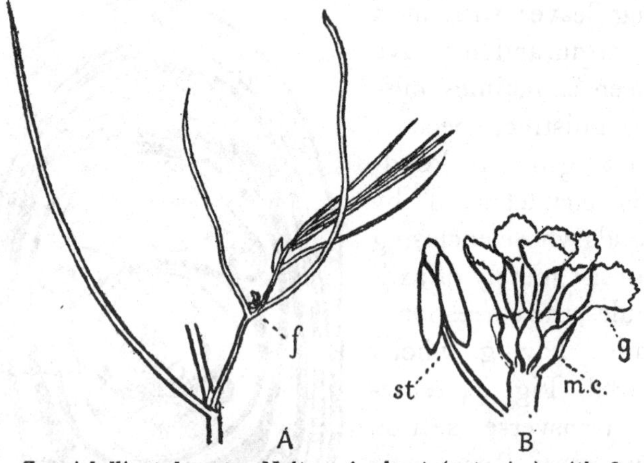

FIG. 45. *Zannichellia polycarpa*, Nolte. *A*, shoot (nat. size) with flowers (*f*). *B*, flowers (enlarged); *st*, stamen; *g*, gynaeceum; *m.c.*, membranous cup. May 25, 1912. [A. A.]

the direction of submerged life. The flowers possess, typically, four stamens, and four free carpels. They appear, at first glance, to possess also four perianth members, but more careful examination reveals that these are, in reality, leaf-like outgrowths from the staminal connective[1]. The spike of *P. natans* is supported above the water by the two floating leaves immediately below it. These are always opposite (cf. Fig. 37, p. 60), although otherwise the leaves are alternate. In some species, e.g. *P. pectinatus*, the spikes, instead of being stiff and erect, are thin and flexible, and float horizontally on the water. In these forms

[1] Information as to the morphology and development of the flower and fruit will be found in Hegelmaier, F. (1870), Schumann, K. (1892), etc.

the flowers are distant, and when mature they are lifted, one by one, a little above the water-surface. In other cases the inflorescences are much reduced—only four flowers being developed in *Potamogeton pusillus*—while in *P. trichoides* the individual flowers are modified, the number of carpels being reduced to one. In *Zannichellia polycarpa* the flowers are unisexual (Fig. 45), a male and female flower (or inflorescence) being found together in one leaf-axil; the male flowers are generally reduced to a single stamen (*st* Fig. 45 *B*), while several carpels with funnel-shaped stigmas (*g*) are grouped together, and enclosed in a membranous cup (*m.c.*). This cup has been interpreted as a spathe enclosing a group of female flowers, each reduced to one carpel. The filament is at first very short, but elongates so as to rise above the pistils at anthesis. The anther dehisces and the pollen grains fall into the open mouths of the cornucopia-shaped stigma, and slide down the stylar canal, whose diameter is almost double that of the pollen grains. The descent of the pollen grains through the water is due to the fact that when they become ripe they are weighted with starch grains[1].

Owing to the air spaces in the pericarp wall, the achenes of some of the Potamogetons float for a time, before becoming waterlogged and sinking. The air-containing tissue of the pericarp in *P. perfoliatus*, and the cuticularised epidermal layer (*o.e.*), are shown in Fig. 46, p. 72.

The fruits of the Pondweeds, after becoming to all appearance ripe, often rest for a considerable period before germination[2], except in the case of *P. densus*, in which the achenes sprout a few days after they fall. But this species is rather remote from the rest of the genus in other respects, such as the opposite arrangement of the leaves, and the absence of the ligule. Sauvageau[3] has shown by experiment that in *P. crispus* it is the hard integument which delays germination; when the embryo is laid bare by the removal of part of the seed coat, sprouting rapidly occurs. The same author observed that when

[1] Roze, E. (1887).

[2] The delayed germination of aquatics in general is considered in Chapter XIX, p. 243.

[3] Sauvageau, C. (1894).

fifty fruits of *P. natans*, which had been gathered in September, 1889, were kept in water at the temperature of the laboratory, none germinated in 1890 or 1891, six germinated in 1892, and thirty in 1893, i.e. after lying dormant for three years and a half.

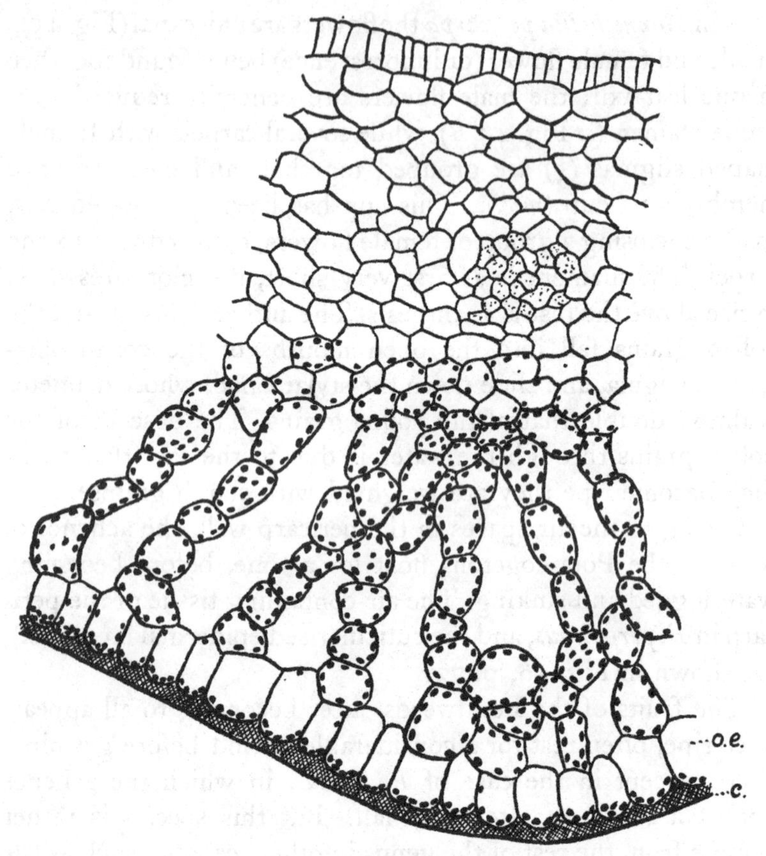

FIG. 46. *Potamogeton perfoliatus*, L. Transverse section of fruit wall to show air spaces in the outer region of the wall, and also the thick outer cell-wall of outer epidermis (*o.e.*). The cross-hatching indicates the non-cuticularised part of the wall: only the outermost surface layer, shown in black, is converted into cuticle (*c*). Chlorophyll grains in epidermis. (× 260.) [A. A.]

The most striking feature of the Potamogetonaceae, as a family, seems to be the remarkable reduction series shown by the vegetative and reproductive organs—the degree of reduction serving in general as a gauge for the degree of completeness with which the aquatic life has been adopted.

CHAPTER VI

THE LIFE-HISTORY OF THE LEMNACEAE[1] AND OF *PISTIA*

EACH of the families with which we have been concerned in the preceding chapters, has shown very great variation in vegetative structure associated with the differing degrees in which its members have adopted the aquatic habit. In the Lemnaceae, which we propose now to discuss, we have, on the other hand, a remarkably sophisticated and uniform group of plants, all of which pass their life floating at or near the surface of the water; the members of the family show, throughout their structure, a high degree of similarity to one another, and a marked difference from other aquatics. The Duckweeds have a very wide range, and occur almost as generally in the Tropics as in the northern countries where we know them so well[2].

In the Lemnaceae the modification of the vegetative body has been carried so far that the usual distinction between stem and leaf is no longer obviously maintained. The Duckweeds are not unique in this disregard of morphological categories—two other groups of water plants, the Utricularias and the Podostemaceae, carry this infringement of botanical conventions to an even more extreme point.

The little green fronds of the Duckweeds produce similar fronds of the second order, and also inflorescences of an extremely reduced type (Fig. 47, p. 74 and Fig. 50, p. 79) from pockets occurring on either side in the basal region. The nature of the fronds has been very variously interpreted. Hegelmaier[3],

[1] Hegelmaier, F. (1868) is still the classic monograph of this group. See also Schleiden, M. J. (1839) and Hegelmaier, F. (1871) and (1885).

[2] Kurz, S. (1867).

[3] Hegelmaier, F. (1868). For another view see Dutailly, G. (1878).

in his monograph of the Lemnaceae, treats them as stem organs which are modified to perform the work of leaves. Engler[1], on the other hand, follows van Horen[2] in interpreting the distal end of the frond as foliar, while the proximal end is axial. Yet a third view is that of Goebel[3] who expresses the opinion that the leaf-like organs of the Lemnaceae are actually leaves, pure and simple. He explains the origin of the lateral shoots of each generation from the base of the preceding one, by assuming that the base of each leaf has the power of functioning as a growing point. Undoubtedly Engler's view—which is based upon a comprehensive study of the Araceae, and a critical examination of *Pistia* and the Lemnaceae—may be accepted as the best founded. The present writer has recently carried Engler's comparison further, and has shown that the buds in the case of *Pistia* arise in minute pockets closely recalling those of the Duckweeds[4].

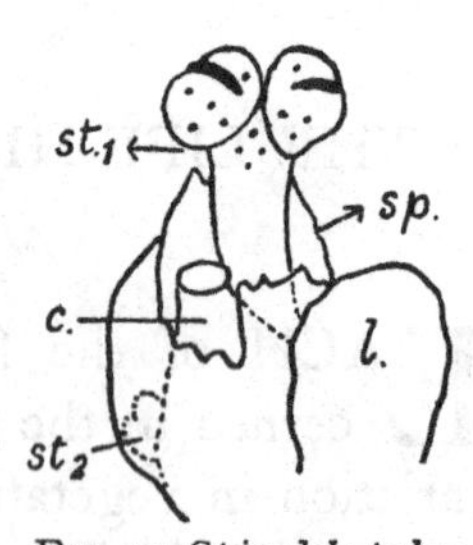

FIG. 47. *Spirodela polyrrhiza*, Schleid. Inflorescence. st_1 and st_2, male flowers reduced to stamens; *c.*, female flower reduced to a gynaeceum; *sp.*, spathe; *l.*, lateral shoot. [Hegelmaier, F. (1871).]

The three genera into which the family is divided—*Spirodela*, *Lemna* and *Wolffia*—are all represented in Britain. *Spirodela polyrrhiza*, Schleid.[5], is the largest member of the Lemnaceae; when it is growing vigorously its fronds attain to about $\frac{2}{5}$ of an inch both in length and breadth. Several roots with conspicuous root-caps hang from the underside of each frond. They are somewhat heavier than water and their tips are the heaviest part. It has been suggested that one of the functions of these roots may be to ensure the equilibrium of the plant[6]. *Spirodela* forms special shoots which outlast the winter.

[1] Engler, A. (1877). [2] Horen, F. van (1869).
[3] Goebel, K. (1891–1893). [4] Arber, A. (1919[4]).
[5] For a description of the very rare flowers of this species see Hegelmaier, F. (1871).
[6] Gasparini, quoted by Hegelmaier, F. (1868); Ludwig, F. in Kirchner, O. von, Loew, E. and Schröter, C. (1908, etc.).

Such turions are of great importance throughout the family, since the flowers are rare and relatively little seed is set. The winter-fronds of *Spirodela* are smaller than the summer ones and almost kidney-shaped. The air spaces in the tissues are reduced, and the cells are packed with starch, with the result that the fronds are heavier than water. The roots remain undeveloped. These winter-buds become detached from the parent frond in the autumn and sink to the bottom of the water. In the spring, a lateral frond begins to grow out; in so doing it absorbs the starch from the parent, and on this account, and also by development of air spaces, the whole body becomes lighter and rises to the surface[1]. The present writer has found that the rising of the winter-buds can be induced, as early as January, as a result of a few days in a warm room, even in a dim light. The time of year at which the turions begin to be formed is variable, and depends on external conditions. It has been shown by van Horen[2] that in shady places they develop very late or even fail altogether, whereas they occur early in bright sunlight, especially if the water is stagnant. Guppy[3], who has made a special study of the habits of the Lemnaceae, mentions that on one occasion he found a large number of plants of *Spirodela polyrrhiza* in a ditch, producing winter-buds, at the beginning of July, to an extent he had never seen before or since; the conditions were precisely those indicated by the previous observer as being favourable to the early occurrence of this phase —namely almost stagnant water which was brilliantly insolated. During the few weeks preceding the observation of the winter-buds, Guppy records that the surface was frequently heated in the day time to 80° Fahr. (nearly 27° C.). It is difficult to understand why conditions so favourable for vegetative growth should initiate turion formation, since in most water plants their production is induced by a state of poor nutrition. Possibly the explanation may lie in the great size of the winter-bud of the Lemnaceae in relation to the entire vegetative body of the

[1] Hegelmaier, F. (1868). [2] Horen, F. van (1869).
[3] Guppy, H. B. (1894[2]).

parent, when compared with the small proportion that the turions of other aquatics bear to the plant producing them. To synthesize enough starch to fill the cells of the winter-bud may be a considerable tax on the parent frond, and may only be possible under conditions peculiarly favourable for photosynthesis.

The commonest British Duckweed is *Lemna minor*, L.[1], which seems to be in some ways the least specialised, among our native species, for its particular mode of life. No definite turions are formed, and the plants are to be found swimming at the surface of the water at almost all seasons. When frozen, the older fronds become water-logged more readily than the younger ones, and they sink to the bottom, dragging down the young laterals with them.

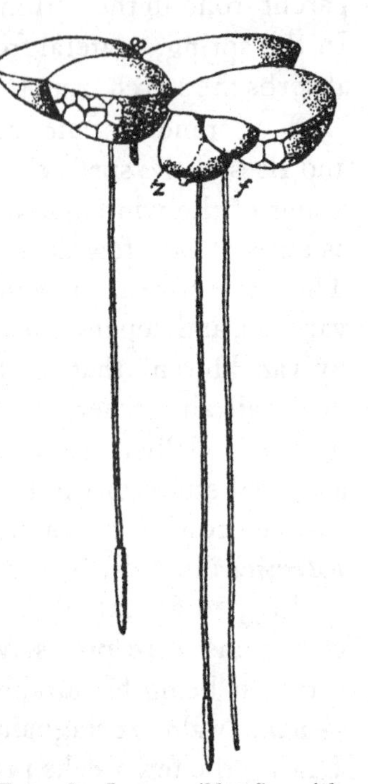

FIG. 48. *Lemna gibba*, L., with fruit, *f*. [Hegelmaier, F. (1868).]

Another species, *Lemna gibba*, L.[2], is notable for having the underside of the frond modified as a spongy aerenchyma—the gibbous form so produced giving the species its name (Fig. 48). The degree of development of the air tissue varies with the external conditions; the fronds are most conspicuously gibbous in running water where the insolation is moderate[3]. At certain periods of the life-

[1] On the flowering of *Lemna minor* see Brongniart, A. (1833) and Kalberlah, A. (1895); on the gametophytes and fertilisation, Caldwell O. W. (1899).

[2] On the flowers and seed of *Lemna gibba* see Micheli, P. A. (1729) and Brongniart, A. (1833); on the germination, Wilson, W. (1830).

[3] Horen, F. van (1869).

history, flat fronds are however produced and we owe to Guppy[1] the elucidation of the part played by the two types of shoot. He observed one hot summer, when *Lemna gibba* flowered profusely in July, that, during August, the gibbous plants gave rise to numerous thin, flat fronds of a dark green hue. These were the turions, and their appearance was accompanied by the death of a large number of the gibbous mother-plants, a result which this author attributes to exhaustion after flowering. Many of the gibbous plants, however, survived and continued to bud off winter-fronds except during the severest weather. Early in February the budding recommenced, but the gibbous character was not displayed until the weather became warmer. This author thinks that for the development of the gibbosity the plants require an average daily maximum temperature at the surface of the water, not much, if at all, under 70° Fahr. (21° C.). After cool summers when *Lemna gibba* does not flower, no flat winter-buds are formed, but the gibbous fronds survive until the next spring. One of the reasons for the relative rarity of *L. gibba*, as compared with *L. minor*, is probably that, as Guppy has shown, it requires a higher temperature than that needed by the Lesser Duckweed, both for initiation of budding in spring and for flowering. Under suitable conditions, however, it shows a wonderful vigour of vegetative growth. It has been recorded, for instance, that an area of water of about half an acre, which was edged on a certain date in June by a border of this plant a few feet wide, nineteen days later was thickly covered with the fronds over almost its entire surface[2].

The surface-living Duckweeds can survive for a time if stranded on the mud by the lowering of the water in which they grow, and in cultivation it has been found possible to establish land forms which can fulfil the whole cycle of normal vegetative development[3]. For instance, *Lemna minor* has been grown for as long as twenty months on wet mud, where it throve and budded at all seasons of the year. Two plants set apart in

[1] Guppy, H. B. (1894²). [2] Hegelmaier, F. (1868).
[3] Guppy, H. B. (1894²).

October had increased under these conditions to fifty in the course of a year. *Spirodela polyrrhiza* can also be cultivated on mud from the winter-buds through the summer phase to the winter-buds again.

The genus *Lemna* contains another British species which is more deeply committed to the water life than either *L. minor* or *L. gibba*. This is *L. trisulca*, L., the Ivy-leaved Duckweed, a submerged plant, floating beneath the surface level[1]. The fronds of *L. trisulca* are longer than those of the other Duckweeds and this elongation may be connected with the tempering of the light due to its passage through a layer of water. Its shoots form very decorative, symmetrical patterns, owing to the circumstance that branches of many different generations remain attached to one another (Fig. 49). This fact is probably to be associated with the relatively sheltered habitat of the Ivy-leaved Duckweed, as compared with *Lemna minor*, *L. gibba*, etc.[2]. These floating species are exposed to all the surface movements of the water—a fact which must encourage detachment. That it is the difference between floating and submerged life that determines the question of the fronds becoming isolated or remaining attached, is confirmed by the fact that the partially surface-floating, fertile fronds of *L. trisulca* (Fig. 50) tend more to separation. In these fertile fronds the basal part, which bears the inflorescence, floats on the surface, but the apical region dips down into the water[3]. The sterile fronds and the submerged part of the fertile fronds agree in having no stomates, whereas the floating part of the fertile frond bears stomates and approaches more closely in structure to the fronds of *Lemna minor* than do the submerged sterile shoots. The very simple vascular strands are dorsiventral with xylem above and phloem below; one vessel and one sieve tube form a characteristic combination[4] (Fig. 51).

[1] Clavaud, A. (1876) puts forward a theory concerning the cause of submergence in this species which seems to be quite unfounded.

[2] Schenck, H. (1885). [3] Hoffmann, J. F. (1840).

[4] Schenck, H. (1886).

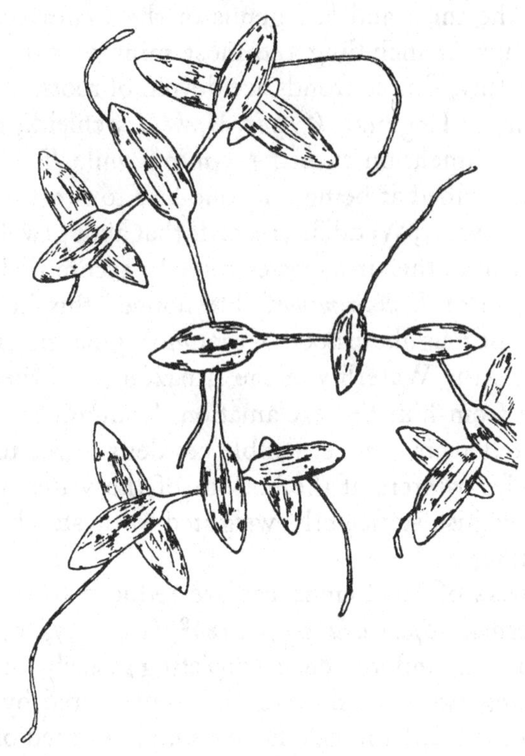

FIG. 49. *Lemna trisulca*, L. Habit drawing. (Slightly enlarged.) [Kirchner, O. von, Loew, E. and Schröter, C. (1908, etc.).]

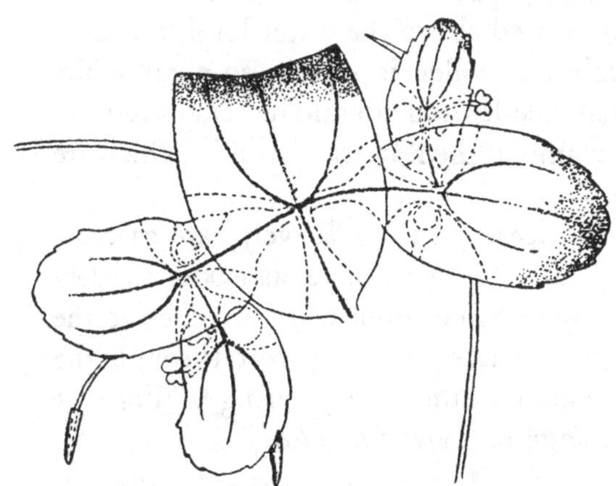

FIG. 50. *Lemna trisulca*, L. Flowering shoot. (Enlarged.) [Hegelmaier, F. (1868).]

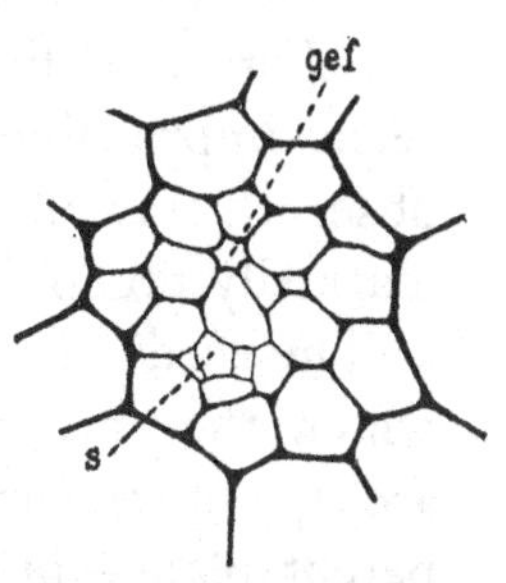

FIG. 51. *Lemna trisulca*, L. T.S. bundle from stalk of frond. One vessel (*gef*) and one sieve tube (*s*) with two companion cells. (×475.) [Schenck, H. (1886).]

Wolffia, the third and last genus of the Lemnaceae, enjoys the distinction of including the most minute of all flowering plants. The tiny, simple fronds are devoid of roots. The species which occurs in England, *Wolffia Michelii*, Schleid., has fronds which in no dimension exceed 1·5 mm., while *W. brasiliensis*, Wedd., is described as being only one-half to two-thirds of this size. Its discoverer, Weddell[1], records that about twelve flowering individuals of this tiny species could be accommodated upon a single frond of *Lemna minor*. He noticed this little *Wolffia* growing in the neighbourhood of that most gigantic of aquatics, *Victoria regia*, the Waterlily of the Amazons, and their propinquity drew from him the exclamation, "Singulière bizarrerie de la nature d'avoir semé ensemble ces deux végétaux!" Our native species winters at the bottom of the water, its minute fronds being just sufficiently weighted with starch grains to induce sinking.

The flowers of the Lemnaceae are reduced to the simplest possible terms. *Spirodela polyrrhiza*[2] (Fig. 47, p. 74), for instance, has an inflorescence consisting merely of a spathe (*sp.*) enclosing two male flowers each represented by a stamen only (st_1 and st_2) and a female flower simply formed of a gynaeceum (*c.*) with one or two ovules. *Lemna minor*[3], and probably other members of the family, appear to be pollinated by insects. The essential organs are raised above the water level, but they are short and stiff, while the pollen is scanty, so anemophily seems improbable. Small beetles and aquatic insects have been observed crawling about among the flowering fronds, which are markedly protandrous.

The seeds of the Lemnaceae, in the relatively rare cases in which they are produced, may germinate as soon as they are ripe in the summer—sometimes even while attached to the parent plant—but in other cases they may rest through the winter and defer germination until the spring[4]. Fig. 52 illustrates the seedling stage of *Lemna trisulca*.

[1] Weddell, H. A. (1849). [2] Hegelmaier, F. (1871).
[3] Ludwig, F. (1881). [4] Hegelmaier, F. (1868).

The extreme reduction and specialisation, which characterise the Lemnaceae, are united with great vigour and vitality. We have already alluded (p. 77) to a special case of the rapid power of vegetative reproduction shown by *Lemna gibba*, and the same capacity characterises other members of the family. Another remarkable trait of the Duckweeds is their power of

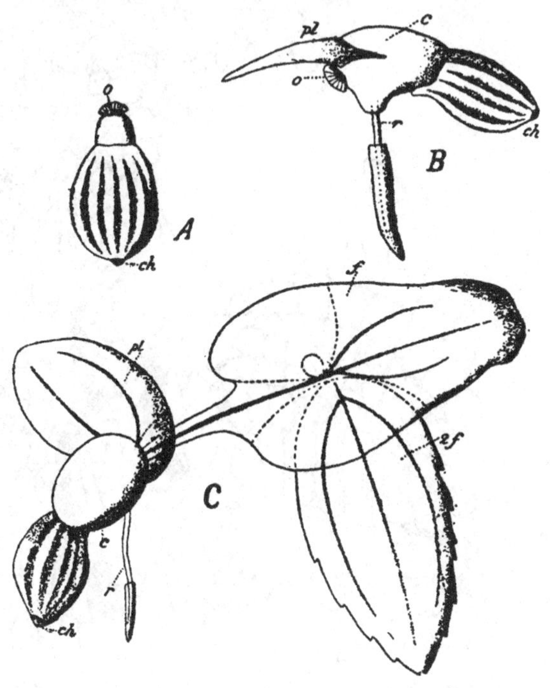

FIG. 52. *Lemna trisulca*, L. Germination. *A*, germinating seed with operculum (*o*) just coming away. *B*, seedling seen from the side. *C*, seedling further developed, seen from above. *ch* = chalaza, *c* = cotyledon, *pl* = plumule, *f* = lateral shoot from plumule, 2 *f* = secondary lateral shoot, *r* = radicle. (Enlarged.) [Hegelmaier, F. (1868).]

living and flourishing in water which is so full of organic impurities that no other Phanerogams can survive in it. If introduced into water with a bad smell, they will purify it until it is a fit habitation for small animals[1].

[1] Ludwig, F. in Kirchner, O. von, Loew, E. and Schröter, C. (1908, etc.); see also p. 287.

The Lemnaceae are generally regarded as related to the Aroids, so it may be well to conclude this chapter by a further reference to *Pistia Stratiotes*, L.[1], the River Lettuce of

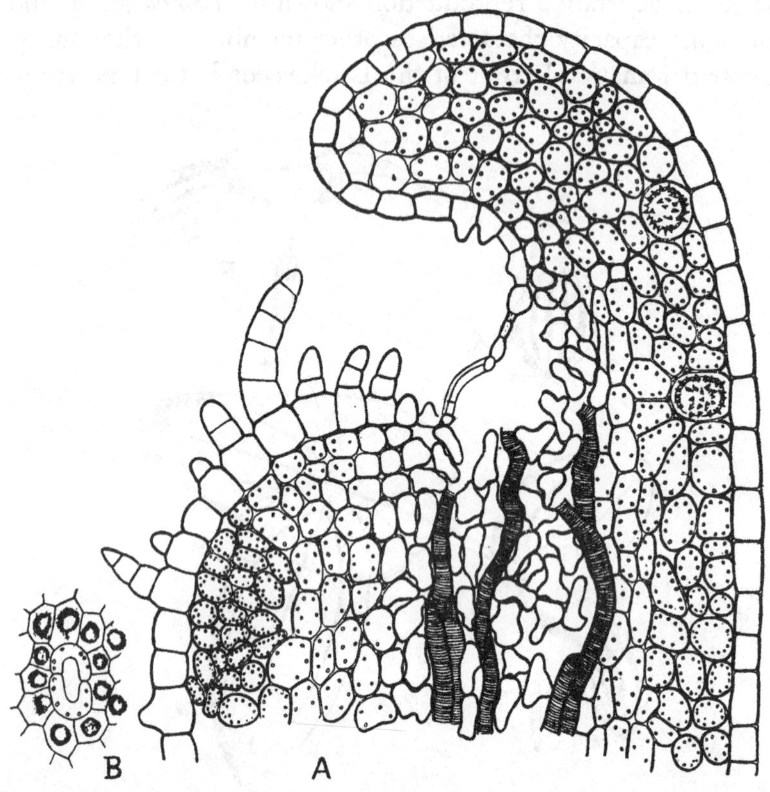

FIG. 53. *Pistia Stratiotes*, L. *A*, radial longitudinal section of leaf apex showing groove into which the water pores open and the space beneath them into which tracheids emerge. *B*, surface view of water pore. [Minden, M. von (1899).]

the Tropics—the member of the Araceae most nearly allied to the Duckweeds. This plant has a floating rosette of leaves, and multiplies by runners from which fresh rosettes arise. The lower side of each sessile leaf bears a swelling, which may reach the size of a pigeon's egg. This swelling consists of spongy air-containing tissues, and serves as a float. The upper

[1] On *Pistia* see Koch, K. (1852), Hofmeister, W. (1858), Engler, A. (1877) and Arber A. (1919[4]).

and lower leaf-surfaces are covered with minute depressed hairs, which prevent the leaves from being wetted[1]. Fig. 53 shows the apical opening, so often found in aquatics, through which water is eliminated from the leaf[2]. Like the Lemnaceae, *Pistia* represents a type which is singularly successful in the matter of vegetative growth. Its reproduction is so rapid that it sometimes chokes water-channels and proves a serious hindrance to navigation[3].

[1] Ito, T. (1899). [2] Minden, M. von (1899). See also p. 267.
[3] This subject is dealt with more fully in Chapter xvii, p. 213.

CHAPTER VII

THE LIFE-HISTORY OF *CERATOPHYLLUM*

EACH of those aquatic families whose life-histories we have hitherto considered, contains a considerable number of species, representing, in the case of the Lemnaceae, three genera, while, in the case of the other groups discussed, the number is much higher, as many as fourteen genera being included, for instance, in the Hydrocharitaceae. The family Ceratophyllaceae, the subject of the present chapter, offers a marked contrast on this point, since it includes only a single genus, containing three species, or, on other interpretations, one alone[1]. *Ceratophyllum*, the Hornwort, is extremely isolated in its structure and habits, so much so that there has been, at various times, the widest diversity of opinion as to the position which should be assigned to the family; the plant, from its taxonomic wanderings, has been opprobriously styled "a vegetable vagabond." The question of its affinities will be discussed in Chapter xxv.

In the genus *Ceratophyllum* the aquatic habit seems to have reached its ultimate expression. The plant not only lives entirely submerged throughout its vegetative life, but even its stigmas do not reach the surface, and the pollen is conveyed to them by the water[2]. The Hornwort is monoecious, the male flowers consisting of a group of stamens enclosed in a perianth of about a dozen members (*p* in Fig. 54 *B*). These stamens, when the flower is mature, become detached—the terminal expansion of the connective acting as a float[3]—and rise to the surface of the water. They then dehisce and the pollen, having a specific gravity very slightly higher than that of water, sinks gently,

[1] Schleiden, M. J. (1837).
[2] Delpino, F. and Ascherson, P. (1871).
[3] Ludwig, F. (1881).

and thus comes into contact with the stigmas[1]. This water-carriage of the pollen is the more striking, since the great majority of aquatic plants show a strong tendency to retain the aerial pollination mechanism of their terrestrial ancestors.

As regards vegetative structure, the most notable feature of the Hornwort is the entire absence of roots. The radicle

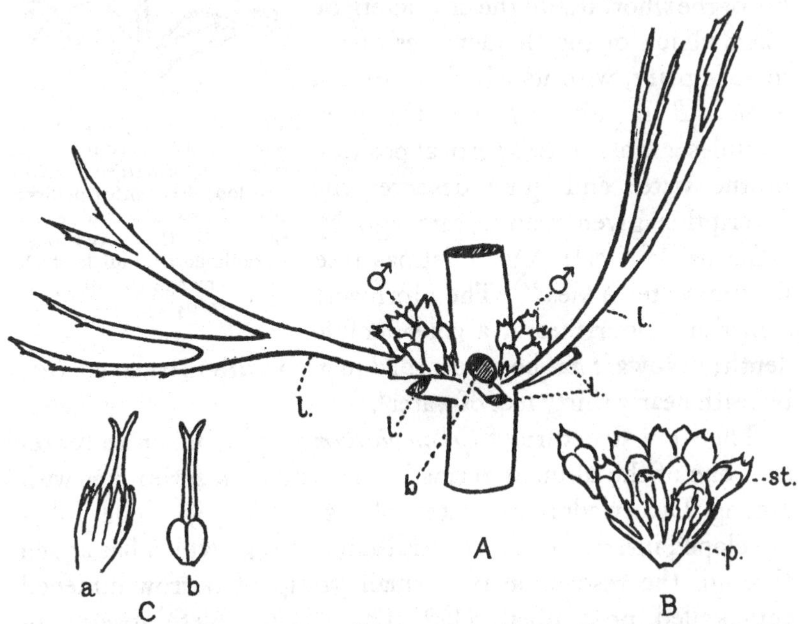

FIG. 54. *Ceratophyllum demersum*, L. *A*, node bearing two male flowers (♂) (Enlarged); a branch (*b*) and all the leaves but two (*l*) have been cut across. *B*, a single male flower on a larger scale; *p*, perianth of about 12 members; *st*, stamens. On the left, a stamen is in the act of being squeezed out. *C*, ♀ flower; *a*, showing perianth, style and stigmas; *b*, with perianth removed showing ovary. The stigma varies from being single to being sometimes much more deeply bifid than in *C*. [A. A.]

never develops beyond a rudimentary stage and no adventitious roots are produced. Fig. 55, p. 86, shows a seedling[2] with its reduced radicle (*r*). The seed germinates at the bottom of the water, the plantlet rising to the surface when it is about three inches long. The leaves of the first pair (*l*) are linear and decussate.

[1] Willdenow, C. L. (1806), Dutailly, G. (1892), Roze, E. (1892), Strasburger, E. (1902).

[2] Guppy, H. B. (1894[1]).

The forked leaves characteristic of the mature plant (*l* in Fig. 54 *A*, p. 85) are not formed immediately; they are preceded by a juvenile type which is simple and linear. It is not until the fourth node above the cotyledonary node that every member of the whorl attains the characteristic form. Each of the slender axes of the mature plant, with its whorls of forked leaves (*B* in Fig. 57, p. 89), often occupies a more or less vertical position in the water and quite deserves the description given many years ago by a German writer[1]: "A Christmas tree for tiny water nixies." The Hornwort sometimes flourishes at a considerable depth; in Iowa it has been recorded to grow with marked success beneath nearly thirty feet of water[2].

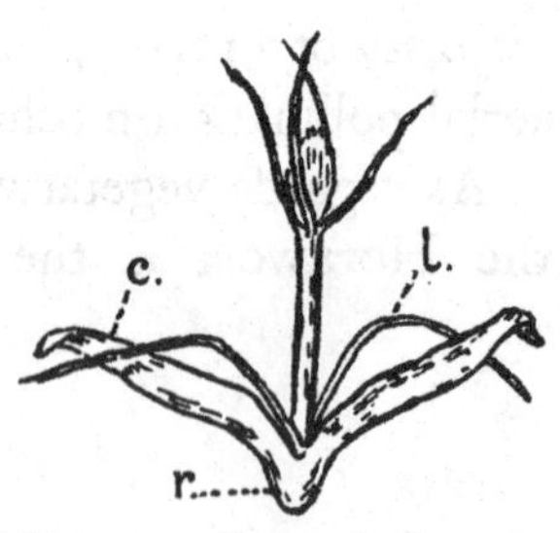

FIG. 55. *Ceratophyllum demersum*, L. Seedling one week old. (Enlarged.) *c*=cotyledon; *l*=member of first pair of leaves which decussate with the cotyledons; *r*=rudimentary radicle which never elongates. [Guppy, H. B. (1894[1]).]

The stem structure of *Ceratophyllum* may be taken to represent one of the ultimate terms in the reduction series met with among Dicotyledonous water plants (Fig. 56). The fully-developed internode has a central axial passage which has arisen through the resorption of a small group of narrow-lumened thin-walled procambial cells[3]. There is complete absence of lignification.

The water content of the plant is very high, representing 88 per cent. of the total weight[4], but as the young parts are cuticularised to a degree unusual in submerged plants, the texture of the shoots is less fragile than one might expect, and collapse does not occur so rapidly in a dry atmosphere as in the case of many hydrophytes. The curious mucilage-containing hairs borne by the leaves, stamens, etc., have been much discussed[5]. They seem to differ from the common mucilage hairs

[1] Schleiden, M. J. (1837). [2] Wylie, R. B. (1912).
[3] Sanio, C. (1865). [4] Schleiden, M. J. (1837).
[5] Göppert, H. R. (1848), Borodin, J. (1870), Strasburger, E. (1902).

of water plants in not excreting any slime, and their special function—if they possess one—remains a mystery.

It is characteristic of the Hornwort to occur sometimes in such great abundance that it drives out nearly all other competitors. It has been described, in the case of a certain Scottish loch, as so luxuriant that a boat could only be rowed through it with difficulty[1]. The present writer has seen it at Roslyn Pits, near Ely, at the beginning of October, in such quantity that the effect, on looking down into the water, was that of gazing into a pure forest of *Ceratophyllum*. The axis at this season of the year

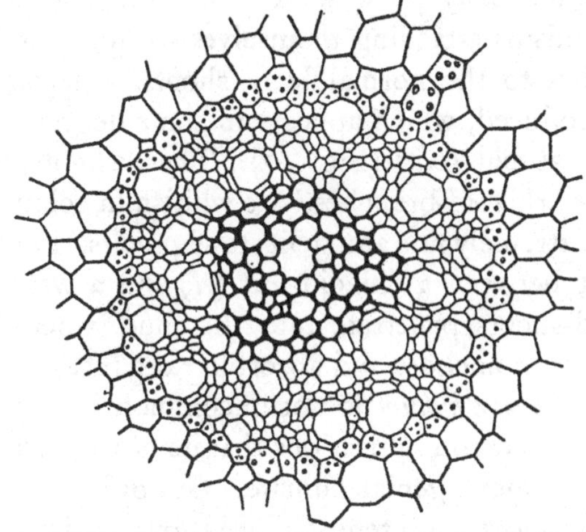

FIG. 56. *Ceratophyllum demersum*, L. Vascular cylinder of stem in T.S. Small xylem space in the centre; xylem parenchyma thickened; phloem zone well developed with large sieve tubes. (× 130.) [Schenck, H. (1886).]

is extremely brittle, snapping asunder at the slightest touch and thus giving rise to countless detached fragments capable of reproducing the plant. The apical regions of the shoots are more crowded with leaves and more deeply green than the rest of the plant, but are scarcely specialised enough to be called winter-buds[2]. During the cold season the stems remain at the

[1] West, G. (1910).

[2] The existence of these winter shoots was noted by Royer, C. (1881–1883); that the plant may vegetate throughout the winter was recorded by Irmisch, T. (1853).

bottom of the water, weighted down with a "living freight of aquatic molluscs, insects and annelids[1]." The young shoots formed in the spring, since they have not had time to become so ballasted, rise erect in the water. The stems of the previous year gradually decay away, and by the flowering time, in June or July, they have practically disappeared. The fact that the Hornwort, which has no surface layer of mucilage, becomes, to so remarkable an extent, an asylum for aquatic animals, may possibly be taken to afford some negative evidence for the theory that the mucilaginous coat, which is almost universal in hydrophytes, may have some value in preventing small foreign organisms attaching themselves to the plant's surface.

In addition to the normal leafy shoots, a second type of branch is produced, which appears in some degree to take the place of the absent roots (Fig. 57)[2]. These shoots, which are described as 'rhizoid-branches,' are whitish in colour and bear leaves with extremely fine and delicate segments. Fig. 58 shows the contrast between a rhizoid-leaf (*A*) and a water-leaf (*B*). The rhizoid-shoots penetrate into the mud, where they presumably serve as anchors and absorbing organs.

Although *Ceratophyllum* is not uncommon in northern latitudes, there are certain indications that its birth-place may have been in some more genial climate. Guppy[3] has shown, for instance, that a very high temperature is required for the maturation of the fruit. He noticed that in the drought of the hot summer of 1893, the ovaries ripened well in a shallow pond where the temperature of the water always rose above 80° Fahr. (27° C.) in the afternoons, and occasionally as high as 95° Fahr. (35° C.), while in the neighbouring waters, which were not so much overheated, no fruits were produced. Curiously enough, even in Fiji the fruit is only matured in the superheated waters of shallow pools, tanks and ditches[4]. Conversely, the vegetative organs cannot endure freezing, even for a period so brief as to be quite harmless to many other aquatics;

[1] Guppy, H. B. (1894[1]).
[2] Glück, H. (1906).
[3] Guppy, H. B. (1894[1]).
[4] Guppy, H. B. (1906).

Guppy found that the shoots were mostly killed by five or six days inclusion in ice.

FIG. 57. *Ceratophyllum demersum*, L. Part of an axis, *A*, which is lying on the soil and bears a normal leaf-whorl, *B*, and a rhizoid penetrating the soil. The lowest leaf-whorl of the rhizoid, *U*, has transition leaves. (Slightly enlarged.) [After Glück, H. (1906), *Wasser- und Sumpfgewächse*, Bd. II, Pl. VI, Fig. 76.]

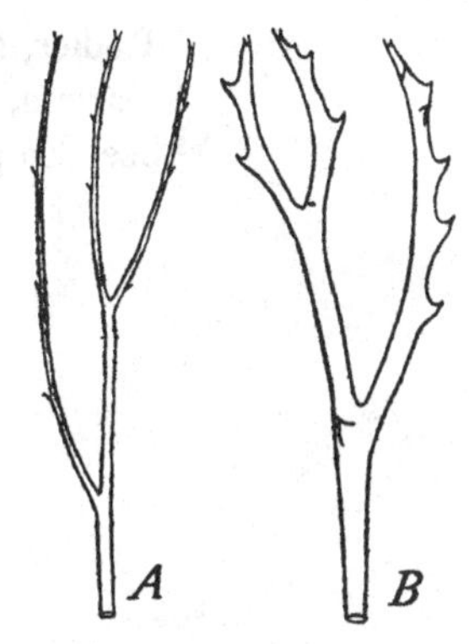

FIG. 58. *Ceratophyllum demersum*, L. *A*, single leaf of a rhizoid. *B*, single leaf of a water shoot (Enlarged.) [After Glück, H. (1906), *Wasser- und Sumpfgewächse*, Bd. II, Figs. 27 *A* and *B*, p. 195.]

The various peculiarities of structure and habit to which we have referred in the preceding pages, are not the only singularities exhibited by *Ceratophyllum*. In 1877 a French observer, Rodier[1], recorded the existence of certain spontaneous movements which characterise the shoots of this plant. He noted that the shoot moved in one direction for six hours, and then returned for another six—then moved for four hours in the opposite direction, and in another four hours returned again to its original position. Darwin[2] drew attention to certain obscurities in Rodier's description, but no more recent work appears to have been done on the subject; the movements of *Ceratophyllum* might repay further investigation[3].

[1] Rodier, E. (1877[1]) and (1877[2]).
[2] Darwin, C. and F. (1880).
[3] See also p. 281.

CHAPTER VIII

THE LIFE-HISTORY OF THE AQUATIC UTRICULARIAS AND OF *ALDROVANDIA*

OF all our native aquatics, the Bladderworts (*Utricularia*) diverge most in their vegetative characters from ordinary terrestrial plants. When not in flower, they live wholly submerged. Roots are entirely absent and the plant consists of an elongated branching axis producing delicate, finely-divided leaves on which small utricles are borne. This is not, however, the only type of vegetative body represented in the genus. Outside Europe there are a number of terrestrial species in which entire leaves of a simple type are produced in addition to bladder-bearing organs. The family to which the genus belongs—Lentibulariaceae—consists chiefly of aquatic and marsh plants; it is probable that the water Utricularias, with which alone we are concerned in this chapter, are the descendants of marsh forms, which, in the course of evolution, have become more and more completely involved in aquatic life[1]. It is impossible to draw a sharp line within the genus between the land and water types; the terrestrial species sometimes produce water forms, and the aquatic species can, to a limited extent, take to life on land. Even among our native Bladderworts, we find that, though *Utricularia vulgaris* cannot live except as a submerged plant, *U. minor* and *U. intermedia* are able, on rare occasions, to produce land forms[2], which are so far adapted to aerial life as to develop stomates—but in this condition they do not flower. The land form of *U. minor* is said to grow as a close moss-like turf.

The little utricles borne by the leaves (Fig. 59, p. 92), which give the Bladderworts their unique appearance, and to which

[1] Goebel, K. (1891–1893).

[2] Glück, H. (1906) and Luetzelburg, P. von (1910).

they owe both their Latin and their English names, are hollow structures with a small apical aperture, closed by a flap serving as

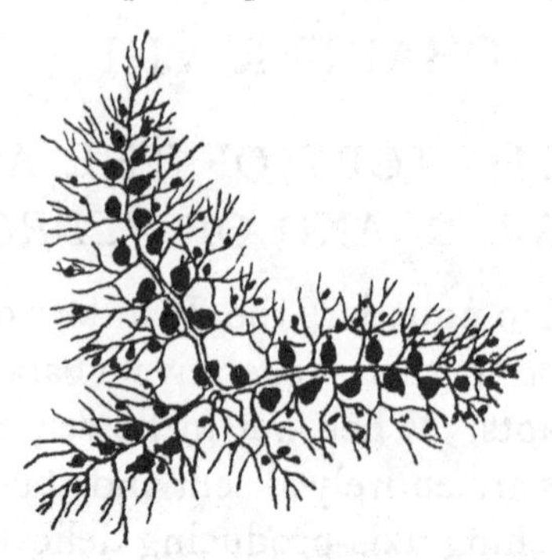

FIG. 59. *Utricularia neglecta*, Lehm. A single trifid leaf with bladders. (Slightly reduced.) [Adapted from Glück, H. (1906), *Wasser- und Sumpfgewächse*, Bd. II, Pl. II, Fig. 15 *b*.]

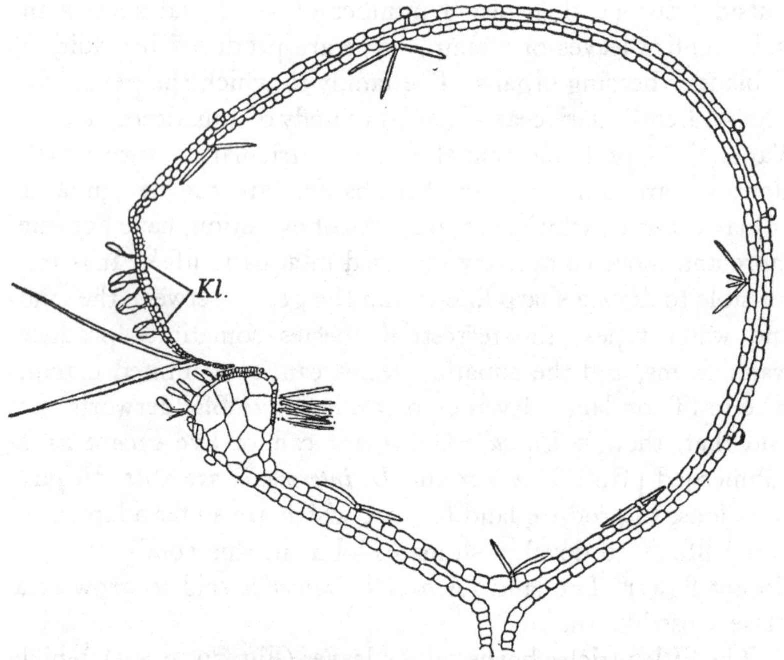

FIG. 60. *Utricularia flexuosa*, Vahl. Longitudinal section through a bladder. (Enlarged.) *Kl.* = valve. [Goebel, K. (1891–1893).]

a valve. Fig. 60 represents a section of the utricle of *U. flexuosa*, a species which plays a part in India corresponding to that of

U. vulgaris in Europe. Darwin[1] describes the valve of *Utricularia neglecta* as attached on all sides to the bladder, excepting by its posterior margin, which is free and forms one edge of the slit-like orifice. This margin is sharp, thin and smooth, and rests on the edge of a rim or collar which projects into the interior of the bladder. The collar obstructs any outward movement, with the result that the valve can only open inwards. The function of the bladders was for a long time in dispute. Certain ingenious but mistaken theorisers regarded the little four-armed hairs (Fig. 61), which occur within the bladders, as root-hairs, and supposed that the bladders existed in order to protect these delicate organs from the direct action of light and the depredations of Crustacea[2]! On a more plausible view, it was maintained that the bladders were to be interpreted as floats, which buoyed up the plant in the water. This idea has been discounted, however, since many terrestrial Utricularias produce large numbers of bladders; moreover it has been shown that the Utricularias do not sink when all the bladders are removed[3].

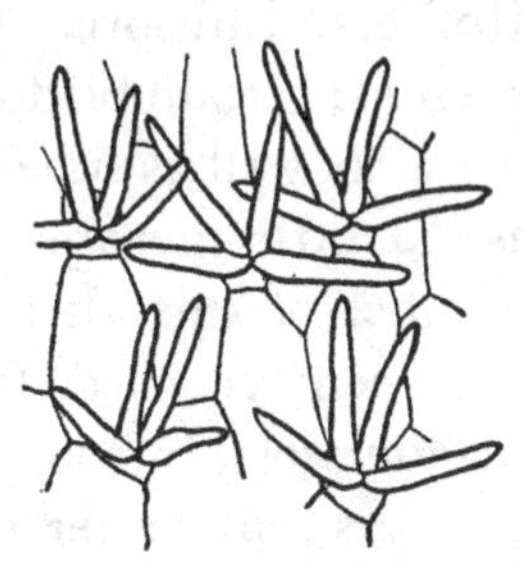

FIG. 61. *Utricularia Bremii*, Heer. Glands from the interior of a bladder. [Meierhofer, H. (1902).]

A third hypothesis now holds the field—namely, that the bladders act as traps for small animals which serve as food for the plant; this theory may now be considered to be fully proved. Before the middle of the last century, Treviranus[4] had recorded the finding of a beetle and some small snails in the bladder of a terrestrial *Utricularia* (*U. Hookeri*) and had suggested the comparison between these organs and the pitchers of *Sarracenia*, *Nepenthes*, and other carnivorous plants. But it was not until 1875 that the fact that our native Utricularias preyed on small animals was definitely proved. In this year Cohn[5]

[1] Darwin, C. (1875).

[2] Crouan (Frères) (1858).

[3] Darwin, C. (1875), Büsgen, M. (1888), Goebel, K. (1889²) and (1891–1893).

[4] Treviranus, L. C. (1848¹).

[5] Cohn, F. (1875). See also Darwin, C. (1875).

showed that in herbarium specimens of *Utricularia vulgaris* the bladders often contained skeletal tissues of Crustacea and insect larvae. He then tried the experiment of putting a living shoot of this plant, which had empty utricles, into water rich in *Cypris*; next morning nearly all the bladders contained Crustacea, swimming about in a restless manner but unable to escape. Rotifers, Infusoria, Rhizopods and other animals were also present; certain bladders containing as many as six living Crustacea, as well as other animals, were described by the observer as "a little menagerie of the microscopic water fauna." The number of animals secured may sometimes be very great. It has been recorded, for instance, that a plant of the Common Bladderwort, introduced into water rich in Daphnidae, in one case was found after 1½ hours to have caught as many as twelve of these little Crustacea in a single bladder[1]. Another plant, which was about 15 cms. long, and bore fifteen fully developed leaves, each with about six bladders, is reckoned to have entrapped at one time as many as 270 individuals of *Chydorus sphaericus*[1]. It is a curious fact that different species of *Utricularia*, even when growing associated in the same water, may, owing to some slight difference of habit, catch quite different animals. In one case Goebel[2] observed *U. intermedia* and *U. vulgaris* growing together, but while *U. intermedia* had caught chiefly *Cypris*, *U. vulgaris* had caught only Copepods. This is to be explained by the fact that *U. intermedia*, being anchored at the bottom of the water, was only able to secure the *Cypris*, which is a creeping form, while the Copepods, because they were free-swimming, were entrapped by the bladders of the unattached *U. vulgaris*. The animals are said to be attracted by edible mucilage secreted by the hairs which grow on the bladders of the Utricularias (Fig. 62), and especially on the valve at the aperture[2].

The observations which we have enumerated and many others which might be cited, leave no room for doubt that the

[1] Büsgen, M. (1888).
[2] Goebel, K. (1891–1893).

Utricularias do, as a matter of fact, catch animals in their utricles, but the questions still remain whether the absorption of organic material actually takes place, and, if so, whether the carnivorous habit is of definite benefit to the plant. The inner epidermis of the bladders is cuticularised except as regards the four-armed hairs (Fig. 61, p. 93) which are thin-walled. These hairs, in the case of a bladder enclosing decaying animals, have been seen to include oil-drops, which may be presumed to be derived from the animal tissues, since the hairs in a bladder which had received no food, showed no such drops[1]. Experimental work has also demonstrated that treatment with ammonium nitrate, etc., produces changes in the hairs which suggest that absorption has occurred[2]. These observations would not be sufficient in themselves to prove that the entrapped animals serve as a source of food for the plant, but a demonstration of this point was supplied by certain comparative cultures of Utricularias growing in water with or without animal life. From the upshot of these experiments it appeared that the plants deprived of animal food only showed about one-half of the growth of those that were allowed to catch their prey in the normal way[3]. A further problem which presents some difficulty is that of the causes which bring about the death and absorption of the entrapped animals. No highly poisonous substance can be present in the bladders, since the imprisoned animals

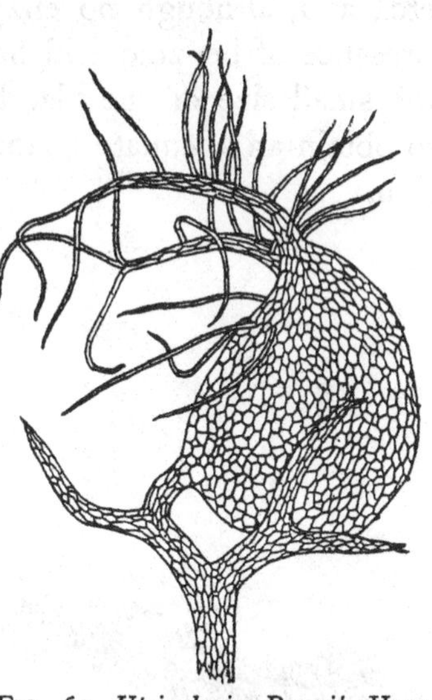

FIG. 62. *Utricularia Bremii*, Heer. Part of leaf with bladder. (Enlarged.) [Meierhofer, H. (1902).]

[1] Goebel, K. (1891–1893). [2] Darwin, C. (1875).

[3] Büsgen, M. (1888). See also Darwin, C. (1888), footnote to p. 365.

may remain alive in them for some days[1]. There is no doubt that the bladders are capable of digesting small animals, algae, etc., and, although no enzyme has yet been recognised, the presence of benzoic acid has been demonstrated[2]. Owing to the small size of the bladders, it must obviously be difficult to obtain an adequate quantity of the secretions for investigation.

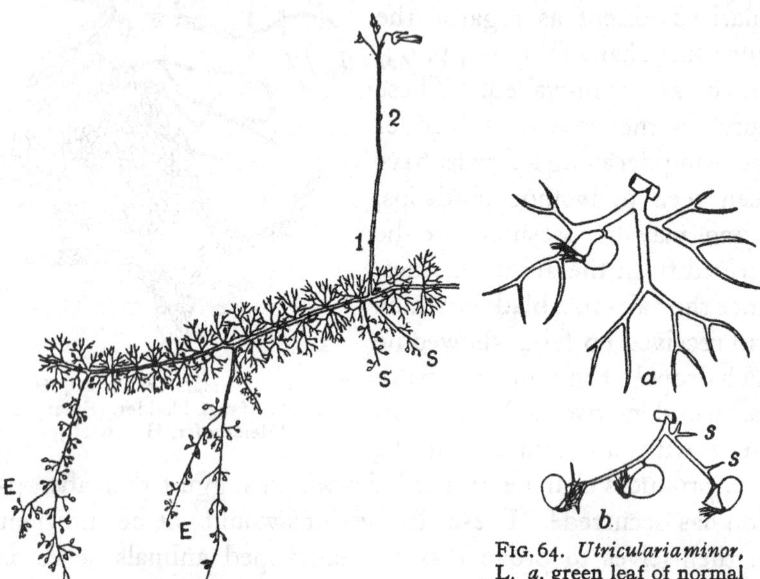

FIG. 63. *Utricularia minor*, L. Part of a shallow-water plant, *E*=earth-shoot. Two branches marked *S* at the base of the inflorescence axis have been cut off. 1 and 2 = bracts on the inflorescence axis. (Reduced.) [Modified from Glück, H. (1906), *Wasser- und Sumpfgewächse*, Bd. II, Pl. II, Fig. 18.]

FIG. 64. *Utricularia minor*, L. *a*, green leaf of normal submerged shoot; *b*, colourless leaf of an earth-shoot. In the latter the leaf segments are reduced to rudiments indicated by *S*. (Enlarged.) [After Glück, H. (1906), *Wasser- und Sumpfgewächse*, Bd. II, Figs. 2 *a* and *b*, p. 42.]

Besides the normal leafy branches, which serve for assimilation and also bear bladders, no less than three modified types of vegetative shoot are borne by certain of the European Utricularias—the 'earth-shoot,' the breathing shoot or 'air-shoot,' and the so-called 'rhizoid[3].'

[1] Cohn, F. (1875). [2] Luetzelburg, P. von (1910).
[3] Goebel, K. (1891–1893) and Glück, H. (1906).

In *Utricularia minor*, *Bremii*, *intermedia*[1] and *ochroleuca*, certain shoots are formed which bear bladders on leaves of a reduced type (*E* in Fig. 63, and *b* in Fig. 64). These branches, which are known as 'earth-shoots,' penetrate the mud at the bottom of the water and apparently serve for purposes of anchorage, and for the absorption of raw food materials. They have retained their power of entrapping small animals, but have substituted the functions characteristic of roots for the assimilatory activities of the water-shoots. The bladders make such efficient hold-fasts that, unless the soil be very soft, it is difficult to pull the earth-shoots out of the substratum without snapping the leaves and leaving the bladders behind. Every transition can be observed between earth- and water-shoots.

The British species of *Utricularia* which produce 'earth-shoots' never show the second form of modification, the 'air-shoot' (*L* in Fig. 65, p. 98), which occurs only in *U. vulgaris* and in the closely allied *U. neglecta*. These curious organs were observed by Pringsheim[2], who did not, however, understand their nature, but called them 'Ranken' (tendrils). It is to Goebel[3] that we owe a very plausible suggestion as to their biological value, and to Glück[4] a definite view as to their morphological status. They are, apparently, reduced inflorescences, and their function is said to be to serve as breathing organs and to connect the submerged vegetative body of the plant with the atmospheric air. In the case of *Utricularia vulgaris*, the air-shoots are fine, whitish, thread-like bodies, some centimetres long. They bear very small undivided leaves, closely appressed to the shoot and with stomates on their outer surfaces. The lower internodes are much elongated. The tips reach the water surface and protrude from it into the air, where the stomates can perform their usual function. The 'air-shoots' are said to occur especially

[1] Benjamin, L. (1848) described *U. intermedia* as 'rooted,' so it is evident that he had observed the 'earth-shoots,' though mistaking their morphological nature.

[2] Pringsheim, N. (1869).

[3] Goebel, K. (1891–1893).

[4] Glück, H. (1906).

when the plants are growing in a thick tangle—that is to say under circumstances in which the oxygen starvation, to which submerged plants are liable, must be particularly acute.

For the third type of modified shoot, the misleading term 'rhizoid' has been used; this name would have been more fitly

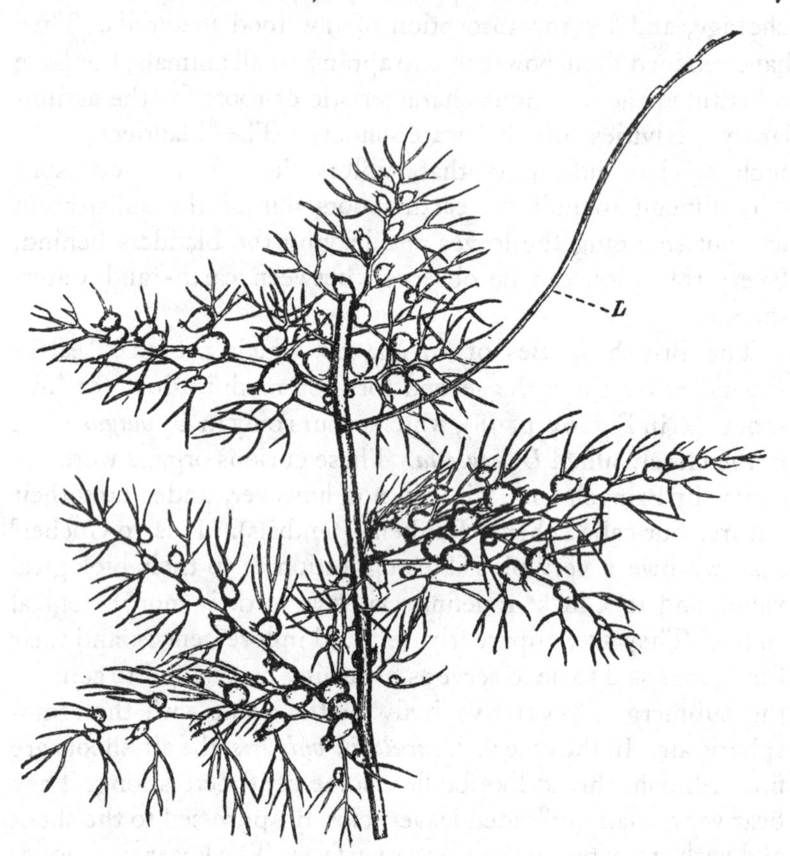

FIG. 65. *Utricularia vulgaris*, L. Part of shoot with bladder-bearing leaves, and an air-shoot (*L*). (Enlarged.) [Goebel, K. (1891–1893).]

applied to the 'earth-shoots,' since in function they approximate to roots, and to the 'rhizoid shoots' of *Ceratophyllum*[1]. The 'rhizoids' are developed at the base of the inflorescence in certain species of *Utricularia* (*R* in Fig. 66). They bear no

[1] See pp. 88 and 89.

bladders, but their leaves are highly glandular and often bent in a claw-like fashion[1]. They are firmer than the ordinary shoots and do not collapse when lifted from the water. Their function is obscure, but it seems possible that they play some part in holding the inflorescence erect. The Utricularias evidently have a strong tendency towards the production of specialised shoots below the aerial part of the flowering axis. Certain extra-European members of the genus (*U. stellaris*, *U. inflexa* and *U. inflata*, Fig. 150, p. 229) have a wreath of air-containing organs surrounding the base of the inflorescence, and undoubtedly serving to keep it erect in the water[2]. A vivid description is given by Spruce[3], in his account of his travels on the Amazons, of a similar arrangement in *U. quinqueradiata*. This is a small species with the usual submersed, finely divided leaves bearing numerous bladders, but the flower-stalk, which is about two inches high, has, midway, a large involucre of five horizontal rays resembling the spokes of a wheel. This floats on the surface and keeps the stalk always erect, and the solitary flower well out of the water, "the whole recalling a floating night-lamp, especially as the large yellow flower may be considered to represent the flame."

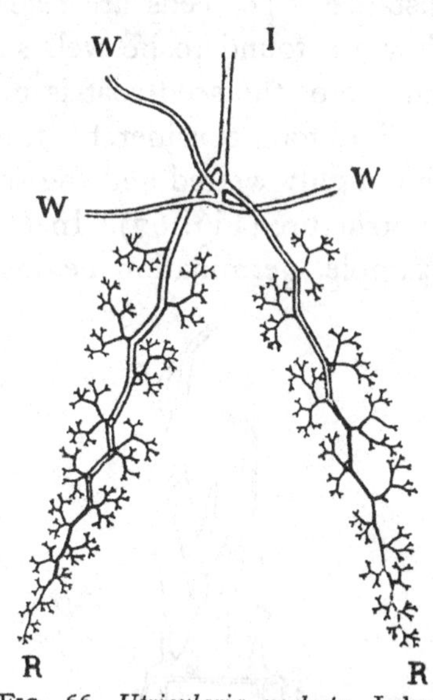

FIG. 66. *Utricularia neglecta*, Lehm. Base of inflorescence axis, *I*, with two 'rhizoids,' *R*. Three water-shoots, *W*, cut away for simplicity. (Slightly reduced.) [After Glück, H. (1906), *Wasser- und Sumpfgewächse*, Bd. II, Pl. IV, Fig. 34 *a*.]

Reproduction by seed appears to be less important among

[1] Goebel, K. (1889[2]).

[2] Benjamin, L. (1848), Treviranus, L. C. (1848[1]) and Wight, R. (1849).

[3] Spruce, R. (1908).

the Utricularias than the method of asexual propagation shortly to be described. In the case of *Utricularia minor*, for instance, ripe seeds are seldom obtained. When they occur, they are found to be well suited to floating on water, as the surface of the seed-coat is pitted and capable of retaining air bubbles for a considerable time[1]. Eventually the testa becomes thoroughly wetted and the seed sinks. The seedling is unique in structure (Fig. 67). In *U. vulgaris*, which may serve as an example, germination begins in spring at the bottom of the

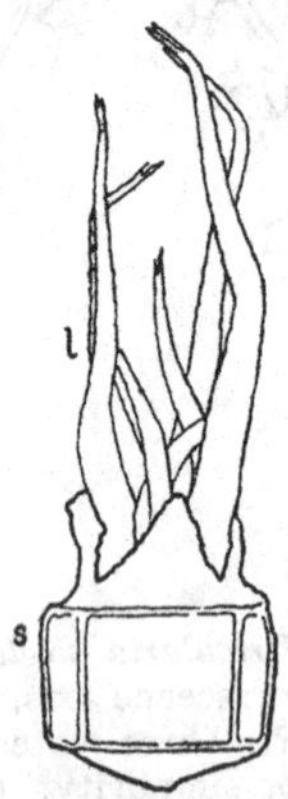

FIG. 67. *Utricularia vulgaris*, L. Geminating seed; *s*, seed coat; *l*, primary leaves. (× about 19.) [Adapted from Kamieński, F. (1877).]

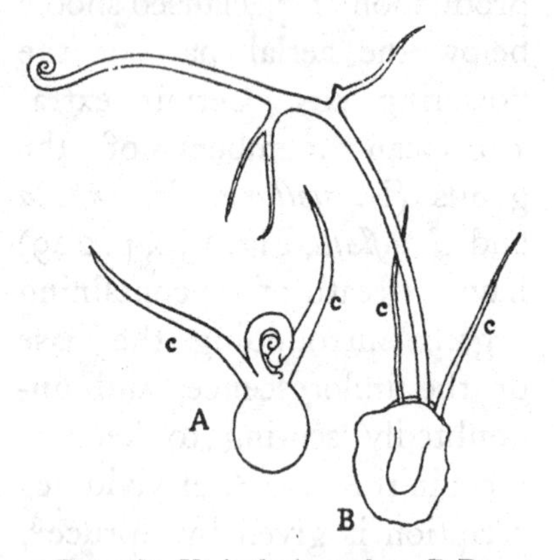

FIG. 68. *Utricularia exoleta*, R.Br. *A* and *B*, stages in germination; *c* = ? cotyledons. In *A* the seed-coat is removed. [Goebel, K. (1891).]

water. The following organs are produced[2]—a number (6–12) of simple primary leaves (*l* in Fig. 67), a bladder, a conical stem apex, from which the main axis develops laterally, and an adventitious shoot (? an air-shoot). No root appears in the seedling, and there is not even any rudiment of this organ in the embryo[3]. In *Utricularia exoleta*[4], a small and simple aquatic form found in Asia and tropical Australia, only two primary leaves (? cotyledons) are formed, but this is perhaps to be interpreted as a case of reduction (Fig. 68).

[1] Meister, F. (1900).
[2] Warming, E. (1874) and Kamieński, F. (1877).
[3] Merz, M. (1897). [4] Goebel, K. (1891).

The Bladderworts are able to reproduce themselves successfully for long periods without having recourse to flowering and fruiting. *Utricularia intermedia*, for instance, was observed in a certain district in Germany to propagate itself for years by vegetative means, when the ditches in which it lived were cleared too frequently to give it an opportunity of flowering[1]. The organs of vegetative reproduction—the turions or winter-buds[2]—are spherical or egg-shaped bodies developed at the ends of the shoots. The case of *Utricularia vulgaris* may be taken as typical. In this species turion formation takes place, in normal circumstances, between the beginning of August and the middle of November. The apical region of the shoot produces a number of reduced leaves separated by highly abbreviated internodes. The concave leaves cover one another in imbricate fashion and are closely packed into a firm ball, clothed with a protective layer of mucilage. When the plant is grown in an aquarium, water-snails are its chief enemies, but the winter-buds, with their coat of hairs and slime, seem immune from the depredations of these creatures[3]. The parent plant sinks to the bottom in the autumn, owing to its tissues becoming water-logged, and carries the turions with it. These, in spite of their firm texture, are lighter than water, and, but for their attachment to the decaying axis, would rise to the surface like pieces of cork. As it is, they remain all through the winter stationary at the bottom, but with their apices directed upwards. In the spring, the turion is at last able to rise to the surface—the parent axis having been reduced by months of rotting to little more than a string-like vascular cylinder, which often adheres persistently to the base of the winter-bud. The axis of the turion elongates with remarkable rapidity, attaining three to six times its original length. The composition of the bud then becomes manifest; a number of bud-scales occur at the base, followed by several transition leaves and then normal foliage leaves which

[1] Schultz, F. (1873).
[2] Benjamin, L. (1848), Glück, H. (1906), etc.
[3] Meister, F. (1900).

receive additions by the apical growth of the germinating turion. The bud-scales resemble reduced foliage leaves, but are specially suited to be protective organs. They are firmer than the other leaves and do not collapse on removal from the water. They are also less subdivided, and bear a more conspicuous development of hairs on their terminal segments—the hairs of the successive leaves amounting, indeed, to a protective felt—so that altogether they form an effective envelope for the bud. In *Utricularia minor*, though the hairs are absent, a similar result is obtained by the leathery texture of the bud-scale and by its form, which is less divided than that of *U. vulgaris*. The contrast between the foliage leaf and bud-scale of *U. minor* is shown in Fig. 69 *a* and *b*. In *U. intermedia* the turion generally becomes free before the winter, and swims among the shore plants instead of spending the dead season at the bottom of the water. The fact that the turion is protected by an especially thick coat of hairs, probably permits it to lead this more exposed existence[1]. Figs. 143 *A* and 143 *B*, p. 220, show the bud-scale and normal leaf of this species.

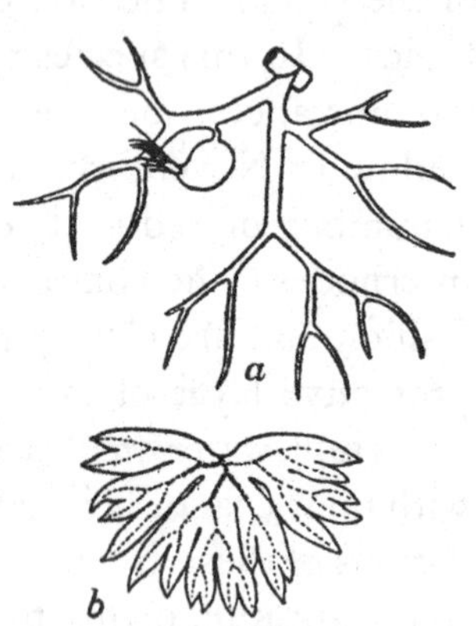

Fig. 69. *Utricularia minor*, L. *a*, normal leaf of the shallow water form, with a bladder; *b*, leaf belonging to a turion. (Enlarged.) [After Glück, H. (1906), *Wasser- und Sumpfgewächse*, Bd. II, Figs. 14*a* and *b*, p. 117.]

Though, under normal conditions, the turions are only formed in the autumn, and carry the plant over the winter season, their formation can be induced at any period of the year by conditions of poor nutrition. In certain experiments made a few years ago[2], some turions of *Utricularia minor*, germinated under starvation conditions on sand, after seventeen days had produced plants 14 cms. long. These were transferred to a culture solution, and after five days, when they had had time to become

[1] Schenck, H. (1885). [2] Luetzelburg, P. von (1910).

vigorous, they were returned to the sand. By the end of twenty-seven days they had formed turions. These were cut off, and the same alternation of sand culture and nutritive solutions was repeated three times. Each time the effect of the starvation culture was to induce the formation of turions, so that the plant went through the entire vegetative cycle, culminating in 'winter' buds, no less than four times between May and the middle of December! The last turions produced were only the size of a pin's head.

In the preceding pages we have, for convenience, used the terms 'shoot' and 'leaf' for descriptive purposes, but it now remains to consider how far current morphological conceptions can be applied to so anomalous a genus as *Utricularia*. There has probably been more controversy about the morphological nature of the different organs of these plants, than about such problems in the case of any other Angiosperm. It is not proposed here to enter into the details of the discussion[1] which seems to have been singularly fruitless. In the upshot, the main point, which emerges from a study of the literature, is that in this genus the distinction habitually drawn by botanists between stem and leaf, breaks down completely. The bladder is probably best interpreted as a modification of part of the "leaf[2]," but even if this be conceded it does not carry us far, since the nature of the "leaf" itself still stands in dispute. By some authors, the entire vegetative body, apart from the inflorescence axis, has been regarded as a root system, while others view it either as wholly axial or as consisting of stem and leaves. A view which has received considerable prominence, is that the entire plant is a much divided leaf[3], but if this be so, it must, as Goebel has pointed out, be admitted that this "leaf" possesses many characters which we are accustomed to

[1] For an historical survey of the literature, see Goebel, K. (1891) and Glück, H. (1906).

[2] Meierhofer, H. (1902). Another interpretation is illustrated in Fig. 72 *B*, p. 106.

[3] Kamieński, F. (1877).

attribute to stems alone, viz. long continued apical growth[1], as well as power of bearing leaves and axillary branches and of developing in more than one plane[2]. The fact that adventitious shoots are produced on the leaves of other Lentibulariaceae is, however, favourable to this view[3]. The unique plasticity of the Utricularias is indicated by the many observations on regeneration phenomena in the genus, which show that almost any part

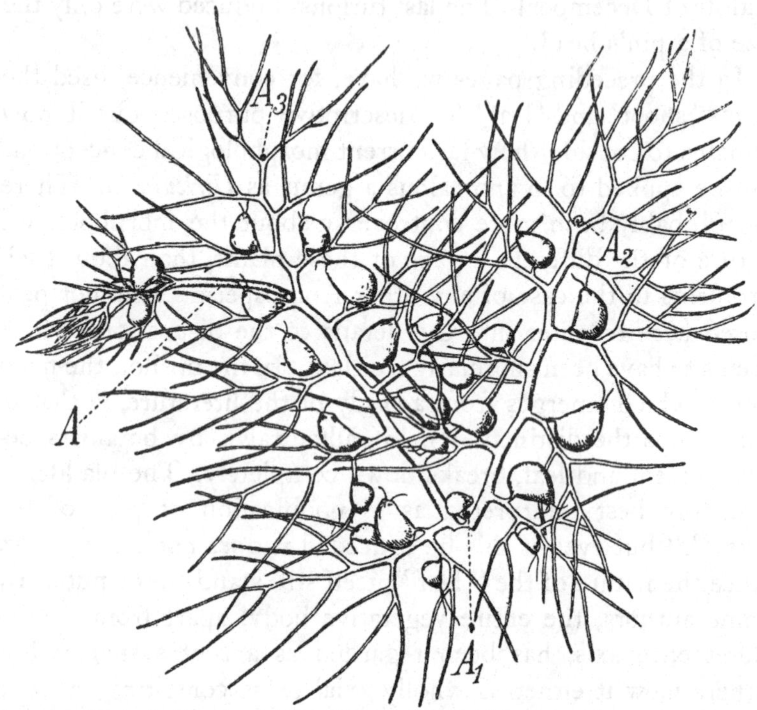

FIG. 70. *Utricularia vulgaris*, L. Detached leaf with four adventitious shoots, A, A_1, A_2, A_3. (Enlarged.) [Goebel, K. (1904).]

of these plants can produce new shoots at will. For instance, in *U. neglecta*, detached leaves, or leaves connected with a dying axis, can produce adventitious shoots which arise endogenously at the points of forking of the leaves, or, more rarely, from the stalks of the bladders[4]. Fig. 70 represents a case in which four

[1] Hovelacque, M. (1888).
[2] Goebel, K. (1891).
[3] Goebel, K. (1904).
[4] Glück, H. (1906).

shoots (A, A_1, A_2, A_3) arose from a leaf of *U. vulgaris*. Again, the inflorescences of various species, if cut off and immersed in a culture solution, have been seen to give rise to lateral shoots from the axils of their scale leaves. These branches may occur in extraordinary abundance: in *Utricularia vulgaris* as many as

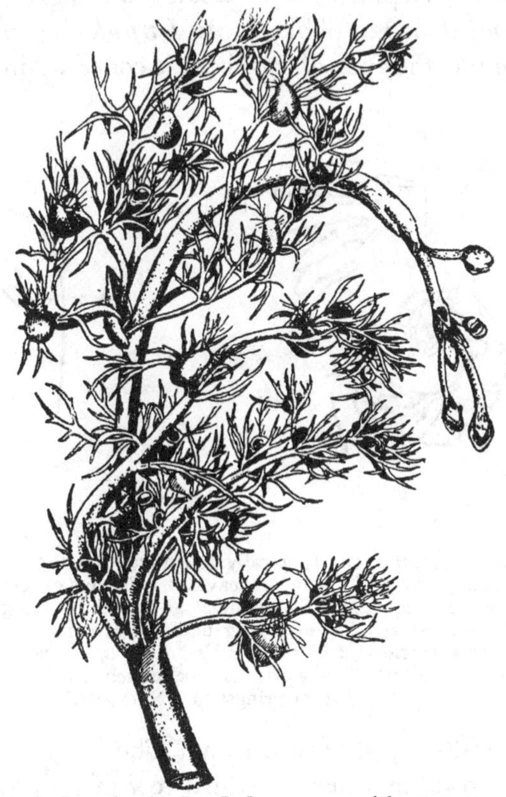

FIG. 71. *Utricularia vulgaris*, L. Inflorescence with numerous lateral shoots arising in axils of scale leaves on inflorescence axis, after 47 days culture under water on peat, and, later, with the addition of a culture solution. (Enlarged.) [Luetzelburg, P. von (1910).]

nineteen lateral shoots have been observed to develop in connexion with one scale[1]; Fig. 71 shows a large number of branches growing from a submerged inflorescence of this species. As illustrations of the numerous abnormalities on

[1] Luetzelburg, P. von (1910).

record, it may be noted that an inflorescence-bract sometimes develops into a water-leaf or even an entire water-shoot, while a bladder rudiment may develop into a water-shoot[1]. In the development of the seedling, the primary leaves may be replaced by stolons[2].

The apical development of the Bladderworts gives little help in interpreting their morphology. In *Utricularia vulgaris* (Fig. 72), for example, the apex of the shoot is coiled up in a singular

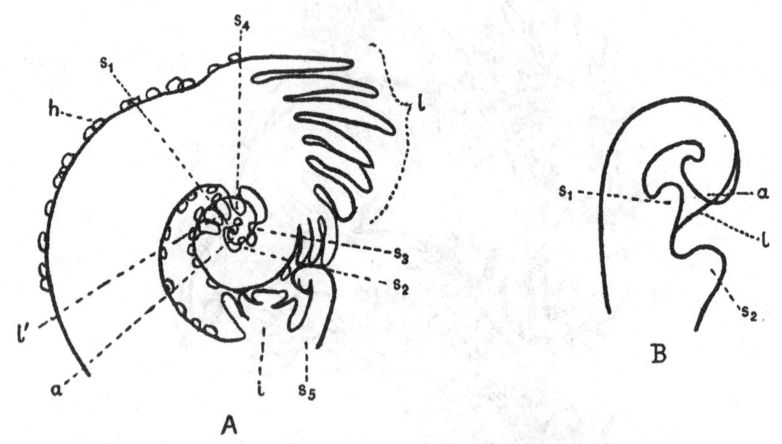

FIG. 72. *Utricularia vulgaris*, L. *A*, spirally coiled end of a shoot, of which *a* is the apex; s_1–s_5, young shoots; *l'*, youngest leaves; *l*, older leaves (between *l* and *l'* some leaves have been removed); *h*, hairs (mucilage glands); *i*, young inflorescence growing from the base of s_5. *B*, developing bladder; *a*, curved apex of shoot; s_1, first shoot, and *l*, single leaf or two leaves fused; *a*, s_1 and *l* fuse to form bladder; s_2 is second shoot which may give rise to a branch or a secondary bladder. [Adapted from Pringsheim, N. (1869).]

way which recalls a young fern frond. The "leaves" (*l*) arise in two lateral rows, and there is a third row of rudiments (s_1–s_5) on the concave face, which give rise to air-shoots. The developing bladders on a leaf are indicated in Fig. 73, while Fig. 72 *B* illustrates that view of the composite origin of the single bladder which regards it as derived from both axial and foliar elements[3].

In general, the only safe conclusion to be drawn from a study

[1] Glück, H. (1906). [2] Goebel, K. (1891).
[3] Pringsheim, N. (1869).

of the available evidence regarding the nature of the organs in the Bladderworts, seems to be that—in the present state of our ignorance—the attempt to fit so elusive a genus into the Procrustean bed of rigid morphology, is doomed to failure. It is probably best, as a purely provisional hypothesis, to accept the view that the vegetative body of the Utricularias partakes of both stem nature and leaf nature. How such a condition can have arisen, historically, from an ancestor possessing well-defined stem and leaf organs, remains one of the unsolved mysteries of phylogeny.

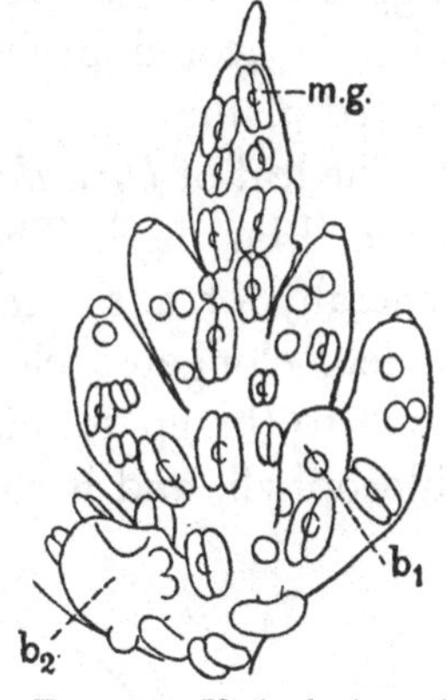

FIG. 73. *Utricularia vulgaris*, L. Developing leaf showing two young bladders b_1 and b_2; *m.g.*, mucilage gland. (Enlarged.) [Meierhofer, H. (1902).]

The anatomy[1] of the water Utricularias, though showing some curious features, is less anomalous than their morphology. In the stem of *U. vulgaris*, the tracheids, of which one or more are present, are placed sub-centrally, and surrounded by little groups of phloem. Some degree of dorsiventrality is given to the structure by the thin-walled character of the small lower sector of the vascular cylinder in which the tracheids lie, while the conjunctive tissue of the rest of the stele, towards the upper side of the axis, is fibrous. The tracheal elements are of the nature of "imperfect vessels," being formed from a file of superposed cells, with imperforate, oblique, separation walls. The incompleteness of the conducting elements is probably to be associated with the relative unimportance of the transpiration stream in a rootless submerged plant. The vascular cylinder is surrounded by an endodermis, and the cortex is lacunar. The structure of the inflorescence-axis differs very markedly from that of the submerged stem; the tracheids form a discontinuous ring enclosing a large central pith containing phloem islands.

[1] Tieghem, P. van (1868) and (1869[1]), Russow, E. (1875), Schenck, H. (1886) and Hovelacque, M. (1888).

The submerged and aerial parts of the axis differ, in fact, so conspicuously in their internal structure that van Tieghem[1] suggested that, if they were submitted separately to an anatomist, he would probably attribute them to distinct and unrelated plants!

The leaf of *Utricularia minor* is typically that of a submerged plant (Fig. 74)[2]. The bundle is extremely small, consisting generally of a single annular tracheid surrounded by thin-walled, elongated elements. The air spaces in the mesophyll reach to the epidermis, which contains the greater part of the chlorophyll, and is the most conspicuous region of the leaf.

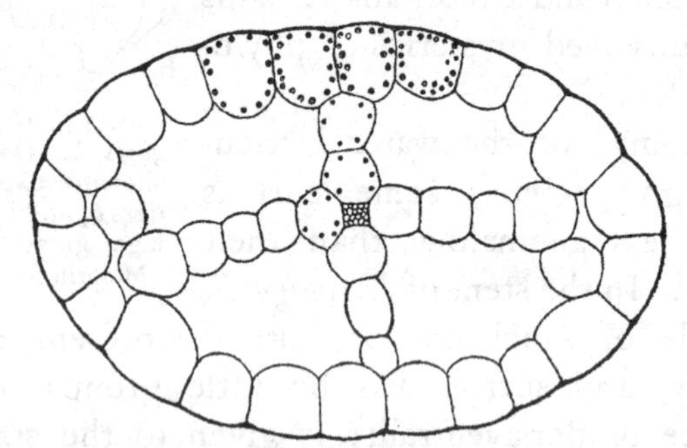

FIG. 74. *Utricularia minor*, L. T.S. lower part of leaf. (× 175.) [Schenck, H. (1886).]

It seems thoroughly in keeping with the uncannily abnormal morphology and the exceptional carnivorous habits of the Utricularias, that they should sometimes locate themselves in odd situations. The oft-quoted case of those Bladderworts which live in association with certain South American Bromeliads, is an instance in point. The leaf rosettes of some Tillandsias form vase-like cavities, which collect and retain water. *Utricularia nelumbifolia* has been described by a traveller in Brazil[3] as only to be found growing in the water which collects in the bottom of the leaves of a large *Tillandsia* occurring on an arid, rocky part of the Organ Mountains, at about 5000 feet above the sea.

[1] Tieghem, P. van (1868). [2] Schenck, H. (1886).
[3] Gardner, G. (1846).

Such a habitat would be impossible for the Bladderwort without the help of the Bromeliad's store of water, while the rich fauna of this water gives it every chance of catching suitable prey[1]. In the observer's own words, the *Utricularia* "propagates itself by runners, which it throws out from the base of the flower stem; this runner is always found directing itself towards the nearest *Tillandsia*, when it inserts its point into the water, and gives origin to a new plant, which in its turn sends out another shoot; in this manner I have seen not less than six plants united."

In British Guiana a similar case has been described[2]. A huge aloe-like Bromeliaceous plant, *Brocchinia cordylinoides*, Baker, grows in the Kaieteur savannah. It may be fourteen feet high, and, in older specimens at least, the crown of leaves is supported on a tall bare stem. Floating in the water retained in the axils of the leaves, is found a beautiful *Utricularia* (*U. Humboldtii*, Schombk.) "with flower stems 3 or 4 feet long, supporting its many splendidly large violet flowers." This form of epiphytism is not obligatory, since in Roraima, although both the Bromeliad and the *Utricularia* occur, the *Utricularia* may live a terrestrial life on marshy ground, instead of being associated with the Bromeliad.

Many of the unusual characteristics of the Utricularias are shared by another flowering plant, extremely remote from them in its affinities—*Aldrovandia vesiculosa*, L., a member of the Droseraceae. This plant has long had a peculiar fascination for botanists, and a detailed memoir upon it by an Italian writer appeared before the middle of the eighteenth century[3]. Like the Bladderworts, *Aldrovandia* is rootless and free-floating, and, but for its flowers, lives entirely submerged. It has a slender axis bearing whorls of leaves; the older internodes and leaf whorls die away successively, as new parts are formed at the apex.

[1] Goebel, K. (1891–1893).
[2] Im Thurn, E. F. and Oliver, D. (1887).
[3] Monti, G. (1747). For an analysis of this paper see Augé de Lassu (1861).

Aldrovandia, like the Bladderworts, is able to form turions; these are the size of a pea and consist of a highly abbreviated axis, which may bear as many as thirty-two leaf whorls. The turions normally sink to the bottom of the water in the autumn, owing to the weight of starch which they contain[1], and rise again in the spring; but it seems that they sometimes fail to reach the surface in the succeeding season, and that the developing plant may even in June be found at the bottom, held there by the remains of the winter-bud[2]. When the turions are kept in an aquarium indoors, it is said that they sometimes fail to sink, but remain floating throughout the winter[3]. In warmer climates these winter-buds are not formed; in Bengal, for instance, the plant is described as vegetating continuously throughout the year[4]. Reproduction by seed also takes place. The flowers are raised above the water, but the young fruits bend down, and the ripening of the seeds takes place beneath the surface[5]. The structure of the embryo recalls the other Droseraceae, the only difference being that the primary root remains rudimentary.

The leaves of *Aldrovandia* are highly peculiar in structure, and serve, like the bladders of *Utricularia*, for catching small animals[6]. The broad petiole terminates in a roughly circular bilobed lamina, and also bears, in its apical region, a number of stiff projections, which at first glance suggest leaflets[7], but are probably only petiolar emergences[8] (Fig. 75). Long sensitive hairs are produced from the upper surface of the lamina in the neighbourhood of the midrib; the touching of these by any passing animal results in the closure of the lobes[9], thus im-

[1] Caspary, R. (1859 and 1862). [2] Maisonneuve, D. de (1859).
[3] Schoenefeld, W. de (1860). [4] Roxburgh, W. (1832).
[5] Caspary, R. (1859 and 1862).
[6] The proof that *Aldrovandia* is carnivorous is due to Cohn, F. (1875), though Delpino, F. (1871) had previously shown that the suffocation of small animals occurs in the leaves.
[7] Cohn, F. (1850). [8] Caspary, R. (1859 and 1862).
[9] Mori, A. (1876) noted that the central region of the leaf was irritable.

prisoning the prey. The sensitiveness of the leaves is greatest at rather high temperatures[1]. The Linnean name, "*vesiculosa*," is an unfortunate one, since it suggests that the leaves form actual bladders, whereas the lobes merely fold together like those of *Dionaea*. Besides the irritable hairs, glands[2] are also present, which apparently secrete a digestive fluid and absorb organic matter[3].

There is good reason to suppose that both *Aldrovandia* and the water Utricularias are descended from terrestrial ancestors

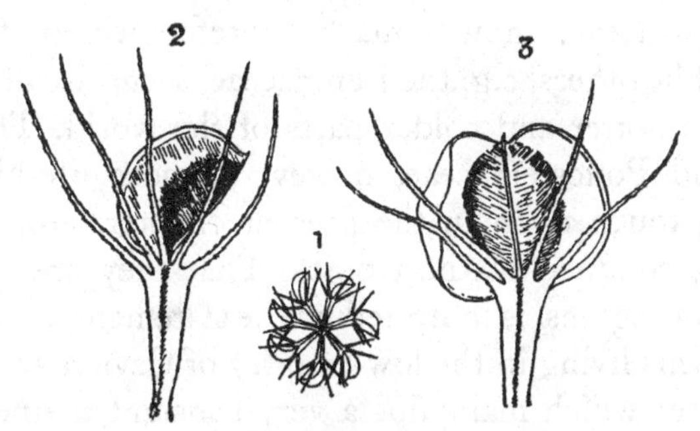

FIG. 75. *Aldrovandia vesiculosa*, L. 1, whorl of leaves (about ½ nat. size); 2 and 3, individual leaves (× 2½ *circa*). Leaf is shown in natural position in 2, and with the lobes open in 3. [Adapted from Caspary, R. (1859).]

which were already carnivorous. *Aldrovandia* is the only aquatic member of the Droseraceae, a family which contains well-known insectivorous types such as the Sundew, while the aquatic Utricularias are associated both with terrestrial carnivorous members of the same genus, and with the insect-catching Pinguiculas, which are not hydrophytes. The habit of consuming animal food has thus not arisen *de novo* in connexion with an aquatic existence, though this mode of life undoubtedly affords unique opportunities to a carnivorous plant[4].

[1] Stein, B. (1874). [2] Fenner, C. A. (1904).

[3] Darwin, C. (1875).

[4] On *Aldrovandia*, in addition to the papers cited in this chapter, see Caspary, R. (1858[4]) and Hausleutner (1850[1]) and (1851).

CHAPTER IX

THE LIFE-HISTORY OF THE TRISTICHACEAE AND PODOSTEMACEAE[1]

ALL the families of aquatics hitherto considered are represented in our own country; some of them, e.g. the Potamogetonaceae, show a marked preference for temperate regions, while others, e.g. the Lemnaceae, seem equally at home in both the hotter and colder parts of the world. The Tristichaceae and Podostemaceae, however, whose life-history we propose to touch upon in the present chapter, are, with rare exceptions, confined to the tropics. That they are essentially plants of hot regions, is indicated by the statement of Dr Willis[2] that the forms living in the low-country of Ceylon and S. India inhabit water which maintains a very constant temperature of 80° F. (27° C.). The two families together form an anomalous group, characterised, as regards their morphology, by remarkable variety, but agreeing, as regards their ecology, in one singular feature—a preference for inhabiting water which flows rapidly or even torrentially over a rocky substratum. This peculiarity, sometimes rendered more noticeable by reason of the striking colour of the plants, has been observed from the earliest time at which Podostemads became known to botanists. The first recognition of a member of this group as the type of a

[1] General accounts of these plants will be found in Gardner, G. (1847), Tulasne, L. R. (1852) and Warming, E. (1881, 1882, 1888 and 1891). They have only recently been divided into these two families (Willis, J. C. 1915[1]), and many authors still refer to them all as Podostemaceae.

[2] Willis, J. C. (1902). This interesting memoir has been largely drawn upon in the present chapter; it contains a bibliography of previous work. See also Willis, J. C. (1914[1]), (1915[1]) and (1915[2]) and Matthiesen, F. (1908).

distinct family, occurred when Aublet[1], nearly a century and a half ago, discovered *Mourera* in rapidly running water in French Guiana. In the case of a certain Venezuelan river, Goebel[2] describes the bed, in places where the water flows quickly, as quite green with a Podostemad, *Marathrum utile*, growing on the stones, and he points out that it flourishes more freely the stronger the current; when the stream is slow it is replaced by Mosses and Algae. Another writer[3] observed *Mourera fluviatilis* in the cataracts of a tributary of the Amazon, growing in such abundance that the rocks, amongst which the waters rushed, were veiled by it, and the colour was so vivid that the river seemed—to use his own expression—"to flow over a carpet of roses." This red hue of the vegetative organs, due to anthocyanin in the surface cells[4], has been noted in many cases. Miss Lister[5], for instance, in her account of the occurrence of a species of *Tristicha* in rapidly flowing water below the first cataract of the Nile—one of the rare records of the appearance of a member of these families outside the tropics—mentions that, when the plant was wet and fresh, the colour was crimson.

The majority of the peculiarities of the Podostemaceae and Tristichaceae are closely related to the nature of their habitat. Life in rushing water—on rocks which are often water-worn to smoothness and into which no roots can penetrate—is obviously impossible except to plants which have a special capacity for clinging to the substratum. In the Tristichaceae[4] (e.g. *Tristicha ramosissima* and *Weddellina squamulosa*) a creeping, thread-like organ is formed, which, though morphologically a root, is dorsiventral in structure, and gives rise to leafy shoots endogenously in acropetal succession. But this thread-like root is not apparently competent to anchor the leafy shoots with the necessary firmness, and additional organs called 'haptera' are formed. They are produced exogenously from the creeping root, and by their positive geotropism and power of flattening them-

[1] Aublet, F. (1775).
[2] Goebel, K. (1891–1893).
[3] Weddell, H. A. (1872).
[4] Willis, J. C. (1902).
[5] Lister, G. (1903).

selves against the substratum, form firm attachment organs. They also secrete a kind of cement which renders their adhesion to the rock very close and permanent. These haptera are found in many Podostemaceae. In *Mourera fluviatilis*, for instance, they are sometimes almost tendril-like[1], while in certain cases they serve as storage organs for reserve carbohydrates[2].

In many of the Podostemaceae the creeping root discards its root characteristics even more completely than in the Tristichaceae, and becomes converted into a thallus, which either follows out every irregularity in the substratum, or, remaining more or less free, develops into all sorts of curious shapes[3]. It still produces secondary shoots bearing leaves, but as the root thallus becomes more important, the secondary shoots become less so, until, in such genera as *Hydrobryum* (Fig. 76), *Farmeria*, *Dicraea* (Fig. 77 and Fig. 79, p. 116), and *Griffithiella* they are much reduced, and assimilation is mainly performed by the thallus. A seedling of *Dicraea stylosa*, with the young thallus (*th.*) developed as a lateral outgrowth from the hypocotyl (*hyp.*), and bearing secondary shoots (*s.s.*) is shown in Fig. 78; the mature plant is represented in Fig. 79, p. 116.

The thallus of the Podostemads is sometimes amazingly polymorphic; its capacity for developing in exceptional forms depends, apparently, on the fact that it is not restricted by a rigid skeletal system, and that nearly all the cells possess the capacity for renewed meristematic activity. *Griffithiella Hookeriana*, for instance, has a thallus which may develop into various shapes recalling different Algae that grow in moving water; one of its forms resembles the basal cup of *Himanthalia lorea*. *Farmeria metzgerioides*, again, recalls *Delesseria Leprieurii*, while *Podostemon subulatus* simulates such an Alga as *Bostrychia Moritziana*, which also grows in rapids. Willis, who draws attention to these cases of simulation, alludes to the great difficulty of interpreting such resemblances between plants far

[1] Went, F. A. F. C. (1910). [2] Matthiesen, F. (1908).
[3] See Willis, J. C. (1902) for further details.

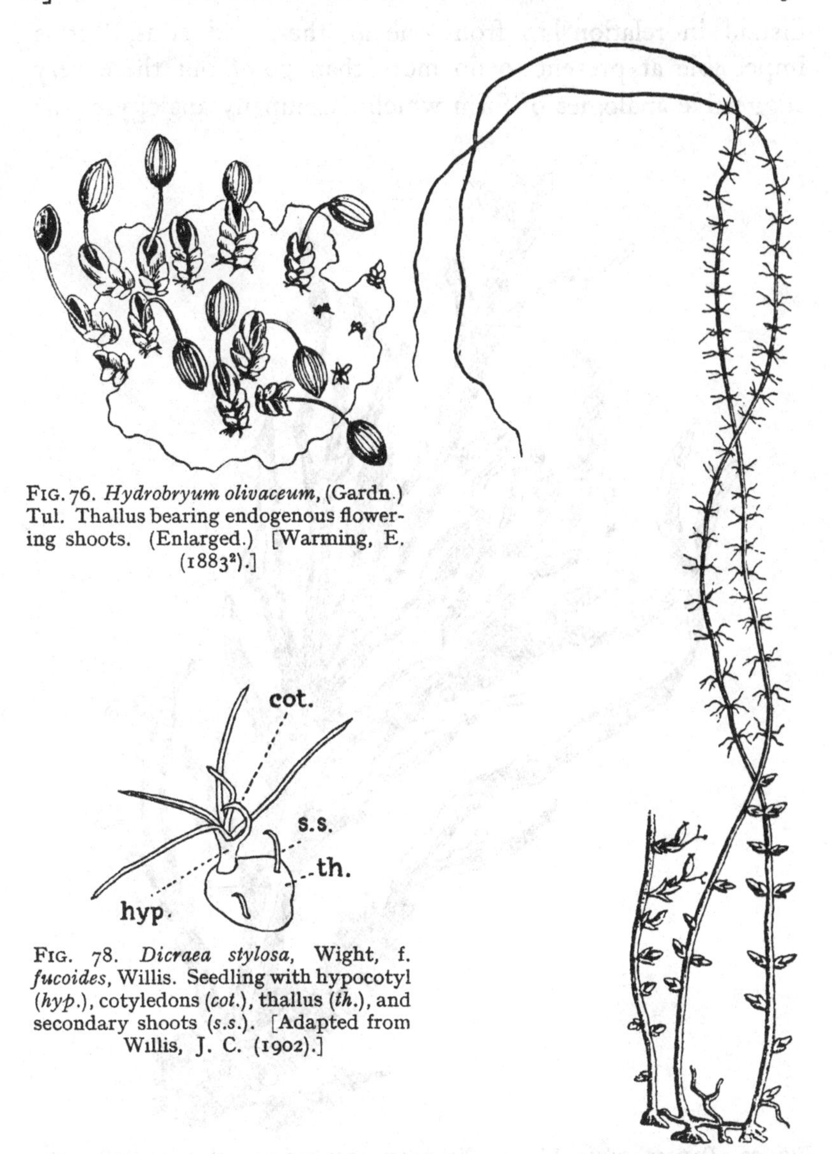

FIG. 76. *Hydrobryum olivaceum*, (Gardn.) Tul. Thallus bearing endogenous flowering shoots. (Enlarged.) [Warming, E. (1883[2]).]

FIG. 78. *Dicraea stylosa*, Wight, f. *fucoides*, Willis. Seedling with hypocotyl (*hyp.*), cotyledons (*cot.*), thallus (*th.*), and secondary shoots (*s.s.*). [Adapted from Willis, J. C. (1902).]

FIG. 77. *Dicraea elongata*, (Gardn.) Tul. Plant with three vertical roots bearing flowers. These float in the water: they spring from a horizontal creeping root. (Nat. size.) [Warming, E. (1883[2]).]

distant in relationship from one another, and adds, "it is impossible at present to do more than point out these very suggestive analogies of form which accompany analogy of the

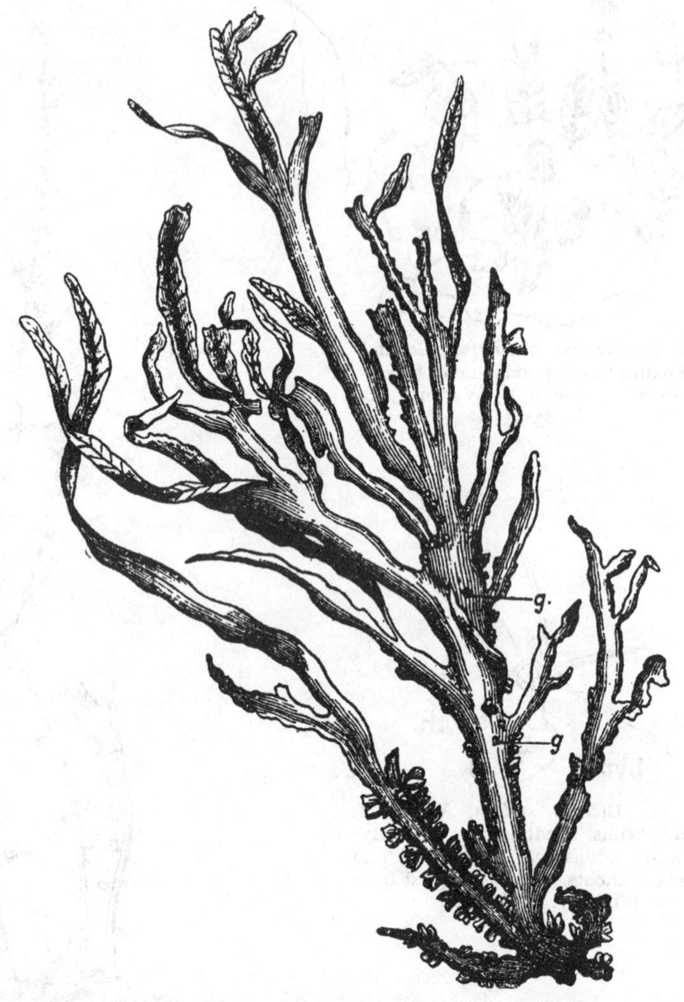

FIG. 79. *Dicraea stylosa*, Wight. Plant somewhat reduced, showing the shoots (*g*, *g*) arising from the band-like root thallus. [Warming, E. (1883[2]).]

conditions of life, and which seem to indicate that an experimental and comparative morphological study of the forms of the Algae and Podostemaceae should be attended with interesting results[1]." Even the Tristichaceae, which do not possess these polymorphic thalli, show "remarkable similarities in morphological features, and in the arrangement and anatomy of the leaves, to many mosses or liverworts, especially to those of wet situations[1]." The specific and varietal names given to various members of these families—such as *bryoides*, *fucoides*, *selaginoides* and *lichenoides*—speak eloquently of their striking resemblance to the lower plants, which the botanists who named them have felt impelled to emphasize[1].

The genus *Lawia* differs from those hitherto mentioned in having a thallus which is not of root nature, but which originates by the fusion of flattened, dorsiventral shoots, while *Castelnavia* also has a shoot thallus. In *Lawia foliosa*[2] the small thallus adheres so closely to the stones that it cannot be separated from them. There are no haptera, but the thallus is attached by hairs. The small simple leaves are without stomates or vascular bundles. They have a midrib of elongated cells, but their structure is altogether more simple than that of the leaves of many Liverworts. In *Lawia zeylanica*[1] the hypocotyl, produced on the germination of the seeds, bends down to the rock and becomes attached to it by unicellular rhizoids from the superficial cells. The hypocotyl then expands and forms a relatively large surface of attachment.

The internal structure of the Podostemads is similar to that of many other submerged plants in reduction of xylem, absence of stomates, and the presence of chlorophyll in the epidermis. On the other hand, a character in which these plants diverge from other hydrophytes is the presence of large quantities of silica in the cells[3]. It seems on the whole most probable that this silica is merely a useless by-product of the plant's metabolism[4]. It has been suggested that it serves as a protection against the attacks of animals[5], but there seems little evidence

[1] Willis, J. C. (1902).

[2] Goebel, K. (1889[3]) and (1891–1893). In Goebel's earlier account this plant is called *Terniola* (*longipes?*).

[3] Goebel, K. (1891–1893).

[4] Matthiesen, F. (1908).

[5] Wächter, W. (1897[1]).

for this view. That these plants, with their large stores of reserve starch, are, as a matter of fact, liable to be preyed upon, is indicated by Im Thurn's[1] observation that, in British Guiana, when the rivers are low, and the rocks which underlie the rapids are partially uncovered, a certain fish (*Pacu myletes*) collects at the falls to feed on the leaves of the Podostemads, which clothe the rocks, and at this time of year come into flower. This fact is so well known that, at this season, large numbers of Indians camp on the sides of the falls, in order to seize the opportunity of shooting the fish.

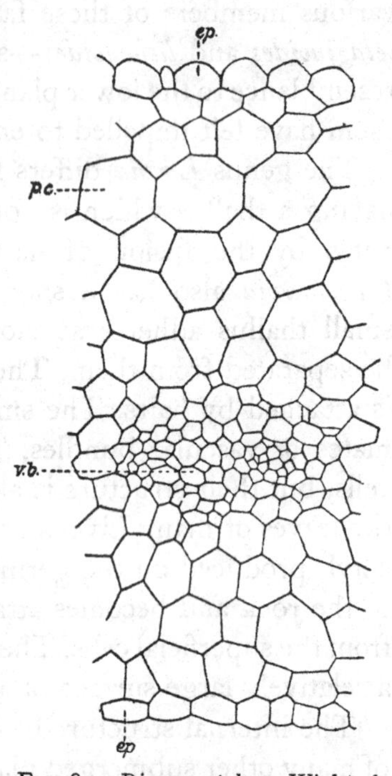

FIG. 80. *Dicraea stylosa*, Wight, f. *fucoides*, Willis. T.S. thallus to show absence of intercellular spaces. *ep.* = epidermis; *p.c.* = parenchymatous cortex; *v.b.* = vascular bundle. (× 150 *circa*.) [Willis, J. C. (1902).]

The most important anatomical peculiarity of the Podostemads is the extreme reduction of the intercellular spaces[2]; in this respect the members of these families contrast most markedly with other water plants (Fig. 80). This feature is probably to be associated with the thorough aeration of the torrential water which they frequent[3]. Certain species, however, possess delicate outgrowths from the surface of the leaves which have been interpreted as "gill-tufts" (Fig. 81). Possibly these structures to some extent compensate for the lack of an internal aerating system.

The water in which the Podostemaceae live is liable to variations in level, and their habit of blooming when the sinking

[1] Im Thurn, E. F. (1883). [2] Warming, E. (1881).
[3] See pp. 256, 257.

of the water exposes them to entire or partial desiccation, has been repeatedly noted by travellers. Barrington Brown[1], in describing his explorations up the Cuyuni River in British

FIG. 81. *Oenone multibranchiata*, Matthiesen. Part of flowering plant showing the numerous "gill-tufts" on the upper surfaces of the leaves. [Matthiesen, F. (1908).]

Guiana, mentions the occurrence of Podostemaceae on the rocks under water where the current runs strongest, and adds,

[1] Brown, C. Barrington (1876).

"These plants bear very pretty flowers at this season of the year [September] as soon as they are left uncovered by the subsiding of the waters after the rainy season, but still kept moist by the wash of the water's edge. One small-leaved species has a little white star-shaped flower, on a short delicate stem, which has a slight perfume and proves an attraction to numerous species of wild bees." Im Thurn[1], again, in his account of the same regions, mentions *Mourera fluviatilis* and *Lacis alata* as growing "on the half-submerged rocks in most of the falls. As the water decreases in the dry season, the tall spikes of bright pink flowers of the former plant rise from their large leaves, the edges of which are cut and curled into the likeness of moss, which lie flat on the rocks; and at the same time and place innumerable tiny pink stars rise an inch or two over the equally moss-like leaves of the *Lacis*."

The vegetative parts of the Podostemads die very quickly when out of their element, and the flowering and seed-setting, both of which take place with the utmost rapidity when the plants are exposed to the air, represent, as it were, their swan-song. In *Lawia zeylanica*, Willis[2] has observed that the enormous amount of starch stored up in the flowering shoots accounts for the great rapidity with which anthesis and seeding take place. In the case of *Rhyncolacis macrocarpa*, Goebel[3] points out that each inflorescence-bud is enclosed in a cavity formed by the connate union of two leaves. These cavities are full of water, so that the life of the flower-stalks is passed in an environment resembling that of ordinary aquatics inhabiting still water; it is thus not surprising that these stalks differ from the other vegetative organs in developing an aerating system, such as is characteristic of water plants in general.

Both entomophily and anemophily occur among the Podostemads. According to Willis, we can trace a series from certain American Tristichaceae with conspicuous, entomophilous

[1] Im Thurn, E. F. (1883). [2] Willis, J. C. (1902).
[3] Goebel, K. (1891–1893).

flowers, to members of the Podostemaceae in which anemophily or autogamy is associated with gradually increasing dorsiventrality. To this subject we shall return in Chapter XXVII. Cleistogamous flowers are also sometimes produced (Fig. 82).

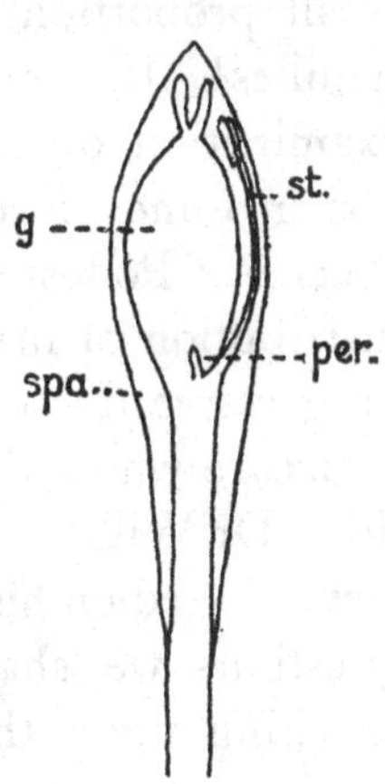

FIG. 82. *Podostemon Barberi*, Willis. Cleistogamic flower in spathe (*spa.*), the front of which is removed to show the gynaeceum (*g*), the single stamen (*st.*) and one of the two thread-like organs representing the perianth or staminodes (*per.*). [Simplified from Willis, J. C. (1902).]

The peduncles of the Podostemaceae contain little water-conducting tissue, and, possibly in correlation with this, the seed-development proves to be of a decidedly xerophilous type[1]—an illustration of the conservatism of the reproductive organs of aquatics and their tendency to retain terrestrial characters. By disappearance of nucellar tissue, a cavity is formed beneath the embryo-sac which, at the time of fertilisation, is filled with fluid. This cavity is bounded by the strongly cuticularised inner wall of the inner integument and the suberised cells of the chalaza. It is open only on the side towards the developing embryo, and is described as "an ideal water reservoir." Mucilage is often present in the neighbouring cells of the inner integument and this may perhaps form an additional protection against loss of water.

The seeds of the Podostemads are often small and numerous. Those of *Rhyncolacis macrocarpa* are about as large as the largest known pollen-grains (e.g. those of *Mirabilis*). The seeds of this species often germinate when caught in some cranny of the parent, so that the old plant may support a number of seedlings. The embryo is strictly rootless, but haptera grow out from the hypocotyl[2].

The morphology of the Tristichaceae and Podostemaceae positively bristles with problems for the botanist, but great caution has to be exercised in dealing with them, since it must

[1] Magnus, W. and Werner, E. (1913).
[2] Goebel, K. (1891–1893).

not be overlooked that the data are still highly incomplete, for, as a recent writer has pointed out, we probably know only a small proportion of the existing species belonging to these families[1]. It was recorded a decade ago, for instance, that the examination of a few kilometres of a river in Venezuela—hitherto unexplored in this respect—yielded no less than four species of Podostemaceae new to science[2]. The extremely local distribution of many forms, their anomalous morphology and progressive dorsiventrality, and the great variety of types of structure which they present, offer every incentive to speculation. Dr Willis has put forward certain far-reaching theoretical views, based on his study of the group, and to these and related questions we shall return in Chapter XXVII, when we are touching upon the problem of Natural Selection.

[1] Went, F. A. F. C. (1910). [2] Matthiesen, F. (1908).

CHAPTER X

THE LIFE-HISTORY OF THE MARINE ANGIOSPERMS

THE small group of Phanerogams inhabiting the sea consists of about thirty species[1] belonging to two families of that Cohort of Monocotyledons known as Helobieae. The Hydrocharitaceae are represented by *Halophila*, *Enhalus* and *Thalassia*, and the Potamogetonaceae by *Zostera*, *Phyllospadix*, *Posidonia*, *Cymodocea* and *Halodule* (*Diplanthera*). The thorough way in which the marine Helobieae have identified themselves with their environment, is shown by the fact that *Cymodocea antarctica* was actually included by Agardh[2] in his *Species Algarum* under the name of "*Amphibolis zosteraefolia*"; in justice to this author it should, however, be mentioned that he lays stress upon the uncertainty of its position "in catena entium." *Zostera marina*, the Grass-wrack, often grows among Seaweeds as if it were one of themselves; in lagoons of the Mediterranean coast it has been observed in association with *Enteromorpha*, *Codium tomentosum*, *Padina pavonia*, *Dictyota dichotoma*, and other Algae[3], while in Danish waters it grows in the midst of varied assemblages of brown, red and green Seaweeds[4]. *Zostera* is even able to descend to considerable depths in the sea; in the Baltic its occurrence at 11 metres from the surface has been recorded[3]. A species of *Phyllospadix* (a genus allied to *Zostera*) is noted for its power of withstanding the violence of the waves; it grows on the Californian coast "in the heaviest surf and on the most exposed ocean shores[5]."

Ascherson, whose work has done much to elucidate this difficult group, pointed out about fifty years ago[6] that the

[1] Sauvageau, C. (1891[1]).
[2] Agardh, C. A. (1821).
[3] Flahault, C. in Kirchner, O von, Loew, E. and Schröter, C. (1908, etc.).
[4] Ostenfeld, C. H. (1908).
[5] Dudley, W. R. (1894).
[6] Ascherson, P. (1867).

Phanerogams inhabiting the sea were, at that time, less well known than most of the higher groups of Algae. These marine Angiosperms often grow in deep water, and botanists have been obliged to depend chiefly on the study of casual fragments washed up by the waves, and thus have been apt to miss the organs of fructification altogether.

The marine Helobieae all show a strong affinity, both as regards vegetative habit and reproductive methods. They all have alternating leaves in two ranks arising from creeping stems. Supple, ribbon-like leaves, sessile, sheathing and capable of following all the undulations of the water, are most characteristic, occurring in *Enhalus*, *Posidonia*, *Phyllospadix*, *Zostera*, etc. Several Halophilas, on the other hand, have broad petiolate leaves with *Potamogeton*-like nervation, while *Cymodocea isoetifolia* is distinguished by awl-shaped succulent leaves[1].

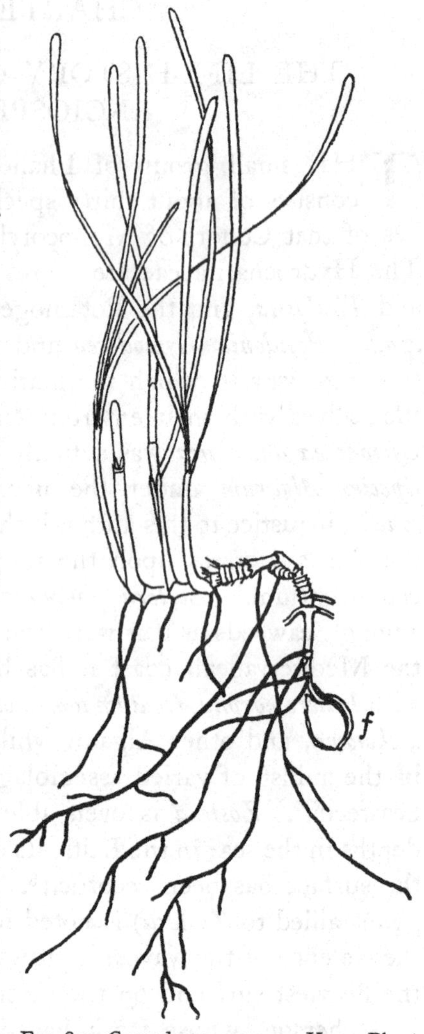

FIG. 83. *Cymodocea aequorea*, Kon. Plant in the middle of the third year of vegetation; *f* = fruit from which plant has grown. (Nat. size.) [Bornet, E. (1864).]

Submerged pollination and Conferva-like pollen are characteristic of all the marine Angiosperms. The thread-like pollen was figured as early as 1792 by the Italian

[1] Ascherson, P. (1867) and Sauvageau, C. (1890[1]) and (1890[3]).

botanist, Cavolini[1], who described it in the case of *Posidonia* as "lanae instar gossipinae." The cases which he records are those of "Zostera oceanica"(= *Posidonia Caulini*, Kon.), "Phucagrostis major" (= *Cymodocea aequorea*, Kon.), and "Phucagrostis minor" (= *Zostera nana*, Roth).

As a typical life-history of one of the marine Potamogetonaceae, that of *Cymodocea aequorea*, Kon. may be briefly outlined. This plant was made the subject of a classic memoir by Bornet[2],

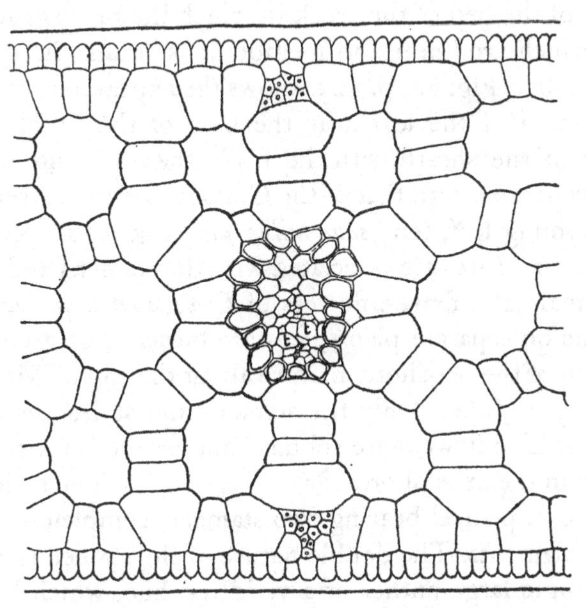

FIG. 84. *Cymodocea aequorea*, Kon. T.S. leaf near base of limb showing median bundle; *t*, *t*, sieve tubes. (×220.) [Sauvageau, C. (1891[1]).]

from which the following account is derived. *Cymodocea aequorea* (Fig. 83) is an herbaceous plant with a creeping stem, which forms submarine meadows after the manner of *Zostera marina*. It occurs in a number of localities in the Mediterranean, growing on muddy sand, in shallow creeks which are not greatly exposed to the shock of the waves. It is a perennial plant, which is in full vegetation from May to October; during the other months it is difficult to detect, for only a few short narrow leaves

[1] Cavolini, F. (1792[1]) and (1792[2]). [2] Bornet, E. (1864).

remain, with their green colour masked by a layer of various animal and vegetable growths. Probably the plant does not attain its full development until the fifth or sixth year, and an individual may live for another six years or more after reaching maturity. The rhizomes are fixed in the soil by long, whitish, fibrous roots, which put out a great quantity of tortuous laterals. The roots form a network, which holds in its meshes the gravel and mud, and thus contributes towards maintaining the stability of the bed of the creek in which the plant grows. The leaves, which are linear and membranous, attain the length of 20 to 30 cms. Fig. 84, p. 125, shows the appearance of a transverse section of the leaf near the base of the limb. At the junction of the sheath with the blade there is a ligule which Bornet compares with that of the Grasses. At the extreme base of each young leaf, ten '*squamulae intravaginales*' occur, and the same structures are associated with the stamens and carpels.

The male and female flowers of *Cymodocea aequorea*, which are borne on separate plants, and are buried 2 or 3 cms. deep in the soil of the sea-shore, mature about the end of May or the beginning of June. Only the stamens and styles emerge into the water. The flowers are solitary, and are borne without any perianth in the axils of ordinary foliage leaves. The male flower consists of a pedicel bearing two stamens, completely fused as to their filaments. The double nature of the stamen is revealed in the single large anther of a vivid red hue, which has eight pollen sacs, and is supplied by two vascular strands. The female flowers are only manifested externally by white, filamentous styles, which emerge in groups of four from the sheaths of certain leaves. Two of these styles correspond to each of the two carpels which constitute the gynaeceum. The ovary is unilocular with one ovule. Until the disappearance of the pollen-mother-cell, the pollen grains are roundish, but at this stage they elongate, without increasing in diameter, until they attain the dimensions of about 2 mm. by $\frac{1}{100}$ mm., thus becoming thread-like. The fruits, which are ripe by August, are flat and oval, being roughly 1 cm. long by 0·5 cm. wide. The endocarp is filled by the embryo with its enlarged hypocotyl, enclosed in a brown membrane. As the fruits develop, mature, and become detached, while still buried in the soil, there is no chance of their becoming disseminated, unless tempests or other accidental causes stir up the sea bottom; this explains the rarity of their occurrence among shore *débris*. Bornet several times found branches bearing two or three generations of fruits.

An Australian species of the same genus, *Cymodocea antarctica*, Endl., exhibits an interesting variant on *C. aequorea* in the matter of the fruit[1]. The plant is annual, or at most biennial, and the germination is viviparous. When the seedling attains a length of 3 to 4 inches, it breaks away from the parent, but carries with it a cup-like body (? the remains of the ovary wall) which has been described as bearing "two unsymmetrical pairs of basket-like spines." The "cup," on account of its relative density, "retains the floating waif in an upright position, and soon proves its ultimate use by acting as a *grappling apparatus*, catching in the tangles of small algae etc." The young plants develop spirally twisted roots, which presumably also serve for anchorage[2].

In *Zostera marina*, L., the Grass-wrack of our shores, the fertile and sterile plants are readily distinguishable from one another, since in the fertile plant the stem is slender, erect, and much branched, while that of the sterile individual is thick, creeping, more luxuriantly leafy, and anchored to the soil by adventitious roots developed in bundles beneath each leaf base[3]. Figs. 85 and 86, p. 128, illustrate the leaf anatomy. The inflorescence, unlike that of *Cymodocea*, consists of a number of male and female flowers, reduced to stamens and carpels and enclosed in a spathe. A French observer[4] has given a vivid description of a successful attempt to observe the actual pollination. Having found a good locality for the purpose, in the month of June, 1872, in his own words, "j'allai m'installer avec mon micro-

[1] Tepper, J. G. O. (1882) and Osborn, T. G. B. (1914).
[2] See p. 205. [3] Grönland, J. (1851). [4] Clavaud, A. (1878).

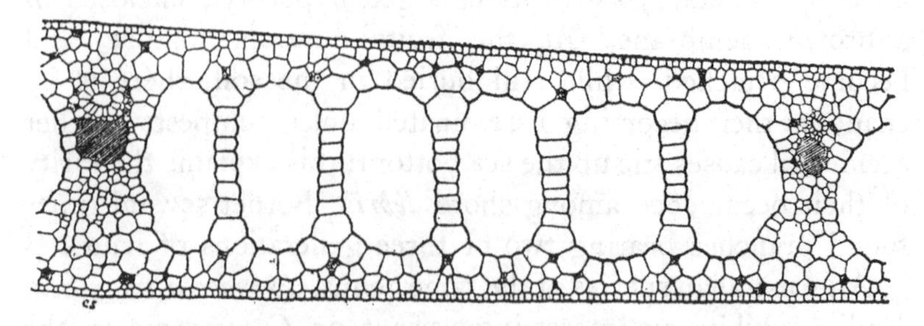

FIG. 85. *Zostera marina*, L. T.S. leaf at base of limb between the median nerve and a lateral nerve, phloem indicated by shading. [Sauvageau, C. (1891[1]).]

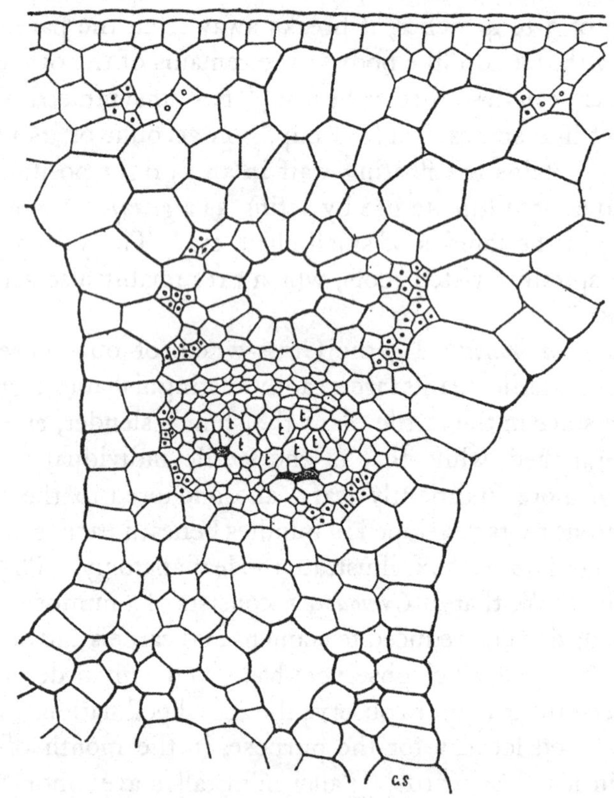

FIG. 86. *Zostera marina*, L. T.S. median bundle of leaf; *t*, sieve tubes. (× 220.) [Sauvageau, C. (1891[1]).]

scope dans la maison d'un ami, à quelques centaines de mètres de la plante, et je résolus de n'en point partir que je n'eusse découvert, si c'était possible, le mode de reproduction de *Zostera*....Il m'importait de ne pas rester plus longtemps dans une incertitude qui commençait à me peser." On a favourable day, hot and absolutely still, he went out in a boat and examined some flowering plants. The three conditions in which the inflorescences were found proved that cross pollination is ensured by protogyny. Some were still enclosed in the spathes, with the anthers intact; others showed stigmatic branches, ready for pollination or recently pollinated, emerging from the spathe, while the stamens were still enclosed and not completely ripe; in others again the stigmatic lobes had all fallen, while the anthers were exposed, and either all empty, or the lower ones empty and the upper ones in the act of dehiscence. The anthers were seen to open, and eject the thread-like pollen which formed a floating cloud. In pollen-grains, which had just been expelled from the anther, an outgrowth was observed at a little distance from one end. When pollinated stigmas were examined, it was noticed that these outgrowths, which were, in fact, young pollen-tubes, were forcing their way into the stylar tissue, between the cells whose walls were becoming mucilaginous and separating from one another. The pollination of *Zostera* is scarcely possible except in still water, as any movement would carry the pollen completely away from the scene of operations.

The best-known genus among the marine Hydrocharitaceae is *Halophila*, three species having been investigated in detail by Bayley Balfour[1] and Holm[2]. Bayley Balfour himself collected his material of *H. ovalis*, (R. Br.) Hook. fil. (*H. ovata*, Gaudich.) and of *H. stipulacea*, (Forsk.) Asch. on the reefs surrounding the island of Rodriguez—east of Mauritius. *H. ovalis* (Fig. 87) grows on spots just uncovered at full ebb tide, while *H. stipulacea* prefers localities where it is always submerged and subjected to a constant current. The rhizomes are creeping, and produce numerous long filiform rootlets

[1] Balfour, I. B. (1879). [2] Holm, T. (1885).

bearing a thick matting of root hairs; this tangle of roots fixes the plant in the sand. The flowers are typically hydrophilous. The filiform styles, which may be 26 mm. long, are receptive throughout their entire length, and, though the individual pollen-grains are not thread-like, the same result is secured by their being united into strings[1]. The seed-coats form an admirable protection for the embryo. The outermost cell-layer is conspicuously thickened on all the walls except that forming the surface of the testa. The next three cell-layers are cuticularised.

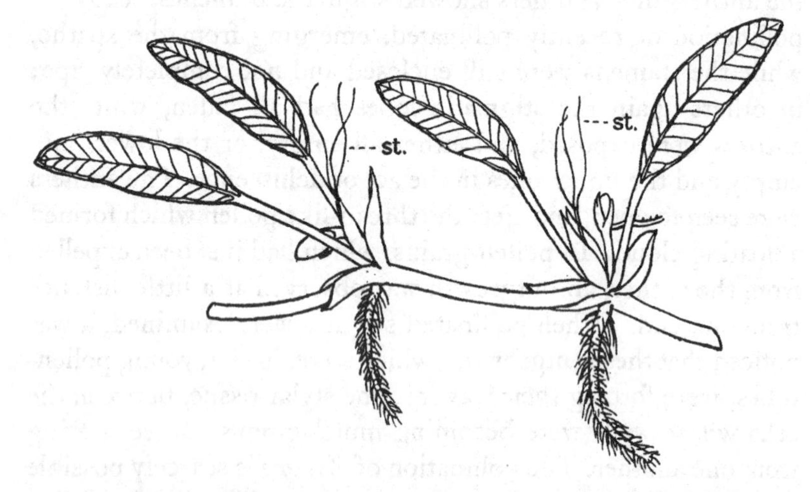

FIG. 87. *Halophila ovalis*, (R.Br.) Hook. fil. Portion of mature plant showing two female flowers in spathes with three thread-like stigmas (*st.*). (Enlarged.) [Balfour, I. B. (1879).]

Since the testa of *Zostera* is similar in structure, it seems not unlikely that in both cases the histological features bear some relation to the mode of life. Bayley Balfour concludes, from the general result of his researches, that *Halophila* forms a link between the Hydrocharitaceae and Potamogetonaceae.

The leaf anatomy of the marine Helobieae has been studied in great detail, partly because these plants are nearly always

[1] The thread-like character of the pollen of *Halophila* was observed by Gaudichaud, C. (1826) who also noticed the same feature in *Cymodocea antarctica*.

collected in a sterile and often fragmentary condition, and it has thus become a matter of importance to systematists to be able to identify them even when no organs of fructification are present. It might have been expected that the examination of the leaves of these plants, which show great similarity in external form and all live completely submerged in a fairly uniform environment, would reveal a monotony of internal structure. But this expectation is far from being realised. Duchartre[1] showed in 1872 that the genera *Cymodocea* (Fig. 84, p. 125) and *Zostera* (Figs. 85 and 86, p. 128) could be distinguished from one another, even in the absence of the flowers and fruit, on anatomical grounds alone. This conclusion was carried much further by Sauvageau[2], who proved, as a result of detailed and critical studies of the anatomy of the marine Phanerogams, that (except among the Halophilas) the anatomy of the leaf gives sufficient data for their exact generic and even specific determination. The variation occurring in the leaf structure is illustrated in Figs. 84, p. 125, 85 and 86, p. 128, 88 and 89, p. 132. Sauvageau pointed out, for instance, that the development of the lignified fibres differs markedly in the three genera, *Enhalus*, *Thalassia* and *Halophila*, and that it is thus impossible to regard this mechanical system merely as an adaptive response to the *milieu*. The differences that are displayed by the different species afford, indeed, another example of the fixity and lack of utility so often observed in specific differences; for it is not conceivable that each of the detailed distinctions between the closely related types of anatomy met with in the leaves of these marine Angiosperms, is to be interpreted as having some definite 'survival value,' though it may be broadly true that some structural variations are more suited to life in a boisterous sea and others to existence in calmer waters.

But though we cannot explain the different types of skeletal system of the leaves on adaptive grounds, there are other leaf-characters which seem definitely related to submerged life. In

[1] Duchartre, P. (1872).

[2] Sauvageau, C. (1890[1]), (1890[2]), (1890[3]) and (1891[1]).

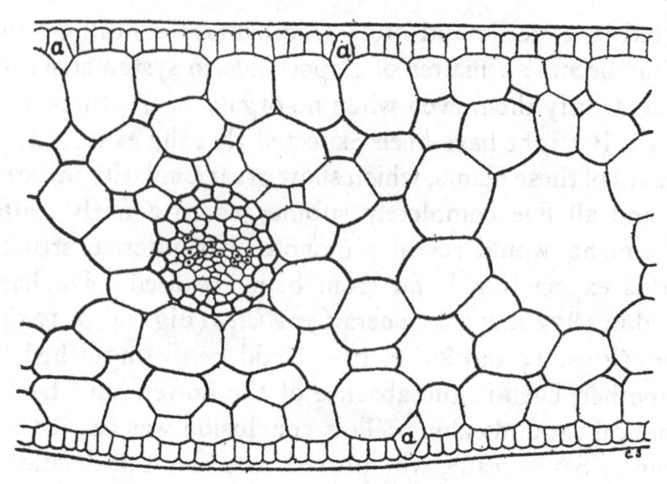

FIG. 88. *Halodule uninervis*, Boiss. T.S. leaf at base of limb; *a, a*, secretory cells. (×220.) [Sauvageau, C. (1891[1]).]

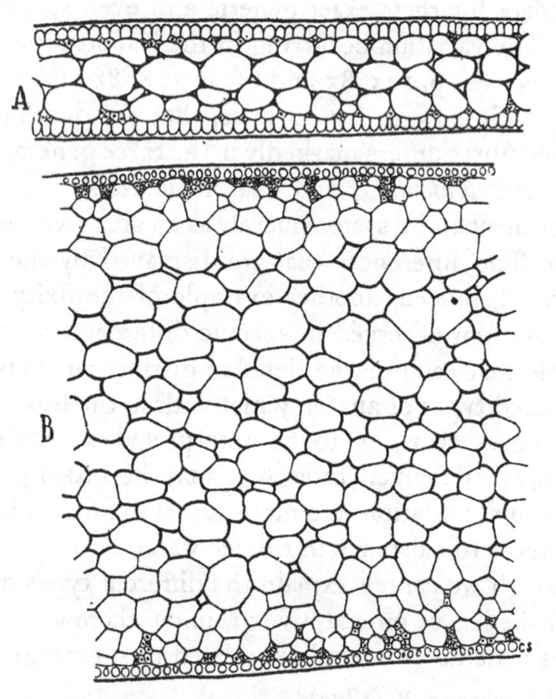

FIG. 89. *Posidonia Caulini*, Kon. T.S. limb of leaf. *A*, 1 cm. from apex; *B*, at base of limb. (×145.) [Sauvageau, C. (1891[1]).]

the marine Potamogetonaceae, the epidermis is characteristically free from stomates and very rich in chlorophyll. Liquid exchange between the plant and the surrounding medium is facilitated by the occurrence of openings at the leaf apices, with which the median nerve is in direct communication. These openings come into existence quite early in the history of the leaf, and are due to the disappearance of the epidermis. In the genus *Posidonia*, again, the fibres of the sheath survive and form a protective covering for the younger leaves. Another small peculiarity, which may be adaptive or may more probably be an indication of community of origin—since it is common to certain genera in the two families under consideration, but is not found elsewhere—is the occurrence of "Flossenzähne" or "dents nageoires" on the leaf margins[1]. These teeth are formed by a peculiar elongation and wall-thickening of the marginal cells.

The fact that a considerable number of Phanerogams live and flourish in the sea, and that yet, on examination, these marine types all prove to be restricted to representatives of two related families, stimulates conjecture as to the origin of this biological group. Both the families to which the flowering plants of the sea belong are typically aquatic, and are widely represented in fresh waters; no marine Angiosperm has a close affinity with any terrestrial plant. These facts suggest that the flowering plants now living in the sea are not the immediate descendants of land plants, but have been derived from ancestors which had already accommodated themselves to life in inland waters. It would seem that, in order to be capable of embarking upon life in the sea, a flowering plant requires four special faculties. These are, firstly, toleration towards a saline medium; secondly, the power of vegetating while wholly submerged; thirdly, the knack of developing a sufficiency of anchoring roots to withstand the wash of waves and tide; and, fourthly, the capacity for hydrophilous pollination, since any aerial method must be doomed to failure, except in halcyon weather in a non-

[1] Ascherson, P. and Graebner, P. (1907).

tidal sea. Both the families to which the marine Angiosperms belong, fulfil these four conditions in the persons of some, at least, of their fresh-water representatives. The existence of such species as *Potamogeton pectinatus* and *Zannichellia palustris*, which inhabit both fresh and brackish waters, and also of *Ruppia* and *Althenia* (Potamogetonaceae), which typically occur in a brackish medium—as well as of *Vallisneria spiralis*[1] (Hydrocharitaceae) and *Callitriche autumnalis*[2] (Callitrichaceae), which are able to tolerate some salt—indicates how the transition from fresh to saline water may have been bridged. The vegetative organs, again, are entirely submerged in such genera as *Elodea* and *Vallisneria* among the Hydrocharitaceae, and *Zannichellia*, *Naias* and many Pondweeds among the Potamogetonaceae. Both families also contain a number of species with well-developed root systems. Finally, floating pollen is carried by water to the stigmas in some species of *Elodea* and *Ruppia*, while actual submerged pollination is found in *Naias* and *Zannichellia*. These families are thus in every respect prepared, as it were, for the evolution of marine members. The reason why other families have not produced any forms adapted to life in the sea, seems to be that, though certain of their species may fulfil some of the conditions which we have enumerated, they fail in others—the one which is most rarely exhibited being a tendency to sub-aquatic pollination. *Myriophyllum spicatum* and *Ranunculus Baudotii*[3], for example, have been observed to live under conditions of slight salinity, but they are handicapped for entry on marine life by the fact that they cannot be cross-pollinated, unless the flowers are raised into the air. *Ceratophyllum* and *Pseudo-callitriche*, on the other hand, owing to their hydrophilous pollination, suggest themselves as possible candidates for marine life, but *Ceratophyllum* lacks roots entirely, and *Pseudo-callitriche* has no rhizome—obstacles that may well prove insuperable. Conceivably in future ages, if the

[1] Chatin, A. (1855[2]). [2] Lebel, E. (1863).
[3] Ostenfeld, C. H. (1908).

evolution of fresh-water plants proceeds on its present lines, a greater number may reach the specialised stage of hydrophilous pollination, and some of these may colonise the sea, thus democratising the narrow and exclusive circle of the Marine Angiosperms[1].

[1] In addition to the papers cited in this chapter the following references may be mentioned:

Ascherson, P. (1870) and (1875); Chrysler, M. A. (1907); Cunnington, H. M. (1912); Delpino, F. (1870); Delpino, F. and Ascherson, P. (1871); Engler, A. (1879); Hofmeister, W. (1852); Magnus, P. (1870[2]) and (1872); Martens, G. von (1824); Sauvageau, C. (1889[3]) and (1891[3]); Solereder, H. (1913); Walsingham, Lord, and Payne-Gallwey, R. (1886); Warming, E. (1871).

PART II

THE VEGETATIVE AND REPRODUCTIVE ORGANS OF WATER PLANTS,

CONSIDERED GENERALLY

"If then the *Anatomy* of *Vegetables* be so useful a *Mean*, we ought not to streighten it; but to force this, as well as the rest, to its utmost Extent. And therefore, first of all, To go through all the *Parts*, with equal care; examining the *Root*, *Trunk*, *Branch*, *Leaf*, *Flower*, *Fruit*, and *Seed*. . . . Together with the *Knife* it will be necessary to joyn the *Microscope*; and to examine all the *Parts*, and every Way, in the use of That. As also, that both Immediate, and Microscopical Inspections, be Compared: since it is certain, That some things, may be demonstrated by Reason and the Eye conjunct, without a Glass, which cannot be discovered by it."

Nehemiah Grew, *The Anatomy of Plants*, 1682.

CHAPTER XI

LEAF TYPES AND HETEROPHYLLY IN AQUATICS

(1) TYPES OF LEAF IN WATER PLANTS

THE types of leaf characteristic of aquatics—excluding those that rise wholly into the air and are thus comparable with the leaves of terrestrial plants—fall into two groups: firstly, those which float on the water surface, and thus preserve contact on the ventral side with the atmosphere and on the dorsal side with the water, and secondly, those which have more completely adopted the water life, since they keep up no direct contact with the atmosphere, but live entirely submerged. The general question of the relation of floating leaves to their environment has been discussed, in connexion with the Water-lilies, on pp. 30–32. There is considerable monotony in the outline and structure of a large proportion of such leaves, associated no doubt with the very definite and uniform physical conditions to which they are subject. Submerged leaves, on the other hand, are characterised by much greater variety. With a number of exceptions, they fall mainly into two groups—those that present a very thin, entire lamina, generally ribbon-shaped but sometimes broad, and those in which the leaf blade is finely subdivided, either by fenestration or dissection. In both these types of leaf, the ratio of surface to volume is higher than is the case in a normal, terrestrial lamina, and many botanists regard their peculiarities as definite adaptations for obtaining from the water an adequate supply of gases in solution. It is generally assumed that the dissected type of leaf is the more efficient form for the purpose. A Russian writer[1] has recently proved, however, that this assumption is scarcely borne out by a critical examination of the facts. By measurements and calculations

[1] Uspenskij, E. E. (1913).

he shows that a cylindrical leaf, in order to have as high a ratio of surface to volume as, for example, the broad, flat leaf of *Potamogeton perfoliatus*, must be only 120μ in diameter, whereas the diameter of the segments of *Myriophyllum spicatum* leaves varies from 220μ to 380μ, and of *Ceratophyllum demersum*, from 600μ to 750μ, while even the ultimate divisions of the leaves of *Ranunculus trichophyllus* reach 190μ. He admits that, apart from the actual ratio of surface to volume, the dissected leaf may possibly have an advantage over the corresponding flat leaf, in tapping a greater volume of the medium[1]; he thinks, however, that though this factor would be of importance in absolutely still water, its significance is much reduced if, as is nearly always the case, movement has to be taken into account. It may be added that the dissected leaf possibly interferes less with its neighbours' light than the undivided type of submerged leaf. From this enquiry and from a general study of submerged leaves, it may perhaps be concluded that both the dissected and flat types of leaf are organs of tolerably equal efficiency for subaqueous gaseous exchange, though the dissected leaf has the advantage of offering less resistance to currents. Which type of leaf a submerged plant shall produce is probably ultimately decided by the general leaf morphology of its terrestrial ancestors, rather than by environmental causes, much as coast scenery is often determined by the forms of the pre-existing land surfaces, rather than by the direct action of the ocean itself.

Among the undissected types of submerged leaf, the ribbon leaf is conspicuous (Fig. 90); it is probably better adapted to resist tearing than, for instance, the large, *Ulva*-like submerged laminae of the Waterlilies. Ribbon leaves are found among many of the marine Angiosperms, such as *Zostera*, which are subjected to the wash of waves and tide. Leaves of this type sometimes grow to a notable length; those of *Sagittaria sagittifolia*, as we have shown in Chapter II, may be more than two yards long, while those of *Vallisneria spiralis* are said to be often a yard or more in length, though hardly a quarter of an inch wide.

[1] Schenck, H. (1885).

Other types of submerged, radical leaf are the small, almost cylindrical leaves of *Lobelia Dortmanna* and *Littorella lacustris* (Fig. 142, p. 218), and the linear serrate leaves of *Stratiotes aloides* (Fig. 32, p. 53), which are too firm and stiff to be called ribbon leaves. In *Lobelia* and *Littorella*, the shortness of the leaves obviates the necessity of pliability to the motion of the water, while in *Stratiotes* the need for flexibility is diminished by the partially free-swimming habit of the plant and its penchant for quiet waters.

FIG. 90. *Sagittaria sagittifolia*, L. Young plant produced from a tuber (*T*) and bearing ribbon leaves only; tuber, with axis and scale leaves, and roots, indicated in solid black. Drift at bank of Cam, May 31, 1911. (Nat. size.) [A. A.]

When the leaves, instead of being radical, are borne on a pliable, elongated stem, the function of flexibility seems to be taken over by the axis and the leaves are generally small and simple, as in the case of *Elodea canadensis*. In *Hippuris vulgaris*, however, the whorled, submerged leaves may reach a considerable length.

The finely divided type of submerged leaf takes two different forms, according to whether the species to which it belongs is Dicotyledonous or Monocotyledonous. There are numerous examples of dissected, submerged leaves among the Dicotyledons, the most familiar case being that of various Batrachian Ranunculi. Among Monocotyledons the submerged leaves are nearly always entire; the characteristic venation of this group does not lend itself readily to the formation of a dissected leaf. As Henslow[1] has pointed out, dissection among Dicotyledons is represented, in the very few equivalent cases among Monocotyledons, by fenestration, which produces a similar result. He adds the ingenious, but probably untenable, suggestion that the fenestration of the aerial leaves of *Tornelia*, *Monstera*, etc., is a character handed down to them from aquatic ancestors. Among the Aponogetons we meet with a slight and irregular perforation of the leaves in *A. Bernerianus*, (Decne.) Hook. fil.[2], while in *A.* (*Ouvirandra*) *fenestralis* the mature leaves are completely reticulate (Fig. 91). According to Mlle Serguéeff[3], who has made a detailed study of the subject, the young leaves are imperforate, the perforations arising at a later stage by destruction of the tissues. When the perforations are formed, a fauna and flora of Flagellates, Rotifers, Bacteria and Algae accumulate in their neighbourhood, without apparently being responsible for their actual initiation; probably

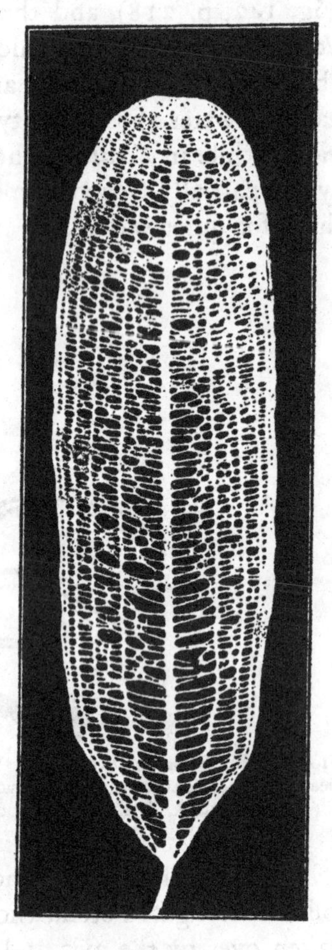

FIG. 91. Perforated leaf of *Aponogeton fenestralis*, Hook. f.=*Ouvirandra fenestralis*, Poir. Lace-plant. [Serguéeff, M. (1907).]

[1] Henslow, G. (1893).
[2] Krause, K. and Engler, A. (1906).
[3] Serguéeff, M. (1907).

they merely make use of the *dèbris* of those cells which are sacrificed to form the perforations.

That the fenestration in *Ouvirandra* may be of some value in connexion with aeration was suggested by Goebel's statement that the tissue of the leaf is poor in intercellular spaces[1]. Mlle Serguéeff, however, shows that Goebel is in error on this point, since numerous lacunae occur in the mesophyll, and she concludes that the main function of fenestration is not aeration, but the reduction of resistance to current. In this connexion it may be recalled that all the imperforate, submerged leaves found among the Aponogetons are thin and ribbon-shaped, thus yielding easily to the movement of the water (e.g. *A. angustifolius*, Ait.)[2]. It may also be significant that *A.* (*Ouvirandra*) *fenestralis*, in its Madagascan home, though it sometimes grows in stagnant water, is also capable of living in torrents. Hansgirg[3] had previously suggested that some forms of submerged leaf might be compared with those of such 'anemophytes' among terrestrial plants as Palms, Bananas, etc., in which the slitting, sub-division and perforation of the leaves are interpreted by some authors as modifications designed to avoid tearing by the wind. But the view that would regard all types of submerged leaf as definite adaptations to water life, probably needs considerable revision. We do not propose to criticise it at this point, since it is included in the broader question of the relation of leaf form to environment, which is better considered in connexion with heterophylly[4].

(2) THE FACTS OF HETEROPHYLLY UNDER NATURAL CONDITIONS[5]

The occurrence of two or more different types of leaf upon one individual, which is so frequently characteristic of water plants, has long attracted the interest of botanists.

[1] Goebel, K. (1891–1893).
[2] Krause, K. and Engler, A. (1906)
[3] Hansgirg, A. (1903).
[4] See Section (3) of this Chapter.
[5] Arber, A. (1919[3]) has been largely incorporated in Sections (2) and (3) of this Chapter.

Lyte's *Herball* (1578) contains a vivid description of heterophylly in the Water Buttercup—a free translation of that given in Dodoens' *Histoire des Plantes* of 1557. Since this description is also noteworthy for its insistence on the influence of external conditions upon the form of the leaves, it may be cited here.

"Amongst the fleeting [floating] herbes, there is also a certayne herbe whiche some call water Lyverworte, at the rootes whereof hang very many hearie strings like rootes, the which doth oftentimes change his uppermost leaves according to the places where as it groweth. That whiche groweth within the water, carrieth, upon slender stalkes, his leaves very small cut, much like the leaves of the common Cammomill, but before they be under the water, and growing above about the toppe of the stalkes, it beareth small rounde leaves, somewhat dented, or unevenly cut about. That kind which groweth out of the water in the borders of diches, hath none other but the small jagged leaves. That whiche groweth adjoyning to the water, and is sometimes drenched or overwhelmed with water, hath also at the top of the stalkes, small rounde leaves, but much more dented than the round leaves of that whiche groweth alwayes in the water."

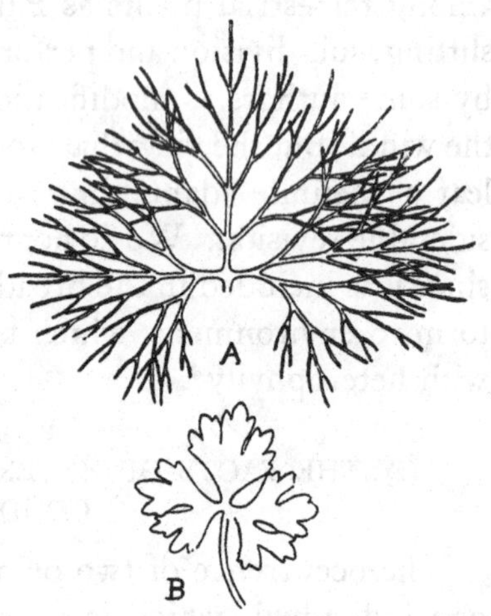

FIG. 92. *Ranunculus Purschii*, Rich. *A*, water leaf (½ nat. size) and *B*, leaf of the land form (reduced). [Goebel, K. (1891–1893).]

The water and land leaves of *Ranunculus Purschii* are illustrated in Fig. 92 *A* and *B*. The heterophylly of the Water Buttercups has been subjected to a great deal of critical investigation. It has been shown that, in the case of *Ranunculus aquatilis*, L.[1], it is impossible to say at an early stage whether a leaf rudiment will produce the floating or submerged form. Up

[1] Askenasy, E. (1870). See also Rossmann, J. (1854).

to a certain point they develop alike and are both deeply subdivided; then the water leaf ceases to change in shape and the segments merely increase in size, while the floating leaf gradually assumes its typical, relatively entire form. In general, the type of leaf produced by the plant can be changed at will by altering the conditions. If a plant that has begun to grow on dry land, be submerged, the new leaves produced by further growth are of the submerged type. The existing leaves, though they cannot alter their form, may, in the basal region which is still capable of growth, develop transitional features as regards the epidermis.

Among species related to *Ranunculus heterophyllus*, Fries, in which floating as well as submerged leaves are usually present, we find some, such as *R. fluitans*, Lamk., in which the floating leaves are rare, while in *R. circinatus*, Sibth. they are unknown. On the other hand, *R. hederaceus*, L. (Fig. 93), which generally grows in shallow ponds and ditches, possesses lobed reniform leaves only, and none that are finely divided and belong to the submerged type.

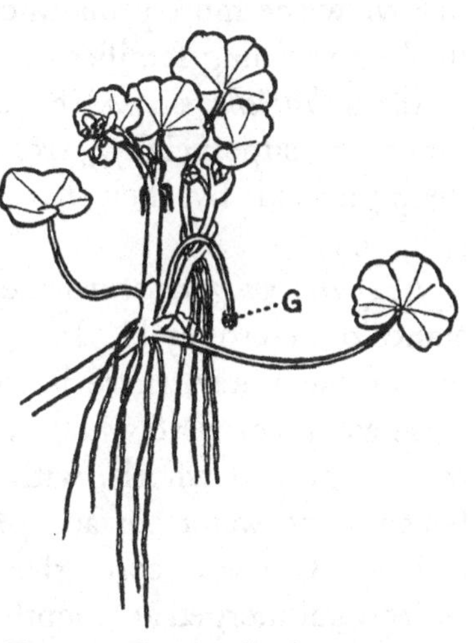

FIG. 93. *Ranunculus hederaceus*, L. An example of a Batrachian *Ranunculus* with undivided leaves. ($\frac{2}{3}$ nat. size.) Shallow pool, Ware Undercliff, March 27, 1912. The gynaeceum, *G*, is bending down to ripen under water. [A. A.]

Heterophylly is not confined to the Batrachian Ranunculi, but is widespread in the genus. *Ranunculus Flammula*, the Lesser Spearwort, though generally terrestrial, may live as a water plant[1], in which case it can develop both submerged and floating leaves. The submerged leaves are not, however, subdivided as in the case of *Ranunculus heterophyllus*. Heterophylly has also

[1] Bailey, C. (1887), West, G. (1910), Glück, H. (1911); references will be found in West, G. (1910) to the earlier writers who observed this form.

been recorded in *R. sceleratus*[1], *R. Lingua*[2] and other species. In *R. sceleratus* the present writer has observed that, in aerial and in floating leaves, stomates occur on both surfaces, but in the case of the floating leaf, the stomates were found to be less numerous on the lower surface than in a leaf growing in air.

The heterophylly of the Nymphaeaceae has been discussed in Chapter III[3], so it is now only necessary to recall that aerial leaves, floating leaves and submerged leaves may occur, the latter belonging either to the *Ulva*-like type of *Nymphaea* and *Castalia*, or the dissected type of *Cabomba*.

Leaving the Ranales, it may be worth while to pass rapidly in review the more pronounced cases of heterophylly met with in the remaining families of Angiosperms.

In *Callitriche verna* the submerged leaves are not very different, superficially, from the floating leaves, but are distinguished by their narrower and more elongated form (Fig. 94).

Hippuris vulgaris furnishes a particularly well-marked instance of heterophylly. In May, when its flowering shoots rise out of the water, there is the sharpest contrast between the close whorls of rigid, short, aerial leaves (*B–D* in Fig. 95) and the submerged whorls, with their long, flaccid leaves, visible beneath the water surface (*A* in Fig. 95; see also Fig. 151, p. 231). Goebel records that he once found *Hippuris* growing entirely submerged at a depth of 3 metres, with leaves 7 cms. or more long[4]. Towards July, when the plant is at its period of maximum activity, the new shoots formed under water, even at a depth of 50 cms., are reported to be of the aerial type and to bear stomates[5]. This statement is of importance in connexion with the problem of the significance of heterophylly, which will

[1] Ascherson, P. (1873), and Karsten, G. (1888).
[2] Roper, F. C. S. (1885).
[3] See pp. 27–29, and Figs. 12 and 14.
[4] Goebel, K. (1891–1893).
[5] Costantin, J. (1886).

be discussed later in the present chapter. When winter comes on, the thin, submerged, stomateless type of leaf is again produced. Fig. 96, p. 148, represents a rather curious case, in which a shoot had reverted to submerged leaves (*a*) after bearing aerial leaves (*c*). It had apparently been beaten down into the water by heavy rains, and this involuntary return to submerged life had induced the production of the submerged type of leaf in the apical region.

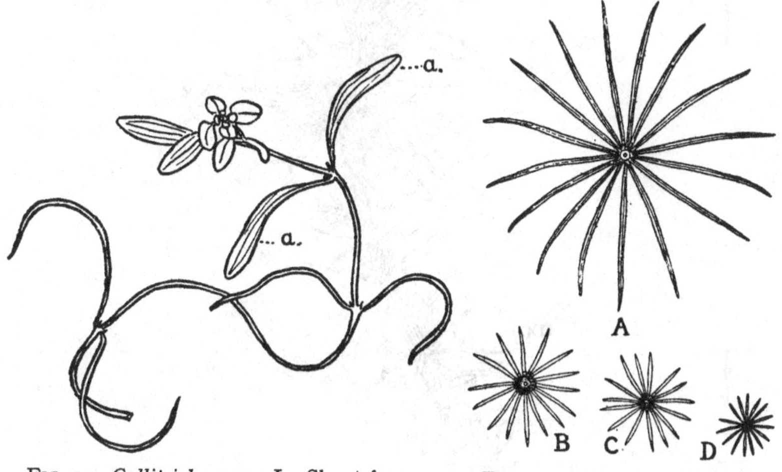

FIG. 94. *Callitriche verna*, L. Shoot from a ditch near the Cam, May 17, 1911, to show the difference between the submerged and floating leaves. The leaves down to, and including, the pair marked *a*, *a* were floating. (Reduced.) [A. A.]

FIG 95. *Hippuris vulgaris*, L. Leaf whorls. ($\frac{2}{5}$ nat. size.) *A*, water leaves; *B–D*, air leaves of land form. *B* and *C* have fruits in the leaf axils. [After Glück, H. (1911), *Wasser- und Sumpfgewächse*, Bd. III, Figs. 23 *a–d*, p. 250.]

Among the Umbelliferae, a differentiation between water leaves and aerial leaves is not at all uncommon. There are several instances even among our native plants. *Sium latifolium* is a very striking case. At the end of May, at Roslyn Pits, Ely, the present writer has seen a quantity of this plant, in a non-flowering condition, bearing three types of leaf—all three sometimes occurring on a single individual (Fig. 97, p. 149). These were—firstly, submerged leaves, either simply-pinnate but deeply incised (Fig. 98, p. 150), or compound-pinnate with

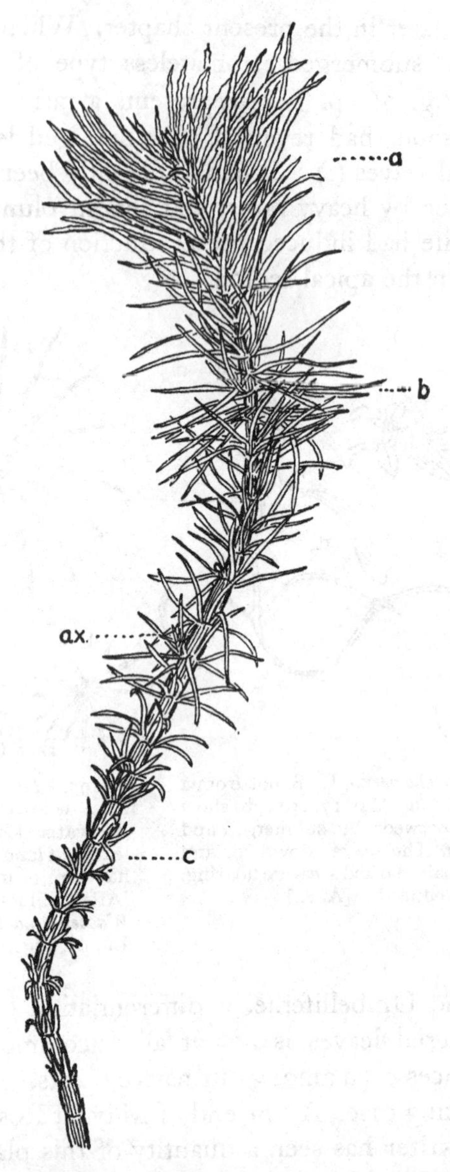

FIG. 96. *Hippuris vulgaris*, L. A shoot which was found lying horizontally in the water, August 17, 1917. It had produced air leaves (*c*), but presumably in very heavy rains, which had terminated a fortnight earlier, it had been beaten down and had produced (*b*) transition leaves and (*a*) water leaves. An axillary shoot (*ax*) bore water leaves. (½ nat. size.) Fig. 96 should be compared with Fig. 151, p. 231, which shows the normal relations of the two leaf types. [A. A.]

linear segments (*a* in Fig. 97): secondly, compound-pinnate air leaves, with each leaflet of the second degree toothed and lobed

FIG. 97. *Sium latifolium*, L. Plant from Roslyn Pits, May 30, 1911, showing three types of leaf: *a*, submerged compound-pinnate leaf with linear segments; *b*, erect air-leaf, compound-pinnate; *c* and *d*, erect air leaves, simply-pinnate. (Reduced.) [A. A.]

(*b* in Fig. 97); and thirdly, air leaves, once pinnate, with the leaflets toothed but not lobed (*c* and *d* in Fig. 97). Some small plants were found bearing the submerged type of leaf alone.

Where the three types were borne together, the simply-pinnate leaves were the latest to be produced, and the submerged leaves the earliest, while the compound-pinnate air leaves were intermediate.

Oenanthe Phellandrium, Lamk. var. *fluviatilis*, Colem.[1] is very common in the Cam near Cambridge. It has graceful, finely cut, pinnate leaves with obcuneate segments, and the plant is generally entirely submerged; a shoot as long as 235 cms. has been recorded. Its identity is liable to be puzzling at first sight, since its aerial axes are comparatively rarely to be found. On one occasion, however, on which the present writer found the plant bearing both submerged and aerial leaves, *Oenanthe Phellandrium* (proper) was noticed, at the same time, growing magnificently in a neighbouring ditch. It had a very stout, lacunate stem, bearing numerous aerial leaves and also a relatively small number of submerged leaves with capillary segments; the abundant lateral roots were lacunate. A comparison of the two plants suggested that *Oenanthe Phellandrium* var. *fluviatilis* is a mutation which has taken more whole-heartedly to water life than the type form of the species.

FIG. 98. *Sium latifolium*, L. Submerged leaf from a plant found at Wicken Fen, June 27, 1914. Less finely divided than leaf *a* in Fig. 97. (Reduced.) [A. A.]

Polygonum amphibium is an example of a hydrophyte which

[1] It is a matter of opinion whether this plant should be regarded as a distinct species or as a variety. See Coleman, W. H. (1844).

can produce either air leaves or water leaves with the utmost facility. The floating leaves and air leaves differ in internal anatomy and in the characters of the epidermis, and also show obvious external differences (Figs. 99 and 100, p. 152); the floating leaves are shiny, leathery and absolutely glabrous, while the air leaves are wrinkled and covered with hairs[1]. The lateral branches from a shoot with floating leaves, or even the end of the branch itself, may rise into the air and develop the characteristics of the land form[2].

Certain Scrophulariaceae are heterophyllous, such as *Ambulia* (*Limnophila*) *hottonoides* and *Hydrotriche hottoniaefolia*. In these cases the submerged leaves are finely divided. Among the Pedaliaceae, *Trapella*[3] has deltoid-rotundate floating leaves and linear-oblong submerged leaves, while *Limnosipanea Spruceana*, of the Rubiaceae, also shows a distinction between water and air leaves[4]. *Bidens Beckii*[5] is an example of a Composite showing heterophylly.

The heterophylly of the Alismaceae and Hydrocharitaceae need not be reconsidered now, since it has been dealt with in Chapters II and IV[6]. Two additional figures may, however, be included here, to illustrate the effect of transferring to water a small terrestrial seedling of *Alisma Plantago* found growing wild (Fig. 101, p. 153). After between two and three months, it had developed into the typical water plant shown in Fig. 102, p. 153.

There are many other cases of heterophylly among the Monocotyledons. Certain Potamogetons, e.g. *P. fluitans*, have air leaves, floating leaves and narrow submerged leaves[7]. *Potamogeton natans* is also a particularly good example; the narrow submerged leaves may attain a length of 50 cms. in running water[5]. The result of planting a land form of *P. natans* in water has been recorded[5]. The aerial leaves soon died, and

[1] Costantin, J. (1886).

[2] Schmidt, E. M. Inaug.-Diss. Bonn, 1879, quoted by Schenck, H. (1885).

[3] Oliver, F. W. (1888).

[4] Hansgirg, A. (1903).

[5] Goebel, K. (1891–1893).

[6] See pp. 9–14, 19–23, 51–52, 57, and Figs. 3–6, 9.

[7] Esenbeck, E. (1914).

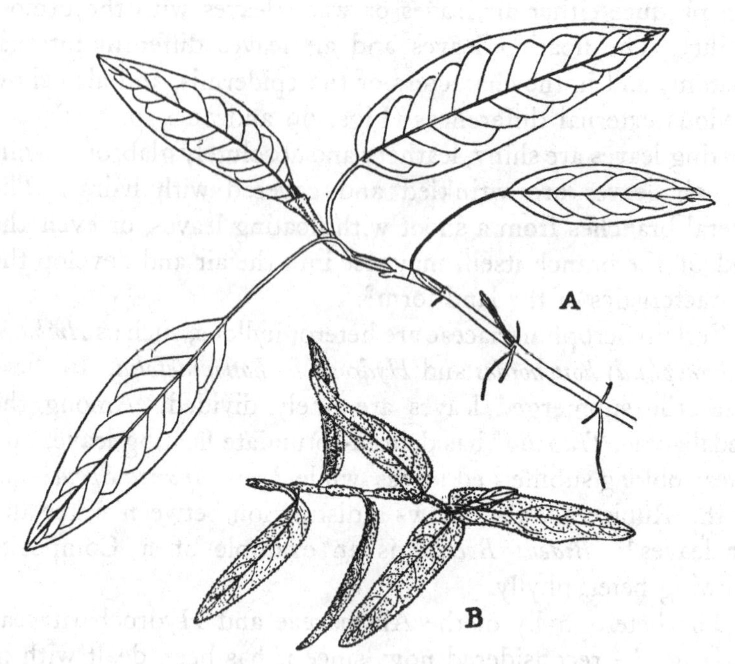

FIG. 99. *Polygonum amphibium*, L. *A*, branch of aquatic plant with floating leaves. *B*, branch of xerophilous plant inhabiting littoral dunes. [Massart, J. (1910).]

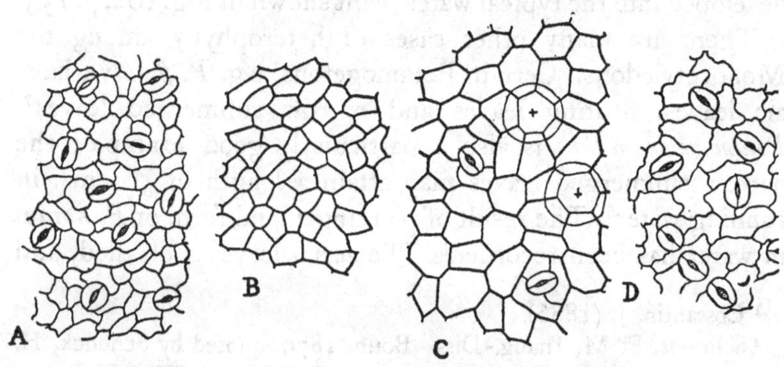

FIG. 100. *Polygonum amphibium*, L. *A*, upper epidermis, and *B*, lower epidermis of floating leaf, cf. Fig. 99 *A*. *C*, upper epidermis, and *D*, lower epidermis of xerophilous leaf, cf. Fig. 99 *B*. The elements marked with a cross are reservoir cells. [Massart, J. (1910).]

the next leaves formed had a smaller blade, a longer stalk, and an upper epidermis with chlorophyll and but few stomates.

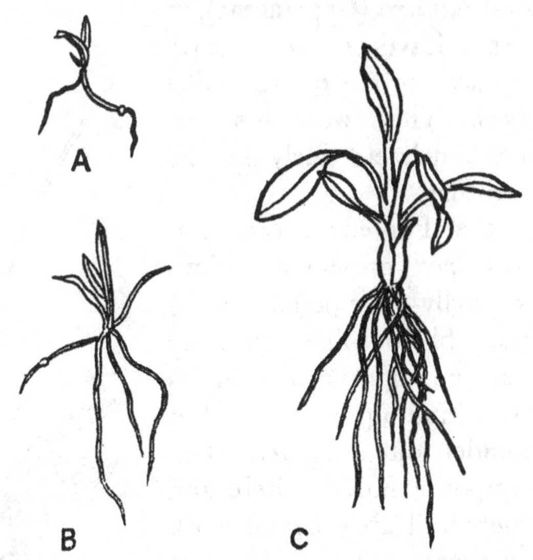

Fig. 101. *Alisma Plantago*, L. Seedlings found growing under the shade of a large *A. Plantago* plant in a dry ditch, May 31, 1911. (Nat. size.) [A. A.]

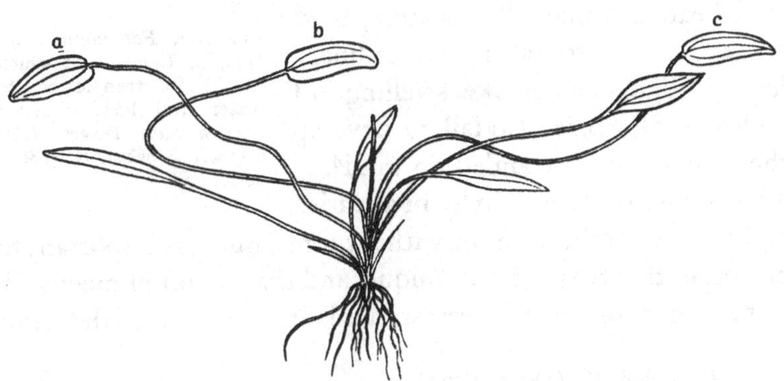

Fig. 102. *Alisma Plantago*, L. One of the seedlings such as those shown in Fig. 101 which had germinated on dry land, but was planted in a pot on May 31, 1911, and submerged in shallow water until August 9, 1911 (two months, nine days). In this time it developed into a typical water form with three floating leaves (*a*, *b*, *c*) and others showing transitions from the submerged type. (Reduced.) [A. A.]

The succeeding leaves were long and simple. Fig. 103, p. 154, illustrates this experiment.

The Pontederiaceae[1] and some of the Aponogetons[2] also have band-shaped, submerged leaves in addition to those that are aerial. *Scirpus lacustris* (Cyperaceae), in which the aerial leaves are very poorly developed, may produce strap-like floating leaves. They were first recorded by Scheuchzer[3] early in the eighteenth century.

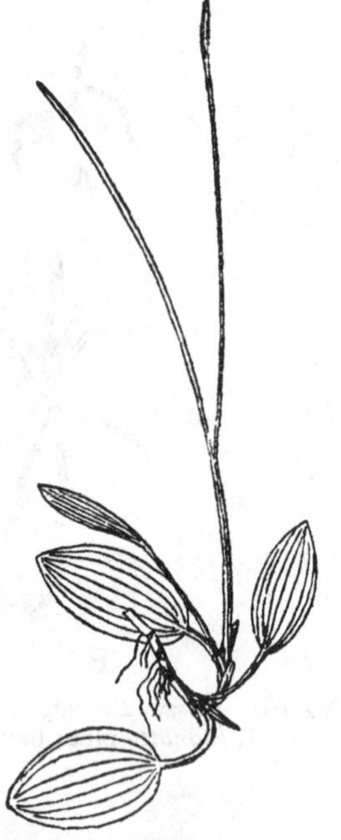

FIG. 103. *Potamogeton natans*, L. Land plant which has been transferred to water and has produced narrow water leaves. (Reduced.) [Goebel, K. (1891–1893).]

Some of the Pontederiaceae, e.g. *Eichhornia crassipes*, present a curious type of heterophylly—the petioles being swollen into bladder-like, floating organs, when the plant grows in its normal free-swimming manner, but becoming slender and elongated when it is thrown upon a muddy shore and takes root there[4]. Experimental work shows that not only a floating life, but full light and low temperature, encourage the spherical form of petiole, while heat, and poor illumination, tend to reduce it to a more ordinary shape[5]. The bladder-like swellings of the leaves of *Pistia* also fail to develop when the plant is stranded on mud[6].

Examples of heterophylly in aquatics might be multiplied almost without limit, but it is important to remember that they are not unique, and that we often meet with the same phenomenon in terrestrial plants. As Nehemiah Grew[7]

[1] Goebel, K. (1891–1893).
[2] Krause, K. and Engler, A. (1906).
[3] Scheuchzerus, J. (1719).
[4] Spruce, R. (1908).
[5] Treviranus, L. C. (1848[1]) and Boresch, K. (1912).
[6] Hansgirg, A. (1903).
[7] Grew, N. (1682).

wrote in the seventeenth century, "there are some, which have *Leaves* (besides the two first *Dissimilar* ones[1]) of Two Kinds or Two distinct *Figures*; as the *Bitter-Sweet*, the common *Little Bell*, *Valerian*, *Lady-Smocks*, and others. For the *Under Leaves* of *Bitter-sweet*, are Entire; the Upper, with two *Lobes*; the Under *Leaves* of the *Little Bell*, like those of *Pancy*; the Upper, like those of *Carnation*, or of *Sweet-William*."

We find parallels to the heterophylly of hydrophytes not only among terrestrial Flowering Plants, but also in the case of the distinct 'youth forms' of Conifers, and even—more remotely —in the *Chantransia* stage of such Algae as *Batrachospermum*.

The conclusion to be drawn from our very brief survey, which only touches the fringe of the subject, is that heterophylly is so widespread that no interpretation can be valid unless the condition be treated broadly as a very general attribute of plant life, rather than as a rare and exceptional phenomenon, for which special and individual explanations will suffice.

(3) THE INTERPRETATION OF HETEROPHYLLY

To the earlier writers, such as Lamarck[2], the problem of heterophylly presented no difficulties. They regarded the submerged or aerial type of leaf as representing a direct response, on the part of the plant, to the medium. The work of the last thirty years, has, however, rendered this simple conception untenable; the theory that now holds the field accords a much less prominent place to adaptation. The first observation that cast doubt upon the idea that leaf form necessarily depended directly on the *milieu*, was that of Costantin[3], who showed that, in the case of *Sagittaria*, the aquatic and aerial leaves were already distinguishable from one another in the submerged bud; he noticed auricles on a leaf which was only 2 to 3 mm. long. In *Ranunculus aquatilis*, also, the leaves destined to be aerial are differentiated in the bud.

[1] I.e. cotyledons. [2] Lamarck, J. B. P. A. (1809).
[3] Costantin, J. (1885²) and (1886).

A large amount of experimental work has been published by various authors on the effect of conditions upon the leaf forms of heterophyllous plants, and, although some of the results are confused and conflicting, a study of the literature seems to justify one general conclusion—namely, that, in many cases, the submerged type of leaf is, in reality, the juvenile form, but can be produced later in the life-history in consequence of poor conditions of nutrition; the air leaf, on the other hand, is the product of the plant in full vigour and maturity. This conclusion, which is primarily due to Goebel[1] and his pupils, is substantiated not only by experiments but by observations in the field.

In many heterophyllous plants, the first leaves produced by a seedling, whether it develops on land or in water, conform, more or less, to the submerged type. This is the case for instance in the Alismaceae. In *Alisma Plantago* (Fig. 101 *A* and *B*, p. 153) and *Sagittaria sagittifolia* (Fig. 90, p. 141), the first leaves produced by the seedling, or the germinating tuber, are ribbon-like, even when the young plant is terrestrial. The formation of this type of leaf can be induced again, even in maturity, by conditions which cause a general weakening of the plant. Costantin[2], thirty-four years ago, recorded that, when the leaves of *Alisma Plantago* were cut off in the process of clearing out a water-course, or in a laboratory experiment, the next leaves produced were ribbon-like, thus representing a regression to the submerged form. More recently, another worker[3] tried the experiment of cutting off the roots of healthy, terrestrial plants of *Sagittaria natans* which bore leaves with differentiated laminae; it was necessary to cut the roots away every week, as they grew again so rapidly. The result of this treatment was that the plants were found to revert to the juvenile stage, the new leaves being band-shaped. When the experimenter ceased to interfere with the roots, the plants again formed leaves with laminae. Other plants, with uninjured roots, grown as water cultures in distilled water, also produced the juvenile leaf form,

[1] Goebel, K. (1896), etc. [2] Costantin, J. (1886)
[3] Wächter, W. (1897[1]).

while those grown in a complete culture solution developed their laminae normally.

The same observer recorded a case in which a plant of *Hydrocleis nymphoides*, Buchenau (Butomaceae), which had been bearing the mature form of leaf, was observed to revert to the ribbon form. On examination it was found that most of the roots had died off. When a fresh crop of roots was produced, the mature type of leaf occurred again.

Another writer[1] demonstrated by a series of experiments upon *Limnobium Boscii* (Hydrocharitaceae) that, in this case also, the heterophylly is not a direct adaptation to land or water life, but that the floating leaves are "Hemmungsbildungen" due to poor nutrition. In *Stratiotes aloides*, also, he showed that the stomateless leaves were primary, and that their production could be induced at later stages by unfavourable conditions[2].

An experiment tried by Goebel[3] on *Sagittaria sagittifolia* indicated that absence of light in this case inhibits the formation of leaves of the aerial type. An observation of Glück's on *Alisma graminifolium*, Ehrh.[4], also points to the same conclusion. But it seems probable that the effect produced in these cases was not due directly to the darkness, but to the state of inadequate nutrition brought about by the lack of light for carbon assimilation.

Among the Potamogetons[5], again, experimental work has shown that reversion to juvenile leaves can be obtained under conditions of poor nutrition. For example, when a land plant of *P. fluitans*, which had been transferred to deep distilled water, had its adventitious roots repeatedly amputated, regression was obtained to the floating type of leaf and then the submerged type (Fig. 104, p. 158). A similar reversion to thin, narrow leaves was brought about, in the case of *P. natans*, by growing the upper internodes of a shoot as a cutting (Fig. 105, p. 159).

Waterlily leaves respond to experimental treatment in just

1 Montesantos, N. (1913).
2 See pp. 51–52.
3 Goebel, K. (1891–1893).
4 See p. 280.
5 Esenbeck, E. (1914).

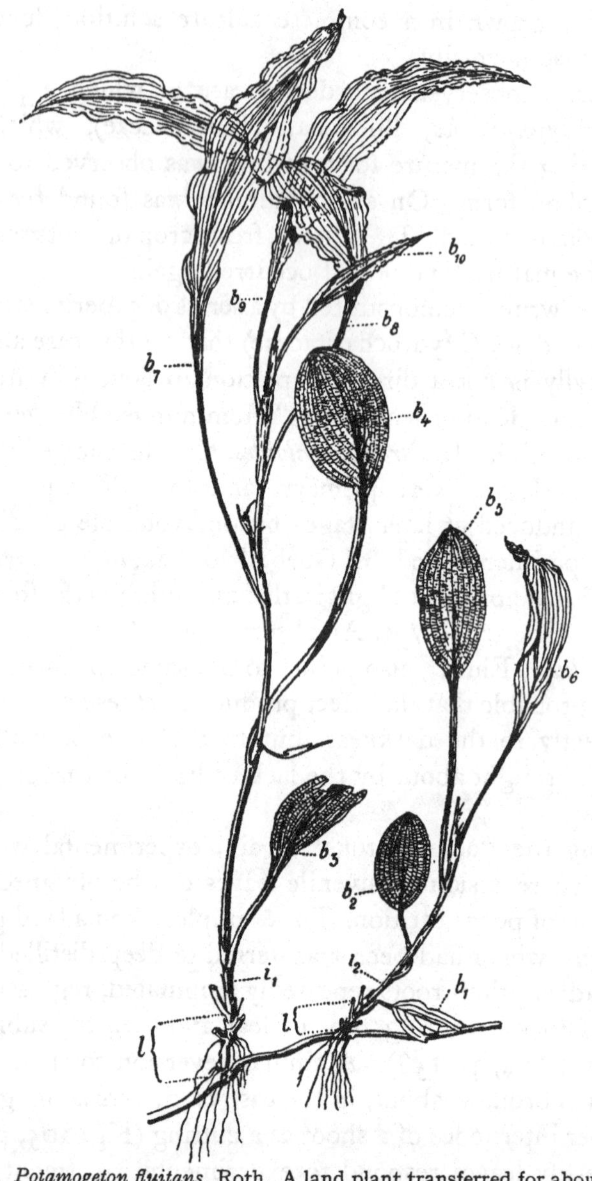

FIG. 104. *Potamogeton fluitans*, Roth. A land plant transferred for about a month to distilled water with no substratum and the adventitious roots repeatedly removed. The internodes marked l were formed during land life; b_1, b_2 and b_3 are the surviving land leaves; the internodes i_1 and i_2 show some elongation as a result of the changed conditions; b_4 and b_5 are floating leaves; b_6–b_{10} are leaves of the submerged type. [Esenbeck, E. (1914).]

the same way as the Monocotyledons already mentioned. In the case of two species of *Castalia*, it has been found possible to induce the mature plants to form submerged leaves, either by removing the floating leaves or by cutting off the roots[1]. This confirms an earlier suggestion, made by an Italian writer[2], that the development of the submerged leaves of *Nymphaea lutea*

FIG. 105. *Potamogeton natans*, L. The uppermost internodes of a normal plant grown as a cutting. One floating leaf (*s*) survives, while the axillary shoots have produced leaves with thin narrow blades, representing a transition between the floating and submerged types. [Esenbeck, E. (1914).]

was due to "un indebolimento o diminuzione di energia vitale." This suggestion has received independent, experimental confirmation from another worker[3], who estimated that a well-developed floating leaf of *Nymphaea lutea* was about eleven times the dry weight of a submerged leaf of the same area.

Another Dicotyledon, *Proserpinaca palustris*, which was in-

[1] Wächter, W. (1897[2]). [2] Arcangeli, G. (1890).
[3] Brand, F. (1894).

vestigated by Burns[1], gave results pointing to the same general conclusion as those observations already quoted. The primitive type of leaf in this plant is always a "water" leaf, but this form of leaf was also produced in the autumn by all the plants, regardless of any external conditions which the experimenter could control. On the other hand, at the time of flowering and in the summer generally, almost every plant, whether growing in water or air, produced the "land" type of leaf—the transition from the "water" to the "land" type taking place earlier on strongly growing than on weak stems. The author considers it evident that the aquatic environment is not the cause of the division of the leaf, nor does it depend on light, temperature, gaseous content of the water or contact stimulus. The only conclusion, which he considers justified by his experiments, is that *Proserpinaca palustris* has two forms—adult and juvenile; under good vegetative conditions, it tends to produce the adult form with the undivided leaf, the flower and the fruit, while, if the vegetative conditions are unfavourably influenced, a reversion can be induced to the primitive form with the submerged type of leaf. These results are consistent with those of McCallum[2], who had dealt with the same species at an earlier date, but his interpretation is slightly different. He is inclined to regard the occurrence of the aquatic form as induced by the checking of transpiration, and by the increased amount of water which hence accumulates in the protoplasm. This explanation is not inconsistent with the more general view that any condition tending to lower the vitality may be responsible for a reversion to the submerged type of leaf.

In nature, the regression to the juvenile type of leaf sometimes occurs, not only in the case of an entire plant subjected to adverse conditions, but also in the case of lateral shoots from an individual which is otherwise producing the mature form of leaf. Goebel[3], for instance, examined an old example of *Eichhornia azurea* (Pontederiaceae) which had wintered as a terrestrial

[1] Burns, G. P. (1904). [2] McCallum, W. B. (1902).
[3] Goebel, K. (1891–1893).

plant in a greenhouse; the leaves were of the mature form—differentiated into sheathing base, petiole and lamina—except in the case of a lateral shoot, which bore the grass-like, simple, leaves which characterise the young plant. Goebel[1] also describes the occurrence of subdivided leaves of the water type on lateral shoots of normal land plants of *Limnophila heterophylla*. A corresponding reversion has been observed in the case of the side branches of plants of *Proserpinaca palustris*[2] developing *in the air* from a plant whose main stem was producing the mature type of leaf; by removing the growing apex of the stem in June, these side branches of the 'water' type were induced to develop.

The interest of these lateral shoots, which show a reversion to an *ontogenetically* earlier type of leaf, is enhanced by the fact that C. and F. Darwin[3] have recorded a case of the occurrence, on lateral shoots, of leaves whose characters are probably *phylogenetically* earlier than those which the species normally exhibits. Their observations related to the sleep habits of the allied genera, *Melilotus* and *Trifolium*. They noticed, in *Melilotus Taurica*, that leaves arising from young shoots, produced on plants which had been cut down and kept in pots during the winter in a greenhouse, slept like those of *Trifolium*, with the central leaflet simply bent upwards, while the leaves on the fully-grown branches of the same plant afterwards slept according to the normal *Melilotus* method, in which the terminal leaflet rotates at night so as to present one lateral edge to the zenith. They suggest that *Melilotus* may be descended from a form which slept like *Trifolium*.

The idea that the 'juvenile' leaves, produced on lateral shoots, may in some cases represent an ancestral type, is consistent with the facts in the case, for instance, of the Alismaceae, provided that the 'phyllode theory' of the Monocotyledonous leaf be accepted in the sense advocated by Henslow and the present writer. According to this theory, which will be dealt

[1] Goebel, K. (1908). [2] Burns, G. P. (1904).
[3] Darwin, C. and F. (1880).

with in some detail in Chapter XXVIII, the ancestral leaf of this family was band-shaped, while the oval or sagittate blade, or 'pseudo-lamina,' is a later development—a secondary expansion of the distal region of the sheath or petiole. The submerged youth-leaves of this family would thus represent a reversion to phylogenetically older forms.

If the interpretation of heterophylly indicated in the present chapter holds good at all widely, the teleological view of the submerged leaf must be considerably modified. The present writer would like to suggest that, for the old conception of heterophylly as *induced* by aquatic life, we should substitute the idea that such a difference between the juvenile and mature forms of leaf as would render the juvenile leaf well suited to life in water, has been in many cases one of the necessary preliminaries to the migration from land to water, and that the aquatic Angiosperms thus include, by a process of sifting[1], those plants whose terrestrial ancestors were endowed with a strong tendency towards heterophylly[2].

[1] Guppy, H. B. (1906) first emphasized the fertile idea that the habitats of plants were determined by their peculiarities of structure, and not *vice versâ*. In relation to the occurrence of plants with buoyant seeds and fruits in water-side stations, he writes, "there are gathered at the margins of rivers and ponds, as well as at the sea-border, most of the British plants that could be assisted in the distribution of their seeds by the agency of water. This great sifting experiment has been the work of the ages, and we here get a glimpse at Nature in the act of selecting a station."

[2] In addition to the references mentioned in this chapter, MacDougal, D. T. (1914) and Shull, G. H. (1905) may also be consulted; the results recorded in these papers emphasize the difficulty and complexity of the problem.

CHAPTER XII

THE ANATOMY OF SUBMERGED LEAVES[1]

THE majority of submerged leaves have certain characters in common, the most obvious of which is their delicacy of structure. On removal from the water they generally collapse rapidly, and in some cases, e.g. *Hippuris vulgaris*, when they are plunged into alcohol the chlorophyll begins visibly to pass into solution almost from the first moment. The general tenderness of the leaves is due to the thinness of the mesophyll and the absence of differentiation between spongy and palisade parenchyma, and also to the relative lack of mechanical elements and the slight development of the cuticle[2]. It is indeed the epidermal characters—such as the reduction of cuticle—which most markedly distinguish submerged from aerial leaves.

It will be remembered that, in general, the epidermal cells of the leaves of Dicotyledons tend to be sinuous in outline, while those of Monocotyledons are more rectangular. But in the case of such a plant as *Callitriche verna* (Fig. 111, p. 170) which has both aerial and submerged leaves, it is found that, though the aerial leaves show the characteristic Dicotyledonous sinuosity in the form of their epidermal cells, the corresponding elements in the submerged leaves have straight walls, and hence approach the Monocotyledonous type. An interesting hypothesis on this subject was put forward long ago by Mer[3]. He drew attention to the fact that the epidermis was the tissue most directly affected by transpiration, and suggested that variations in that function might exercise an influence upon the form of the epidermal cells. According to his view, when transpiration is

[1] For a comprehensive account of this subject see Schenck, H. (1886), which has been largely drawn upon in the present chapter.

[2] A cuticle, though thin, seems to be invariably present. See Géneau de Lamarlière, L. (1906).

[3] Mer, É. (1880[1]).

feeble, as in the case of submerged plants, the epidermal cells are kept in a constant state of turgescence, and hence their growth takes a uniform course resulting in regularity of form. But, on the other hand, when transpiration is active, as in land life, the current is subject to great variations which react upon the form of the epidermal cells and produce sinuosity. It is scarcely possible to submit such a theory to direct proof, but it seems to the present writer that it is at least consistent with the fact, established at a much later date than Mer's work, that Monocotyledons with their rectangular epidermal cells, are in general, though with many exceptions, 'sugar-leaved' and weak transpirers, while Dicotyledons, with their epidermal cells often resembling a Chinese puzzle, are 'starch-leaved' and strong transpirers[1].

The epidermal cells of submerged leaves differ from those of air leaves not only in form but also in contents. Chlorophyll grains, which are generally described as absent from the epidermis of terrestrial plants, are often present in great abundance in this tissue in submerged leaves[2]. Treviranus[3], nearly a century ago, alluded to the lack of distinctively epidermal characters—or, to use his own expression, the "absence of an epidermis"—in the case of the lower surface of the leaf of *Potamogeton crispus*, while Brongniart[4], a few years later, observed the presence of chlorophyll in the leaf epidermis of *P. lucens*. Subsequently, epidermal chlorophyll has been observed widely among aquatic plants[5], though there are certain exceptions, such as *Callitriche*[6]. In some cases, e.g. *Zostera*, *Cymodocea*, *Posidonia*[7], the epidermis is actually the part of the leaf richest in green corpuscles. The presence of chloroplasts does not constitute, however, so absolute a difference from land plants as is sometimes assumed, since it has been shown that chlorophyll grains can be found in the epidermis of the green organs of the

[1] Stahl, E. (1900). [2] Schenck, H. (1886).
[3] Treviranus, L. C. (1821). [4] Brongniart, A. (1834).
[5] Chatin, A. (1855[1]), etc. [6] Schenck, H. (1886).
[7] Sauvageau, C. (1890[1]).

majority of terrestrial Dicotyledons, though they are generally absent in the case of terrestrial Monocotyledons[1]. They are usually to be observed only in the lower epidermis of the leaf, but it seems probable that this is due to the destructive action of sunlight upon the chlorophyll in the upper epidermis. In support of this view it may be mentioned that, in diffused light, chlorophyll occurs in the upper epidermis of the leaves of *Bellis perennis*, whereas under normal conditions there is chlorophyll only in the lower epidermis. The presence of green plastids in the epidermis of submerged plants may thus be regarded as representing merely the elaboration of a character already existing in terrestrial plants, which finds favourable opportunities for development in the relatively dim illumination which submerged plants receive.

The statement, frequently made, that stomates are absent from submerged leaves, and from the lower surface of floating

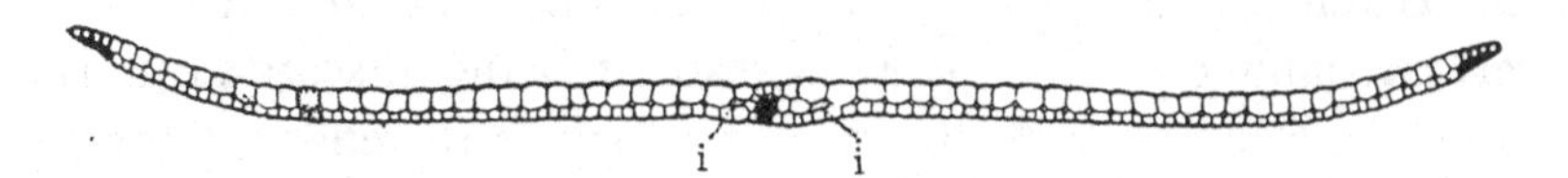

FIG. 106. *Elodea canadensis*, Michx. T.S. leaf; *i*, intercellular air channels. [Schenck, H. (1886).]

leaves, needs considerable qualification[2]. It is, indeed, broadly true that stomates are much less frequent in submerged than in terrestrial leaves, and, moreover, in certain water plants, such as *Elodea* (Fig. 106), *Vallisneria*, *Thalassia*, and other Hydrocharitaceae which always live entirely submerged, stomates never occur[3]. Among the Cryptogams, *Isoetes lacustris* is entirely free from stomates, and Goebel[4] even found that it failed to produce any when grown for two years as a land plant. Submerged leaves in general are not only poor in stomates but also in hairs; it has been suggested by Mer[5] that this—like the

[1] Stöhr, A. (1879).
[2] Costantin, J. (1885[1]). See also Porsch, O. (1903) for citations of a large number of cases in which the occurrence of stomates on submerged organs is mentioned in the literature.
[3] Solereder, H. (1913).
[4] Goebel, K. (1891–1893).
[5] Mer, É. (1880[1]) and (1882[1]).

form of the epidermal cells—may be correlated with the feebleness and uniformity of the transpiration stream. He supposed that the active and variable flow of sap in land plants might bring about the accumulation of nutriment at certain points of the epidermis, thus favouring localised cell-multiplication and the production of hairs and stomates. It seems possible to the present writer that this suggestion contains an element of truth. But on the other hand it must be remembered that stomates have been observed in a large number of submerged leaves, such as those of *Lobelia Dortmanna*[1], *Villarsia ovata*[2] and *Pontederia cordata*[2], and on the lower surfaces of certain floating leaves, such as *Limnocharis Humboldtii*[3] and *Hydrocharis Morsus-ranae*[4]. Porsch[5], who has considered the subject comprehensively, concludes that the stomatal apparatus must have been gradually evolved over a long period of time, so that its characters have become fixed with great tenacity; for, in cases where its existence must be not only superfluous, but attended by a certain danger to the plant, instead of being discarded, it is often modified secondarily in such a way as to render it functionless. He shows that, in the case of submerged plants which retain their stomates, four different modifications are found, each of which must have the result of preventing water entering the tissues through the aperture between the guard cells:

(1) The guard cells may close on submergence, even in full illumination, e.g. *Callitriche verna* and *Hippuris vulgaris*.

(2) The aperture may be permanently closed, as in the case of *Potamogeton natans* (Fig. 107 *B*), in which the whole stomatal apparatus remains roofed in with cuticle.

(3) The development of each stomate may actually cease at an early stage. This is rare, but such abortive stomates are found in the submerged parts of a species of *Oenanthe*[6].

[1] Armand, L. (1912). [2] Costantin, J. (1885[1]).
[3] Duchartre in discussion following Chatin, A. (1855[1]).
[4] Goebel, K. (1891–1893). [5] Porsch, O. (1903) and (1905).
[6] Porsch uses the specific name "*Oenanthe aquatilis*, L."; he is probably referring to *Oe. Phellandrium*, Lamk. var. *fluviatilis*, Colem.

(4) The stomates may develop normally, but the guard cells remain pressed together with their cuticular ridges interlocked, e.g. *Calla palustris* (Fig. 107 *A*).

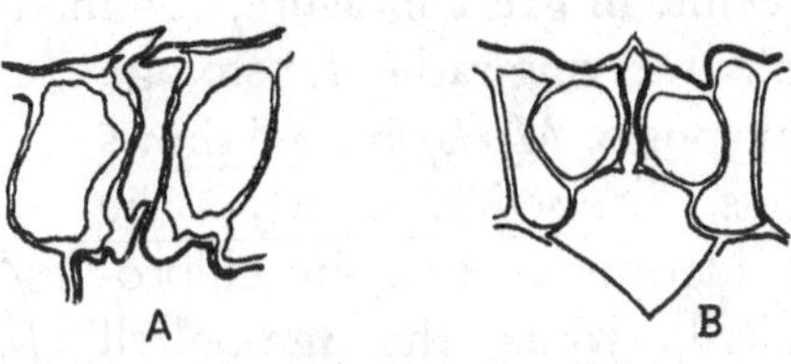

FIG. 107. A. *Calla palustris*, L. T.S. stomate in submerged leaf stalk; the thickening bands fit closely together. B. *Potamogeton natans*, L. T.S. submerged stomate from leaf stalk of floating leaf. This stomate is entirely roofed in with cuticle. [Porsch, O. (1905).]

In addition to ordinary stomates, which, in submerged life, are incapable of exercising their normal function, submerged leaves also very commonly bear water stomates, which are probably of importance in keeping up the 'transpiration' stream by exudation[1]. A longitudinal section passing through the water pores of *Pistia Stratiotes* is shown in Fig. 53, p. 82, while the apical opening of *Potamogeton densus*—in which the tracheids communicate directly with the exterior without the intervention of water stomates—is represented in Fig. 108.

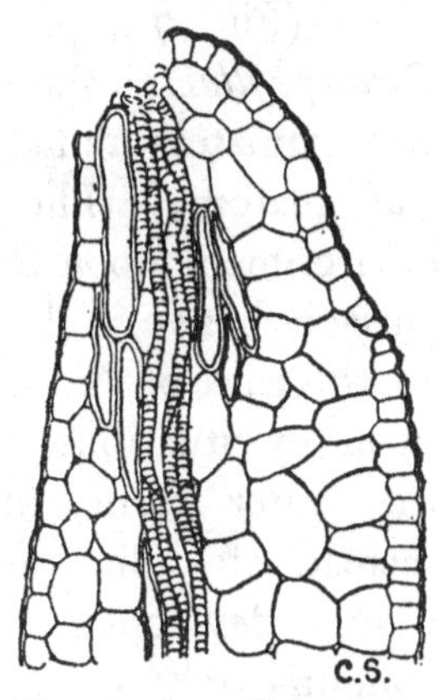

FIG. 108. *Potamogeton densus*, L. L.S. apex of leaf passing through the median nerve and showing the apical opening. (Upper surface of leaf to right hand.) (× 220.) [Sauvageau, C. (1891[1]).]

The aerating system of submerged leaves is a very conspicuous feature. The mesophyll of such subcylindrical radical leaves as those of *Littorella* and *Lobelia Dortmanna* is traversed from end to end by air passages, interrupted only by porous diaphragms, and the same feature is markedly developed in the elongated petioles of such leaves as *Sagittaria* (Fig. 8, p. 19) These diaphragms form *points d'appui* for the secondary nerves connecting the longitudinal bundles[2].

The mesophyll of submerged leaves shows, as has been already indicated, little sign of differentiation into palisade and

[1] This subject is considered more fully in Chapter XXI.

[2] Duval-Jouve, J. (1872).

spongy parenchyma. In many cases the assimilatory activity seems, in great measure, confined to the epidermis, the mesophyll serving rather for storage purposes. *Myriophyllum*[1] shows this distinction clearly; the epidermis is rich in chlorophyll, while the mesophyll contains large starch grains (Fig. 109). This leaf is a good example of the subdivided, submerged type, each limb of which exhibits a tendency to a radial arrangement of the tissues. *Littorella*, *Utricularia minor* (Fig. 74, p. 108) and *Ceratophyllum* all show the same approximately radial type of leaf anatomy. The effect of environment upon this kind of leaf, is illustrated by a comparison between the land and water forms of *Myriophyllum*. In the case of *M. alternifolium*, the land form, when growing in sunny situations, has shorter and thicker leaf segments than the water form; they are also dorsiventral and elliptical, instead of radial and cylindrical, while the xylem is more highly developed than in the water form. The epidermis contains only a few small chlorophyll grains, and stomates occur. The epidermal cells also have the sinuous outline which is lacking in the water form. The absence of marked dorsiventrality in the leaves of many submerged plants, such as *Myriophyllum*, may in part be attributed to the fact that they are perpetually being moved about by water currents, and thus they do not retain any constant position in relation to the incident light.

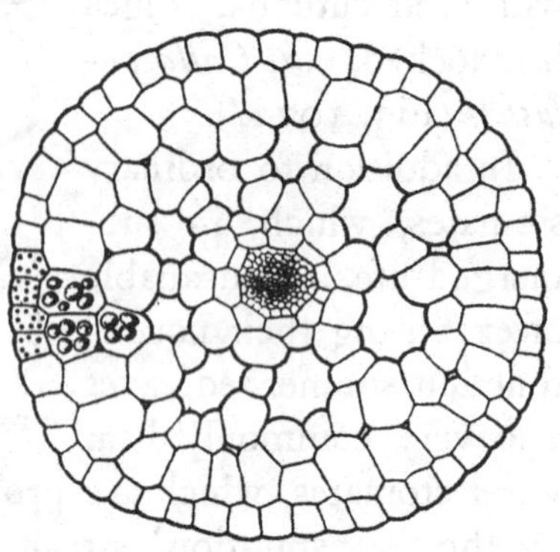

FIG. 109. *Myriophyllum spicatum*, L. T.S. through a segment of the leaf of the water form. The epidermis contains chloroplasts and the mesophyll is laden with large starch grains, only indicated in a few cells. (×157.) [Schenck, H. (1886).]

The very young submerged leaves of *Myriophyllum verticillatum* and *M. spicatum* show a peculiarity which has repeatedly

[1] Schenck, H. (1886).

attracted the attention of botanists[1]—the occurrence, namely, of little colourless cellular plates, arising generally at the apex and base of each lobe, but sometimes elsewhere (Fig. 110 *A* and *B*, p. 170). The cells at their base (*c* in Fig. 110 *B*) become corky at an early stage, and the plates drop off. They are probably best interpreted as caducous trichomes; their function, if they have one, is quite unknown.

As examples of the flat, non-radial type of submerged leaf, *Callitriche*, *Elodea* and *Alisma* may be mentioned. In Fig. 111, p. 170, the contrast between the aquatic and aerial leaf of *Callitriche verna* is indicated. The water leaf is thin, but still retains some mesophyll; the outlines of the epidermal cells in the two forms show the distinguishing characters to which reference has already been made. *Callitriche autumnalis*[2], which lives and flowers completely submerged, has a thinner leaf. The leaf of *Hottonia* resembles that of *Callitriche*. The ribbon-leaf of *Alisma Plantago* shows a slightly different type of structure. The chlorophyll-containing epidermis forms the essential part of the leaf, and the large air passages are bounded by it. There is one main bundle, accompanied by two tiny laterals placed close to the margins. In *Elodea canadensis* (Fig. 106, p. 165) we reach almost the ultimate phase in reduction of the mesophyll, for here the entire assimilating tissue is reduced to the two epidermal layers. The extremely delicate leaf is strengthened by some fibrous cells. Supporting sclerenchyma is characteristic of a certain number of submerged leaves such as those of the Potamogetons (e.g. Fig. 38, p. 61).

There is a strong tendency, in submerged leaves, to the reduction of the tracheal system. Among the Hydrocharitaceae, for instance, though typical spiral tracheids occur in the submerged leaves of *Stratiotes*, the leaves of a number of other genera show either no tracheids at all, or else more or less ephemeral elements with annular thickenings, e.g. *Elodea*, *Halophila*, *Vallisneria* and *Thalassia*[3].

[1] Irmisch, T. (1859[1]), Borodin, J. (1870), Magnus, P. (1871), and Perrot, É. (1900).

[2] Hegelmaier, F. (1864).

[3] Solereder, H. (1913).

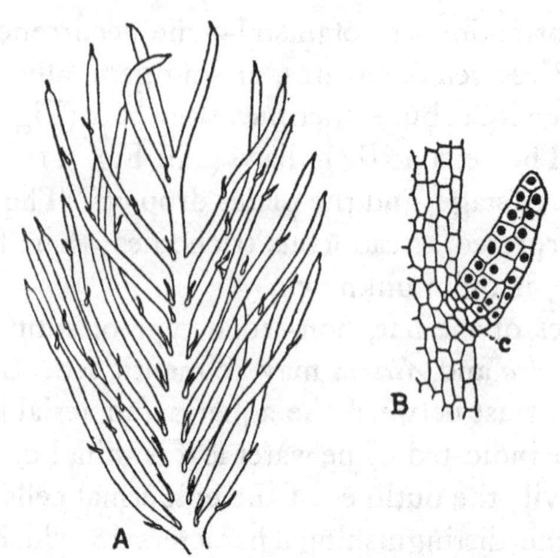

FIG. 110. Trichomes of *Myriophyllum verticillatum*, L. *A*, diagram of a young leaf showing the arrangement of the trichomes. *B*, a single multicellular caducous trichome at leaf margin with corky cells, *c*, at its base. [Perrot, É. (1900).]

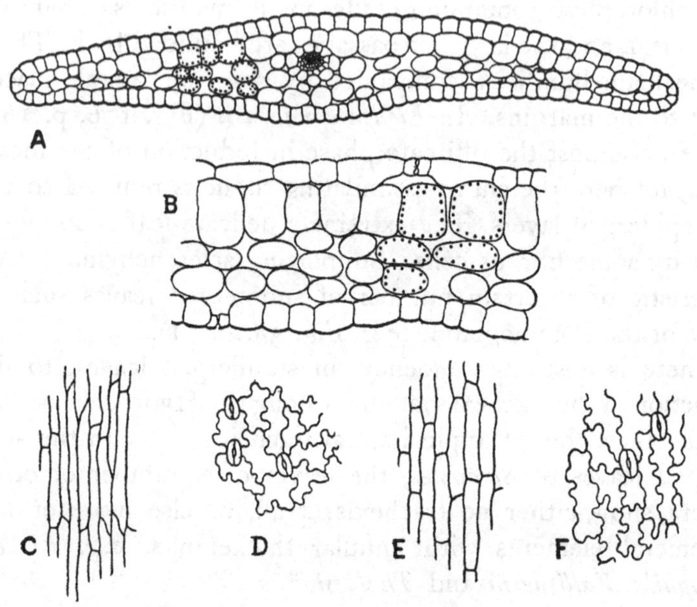

FIG. 111. *Callitriche verna*, L. *A*, T.S. submerged leaf, × 80; *B*, T.S. leaf of land form, × 147; *C*, upper epidermis of submerged leaf, × 92; *D*, upper epidermis of land leaf, × 88; *E*, lower epidermis of submerged leaf, × 92; *F*, lower epidermis of land leaf, × 88. [Schenck, H. (1886).]

A consideration of the structure of submerged leaves opens up a series of perplexing theoretical problems. The idea that the submerged type of leaf arises as an adaptive response to the *milieu*, proves on examination altogether inadequate. The general form of these leaves seems attributable to poor nutrition, while certain characters—thinness, lack of differentiation of spongy and palisade parenchyma, and presence of chlorophyll in the epidermis—are also common, in some degree, to terrestrial plants growing in the shade, and seem intimately connected with lack of sunlight[1]. We may perhaps suppose that the dimness of the light which reaches a plant living below the surface of the water may be directly responsible for these characters; the green pigment, for instance, may be present in the epidermis simply because the leaf is not exposed to direct sunlight, which in the case of terrestrial plants destroys the chlorophyll in the epidermis as fast as it is formed[2]. Now there is little doubt that a thin leaf with an epidermis rich in chlorophyll is particularly well adapted for the assimilation of dissolved carbon dioxide; how then are we to account for the singular coincidence that characters arising in this fortuitous and mechanical fashion prove definitely advantageous to the plant? It is perhaps conceivable that it is the very fact that terrestrial plants under conditions of poor illumination tend to develop this type of leaf, which has rendered possible the assumption of the submerged habit, and that it is those plants whose leaves happened under such conditions to develop on the lines particularly suited to water life, which have accomplished the transformation into thorough-paced aquatics.

[1] Schenck, H. (1885). [2] Stöhr, A. (1879).

CHAPTER XIII

THE MORPHOLOGY AND VASCULAR ANATOMY OF AQUATIC STEMS[1]

THE stems of plants that pass the greater part of their vegetative life entirely submerged, fall in general into two categories. The less common type is the abbreviated axis bearing a tuft of long narrow leaves (e.g. *Stratiotes*, Fig. 31, p. 49 and Fig. 32, p. 53) while, on the other hand, the majority of submerged plants are characterised by thin, elongated, branched stems rising wholly or partially into the water, clothed with leaves and often capable of rooting at the nodes (e.g. *Potamogeton*, Fig. 37, p. 60 and *Myriophyllum*, Fig. 144, p. 221). Owing to the high specific gravity of the water, and the lightness of the stems, due to the air in the intercellular spaces, each axis is to a large extent relieved of the task of supporting the weight of its branches. In consequence there seems to be no impulse to the relatively strong development of a single main axis, and, in conformity with this, the general system is often sympodial (e.g. *Hippuris*, Fig. 112). The plant frequently grows actively in front while it dies away behind, and may thus be regarded, to use Schenck's expression, as being in a state of perpetual youth. The older regions tend to become infested with a flora of epiphytic Algae and Fungi, among which a microscopic fauna makes its appearance. This is an obvious disadvantage, since no leaf thus laden can perform its functions successfully. Possibly the rapid growth of fresh leafy shoots at the apex serves as a compensation for a loss of activity in the older regions, traceable to this cause.

The vascular system of submerged stems shows certain modifications upon the terrestrial type, the most striking differ-

[1] For a detailed treatment of this subject see Schenck, H. (1886), which has been largely drawn upon in the present chapter.

ence being that the xylem tends to be reduced in amount, while the lignification is often very poor. Spiral or annular vessels, when present in the neighbourhood of the growing apex, may, in some instances, be completely destroyed by the elongation of the internodes, and may survive only at the nodes, e.g. *Potamogeton lucens*[1], *Zannichellia palustris*[1], *Althenia filiformis*[2], etc., while in the case of *Elodea canadensis*[1] the tracheal thickenings do not even persist in the nodal tissues. *Ceratophyllum* (Fig. 56, p. 87) is an example of a further degree of reduction, since here lignification is entirely lacking, even in the apical region. This loss of lignification has been sometimes regarded as a corroboration of the widely-held view that the transpiration stream has no existence in submerged plants. But, as we shall show in Chapter XXI, the idea that such a current is absent in these plants, seems often to have been accepted on totally inadequate grounds. In

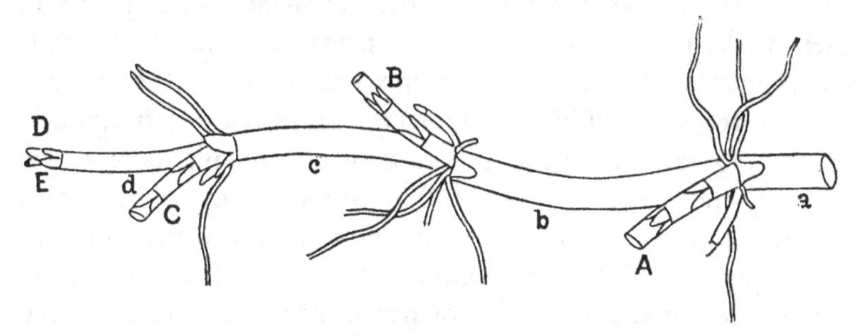

FIG. 112. *Hippuris vulgaris*, L. Diagram of the horizontal rhizome as seen from above to show sympodial growth; *a–A*; *b–B*; *c–C*; *d–D, E*, represent successive axes. [Irmisch, T. (1854).]

this connexion it appears to the present writer that, when xylem and the part which it plays in water-conduction is being considered, too much stress is often laid—almost unconsciously perhaps—on the question of lignification. It seems sometimes to be assumed that the functional importance of the xylem is proportional to its degree of lignification; an idea which may perhaps be interpreted partly as a hypnotic impression conveyed to the botanist's mind by the vividness of the xylem in

[1] Caspary, R. (1858[2]). [2] Prillieux, E. (1864).

stained sections, and partly as a survival from the old days of the 'imbibition theory,' when the ascent of water was supposed to be due to some mysterious property peculiar to the lignified membrane. But it is now universally recognised that water travels in the cavities of the vessels and tracheids rather than in the walls. What part then does lignification play in the ascent of water? It must be remembered that the water-conducting elements are dead and empty, and that in terrestrial plants they often contain air, which is more or less rarefied, and is thus at low pressure. These dead elements are generally in contact with turgid living cells, which exert a strong pressure against their walls. From the point of view of the ascent of water, the only function of the lignified walls of vessels and tracheids appears to be to prevent their being crushed by the neighbouring living elements. The way in which tyloses force themselves into vessels through the defenceless, thin places in their walls, gives some idea of the pressure which living cells are prepared to exert. In hydrophytes, however, the circumstances are very different. The vessels, instead of frequently containing rarefied air, as in the case of land plants, are presumably more continuously full of liquid, and are therefore less liable to be crushed and obliterated by the surrounding living elements. The conduction of water is not, in their case, conditioned by the possession of armoured walls. There is every reason to suppose that the non-lignified conducting elements of a submerged plant may be as effective in raising water as the woody vessels of a terrestrial tree; that water does, as a matter of fact, travel freely in the non-lignified xylem spaces of the submerged Potamogetons has been shown by experiment[1].

Elongated, submerged stems, unless they grow in perfectly still water, must be subjected to some amount of tension from currents. It is probably more than a mere coincidence that the vascular system of aquatics is so often condensed into a central strand, recalling the central cylinder of roots and of climbing stems, both of which are organs subjected to pulling forces.

[1] Hochreutiner, G. (1896); see pp. 261–263, Chapter XXI.

The central strand, even when extremely simple as in the case of *Callitriche*[1], the Hydrilleae, *Aldrovandia*, *Naias*, *Hippuris*[2], etc., is not a single bundle, but represents an entire vascular system, in which the strands are not differentiated as individuals. That the xylem reduction, to which we have already referred, is not itself the cause of the union of the single bundles into an axial strand, may be deduced from a comparison with the stems of colourless saprophytes or parasites. In such plants there is little transpiration and no assimilation and the xylem is proportionately reduced. But the simplified bundles retain their ancestral position and do not fuse into an axial strand[3].

Among the Dicotyledons there are certain hydrophytes, e.g. the Water Buttercups (Fig. 113, p. 176), in which the bundles remain perfectly separate, but in the majority some degree of condensation may be observed. The Potamogetons (Fig. 39, p. 62 and Fig. 40, p. 64) provide an exceptionally interesting series illustrating, within a single Monocotyledonous genus, stages in the concentration of the vascular cylinder. It must suffice here to draw attention to a few other typical examples, showing various grades in the reduction of the vascular system.

In *Peplis Portula* there is a well-marked axial strand, in which individual bundles can no longer be distinguished. In transverse section, an external ring of disconnected phloem groups is seen to enclose a ring of xylem, consisting of short radial rows of vessels separated by rows of parenchyma. The internal phloem characteristic of the Lythraceae is developed within the xylem, and a pith is formed. A cambial layer occurs, but does little work.

The next stage of reduction may be illustrated by the stem of *Callitriche* (Fig. 114 *A* and *B*, p. 176) which shows in transverse section a small ring of xylem surrounded by phloem; there is no cambium. In the water forms (Fig. 114 *B*) the pith is resorbed at an early stage and is represented by a space.

Hippuris has travelled still further upon the road of speciali-

[1] Hegelmaier, F. (1864). [2] Sanio, C. (1865).
[3] Schenck, H. (1886).

sation. The vascular tissue is concentrated into a definite cylinder, with external phloem and internal xylem, enclosing what seems at first sight to be a pith. But Sanio[1], who described the

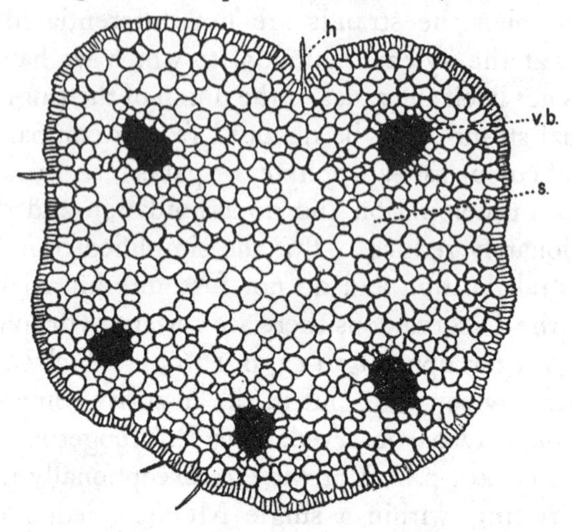

FIG. 113. *Ranunculus trichophyllus*, Chaix. T.S. young stem to show the numerous air spaces, *s*, in the ground tissue. *v.b.* = vascular bundle; *h* = hair. (× 47.) [A. A.]

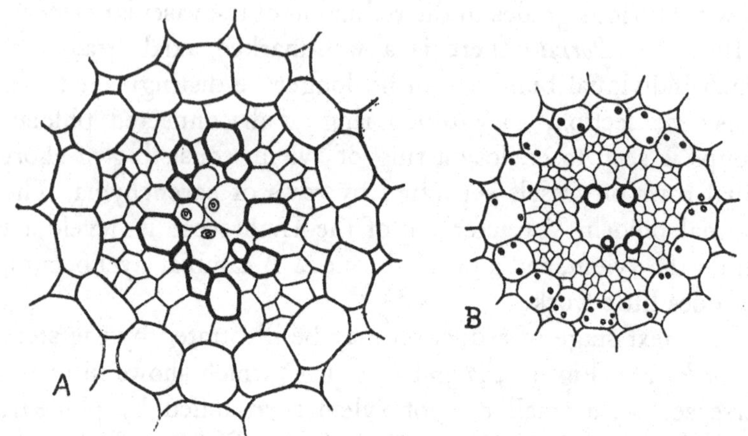

FIG. 114. *Callitriche stagnalis*, Scop. Central cylinder of stem. *A*, land form. (× 475.) *B*, water form. (× 290.) [Schenck, H. (1886).]

anatomy of the stem, demonstrated that the central region, which, if the mature structure alone were examined, would

[1] Sanio, C. (1865).

certainly be regarded as pith, is in reality to be interpreted as xylem parenchyma. He described the occurrence of a number of cauline tracheal elements in the 'pith' region of the embryonic vascular cylinder near the growing point. These cauline elements were found by Sanio to be ephemeral and impersistent; he observed their first appearance at levels above the entry of the first lignified leaf traces. This account appeared to the present writer so singular, that she repeated Sanio's observations in order to see whether the application of microtome methods, by which the history of the tissue in question could be traced element by element, would confirm or refute his conclusions. The result was in all essentials to confirm Sanio's description; the accuracy of his work is indeed remarkable, when it is considered that he was obliged to rely entirely on hand sections for the interpretation of this delicate piece of apical structure. In one stem-apex examined by the present writer, the first cauline xylem element appeared when the stele was only 0·08 mm. in diameter (Fig. 115 *A*, p. 178). This harmonises with Sanio's statement that in one preparation he observed the first cauline element when the cylinder was about 0·1 mm. across. The cauline elements gradually increased (Fig. 115 *B*) and persisted for a distance of a few millimetres from the apex, becoming gradually less lignified and thinner-walled until they finally disappeared. At the level at which the first lignified leaf trace began to pass in towards the stele (Fig. 115 *B*), there were twenty-one cauline tracheal elements. At a slightly lower level, at which the tracheids belonging to eight leaf traces (*L*) had entered and taken up a position at the periphery of the stele, twenty-one cauline elements could still be identified (Fig. 115 *C*). In this particular case, they were found to be just finally vanishing at the level at which the seventh set of lignified leaf traces (counting from the apex) entered the stele; at this level the stele was only 0·2 mm. in diameter. However, a few of the outermost cauline elements were more persistent than the rest, and either themselves became part of the xylem ring, or fused with the leaf traces as they entered. That the lignified elements in the 'pith' are actually xylem, and not merely altered pith cells, is indicated by their possession of typical tracheal thickenings, and also by their occasionally identifying themselves, as just mentioned, with the xylem ring.

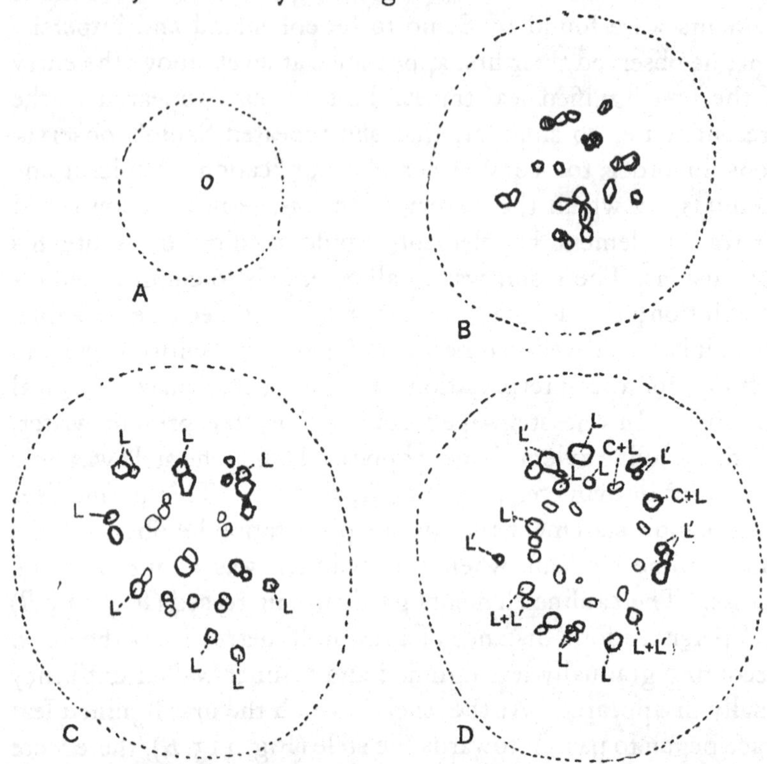

FIG. 115. *Hippuris vulgaris*, L. Series of transverse sections of stele of a stem near apex to show relation of cauline and leaf trace xylem; the dotted line in each case represents the periphery of the stele. (× 280 *circa*.) *A*, appearance of first cauline element when stele is 0·08 mm. in diameter. *B*, level at which first lignified leaf trace begins to pass in towards the stele, which contains 21 cauline xylem elements, but no leaf traces. *C*, the level at which eight lignified leaf traces (*L*) have taken up a position at the periphery of the stele, in which 21 cauline elements can still be counted. *D*, a lower level at which traces (*L'*) from a second node have entered. Fusion of traces from the two nodes or of cauline elements with either is indicated by (*L* + *L'*), (*C* + *L*), etc. [A. A.]

Myriophyllum (Figs. 116 and 117) closely resembles *Hippuris* in vascular anatomy and has the same cauline tracheal elements in the pith, but the xylem is more reduced[1].

[1] Vöchting, H. (1872).

Ceratophyllum (Fig. 56, p. 87), as we have already shown, may be regarded as representing the extremest stage in the simplification characteristic of the stem-anatomy of Dicotyledonous water plants. There is a central duct, surrounded by

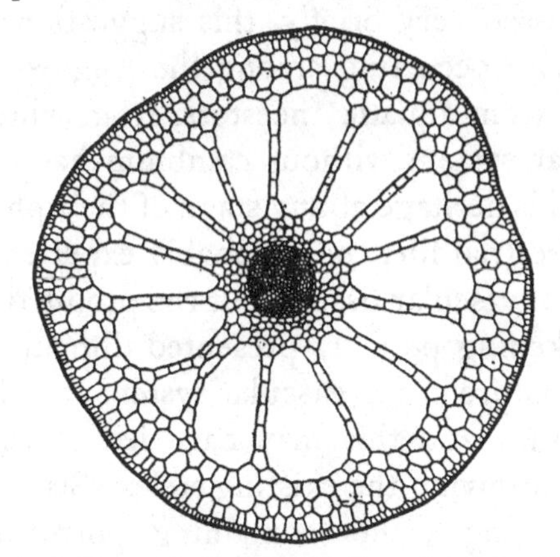

FIG. 116. *Myriophyllum spicatum*, L. T. S. moderately old axis. (× 30.) [Vöchting, H. (1872).]

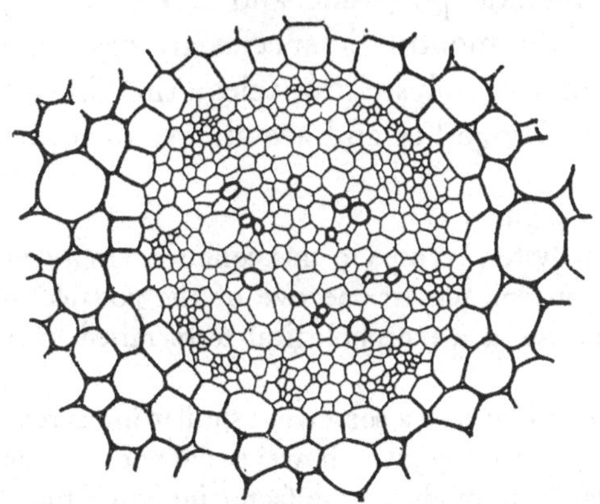

FIG. 117. *Myriophyllum spicatum*, L. T.S. stele of young axis showing the scattered internal vessels and eight phloem groups near the periphery of the stele. (×215.) [Vöchting, H. (1872).]

elements whose walls are somewhat thickened, but consist of cellulose only[1]. These thick-walled cells are again surrounded by a broad zone of phloem[2].

In connexion with the strong tendency shown by aquatic

[1] Sanio, C. (1865). [2] Schenck, H. (1886).

plants towards the condensation of the vascular system to a single strand, devoid of secondary thickening, and in which individual bundles cannot be distinguished, an interesting suggestion, put forward some years ago by Scott[1], may be considered. Expressed very briefly, this suggestion was that the cases of polystely[2] occurring among the Angiosperms may be due to descent from aquatic ancestors, from which a reduced type of vascular system without cambium has been derived. If plants with this heritage at any stage of their phyletic history returned to terrestrial life, they probably experienced the need for an increase of vascular tissue; but the production of normal secondary thickening possibly presented difficulties, owing to the condensed nature of the vascular system and the loss of the cambial apparatus, and this may have led to the alternative expedient of multiplying the existing steles. Scott refers to two genera of flowering plants containing polystelic species—*Auricula* (Primulaceae) and *Gunnera*[3] (Haloragaceae). Both these genera include polystelic and monostelic species. The single steles of the monostelic species are exactly like the individual steles of the polystelic species; they have the vascular bundles crowded together and are almost devoid of pith and

[1] Scott, D. H. (1891).

[2] The word 'polystely' is used in this connexion in a descriptive sense, as a matter of convenience, irrespective of the possible validity of the objections to its use as a morphological term raised by Jeffrey, E. C. (1899).

[3] For the case of *Gunnera* a somewhat similar interpretation had been proposed in 1875 by Russow, who however did not perceive that a return from water to land life might be the factor initiating the polystelic condition. He suggested that the Gunneras were descended from ancestors whose vascular system had been condensed into a single central strand, and that in the course of generations this form of stele might have become so far stereotyped that it could no longer separate into its original constituents (collateral vascular bundles) when a more elaborate conducting system was required; it thus adopted the alternative of branching, and reproducing its structural peculiarities in each branch. (Russow, E. (1875).)

secondary thickening. Both *Auricula* and *Gunnera* have near relatives which are aquatic in habit. The reduced aquatic stele of the submerged stem of *Hottonia* has much in common with an individual stele of *Auricula*. This comparison between *Hottonia* and *Auricula* has had its force greatly increased by Prankerd's[1] subsequent discovery of a transient polystelic phase in *Hottonia palustris* in the base of the inflorescence axis—that is to say, in the region of transition from an aquatic to an aerial type of stem.

It was observed by Scott that the stele of *Myriophyllum* or *Hippuris* agrees closely in structure with that of the monostelic Gunneras, or with a single stele from one of the polystelic species. The comparison of the stele of *Myriophyllum* with that of the Gunneras has been fully confirmed by more recent work[2]. In the case of *Gunnera*—assuming a descent from an aquatic ancestor—it is easy to realise how acute the need for increased vascular tissue in the rhizome must have become when the present type of habit was acquired, since the leaves grow in some cases to an enormous size. Darwin[3], in the *Voyage of the Beagle*, describing the occurrence of *Gunnera scabra* on the Island of Tanqui, off Chili, remarks—"I measured one [leaf] which was nearly eight feet in diameter, and therefore no less than twenty-four in circumference! The stalk is rather more than a yard high, and each plant sends out four or five of these enormous leaves, presenting together a very noble appearance."

It should be noted that Scott had overlooked one previous record of polystely due to Dangeard and Barbé[4]—that of the occurrence of four or five steles in the axis of *Pinguicula vulgaris*. But this case introduces no difficulty so far as Scott's hypothesis is concerned, for *Pinguicula* is related to *Utricularia*

[1] Prankerd, T. L. (1911).

[2] Schindler, A. K. (1904). This author takes the view that *Hippuris* does not belong to the same cycle of affinity as *Gunnera* and *Myriophyllum*.

[3] Darwin, C. (1890). [4] Dangeard, P. A. and Barbé, C. (1887).

with its numerous aquatic species. Further instances of polystely have been subsequently discovered among the Nymphaeaceae[1]. Though the anomalous structures met with in this family cannot perhaps be explained on quite the same lines as those of *Auricula* and *Gunnera*, their existence does not invalidate Scott's view; they are of interest as furnishing another example of the tendency towards the development of distinct steles or vascular zones in aquatic plants in which secondary increase in thickness is lacking.

The present writer would like to suggest that there is possibly some significance in the fact that nearly all the known cases of polystely in Angiosperms occur in plants whose main vegetative axis takes the form of a rhizome. This organ, not being subjected to the same mechanical strains as an erect stem which has to support leaves and branches, is not so irrevocably committed to the 'continuous cylinder' type of vascular system, which is the best form of structure for withstanding bending forces. That the polystelic type of anatomy does not make for strength, is indicated by the recent observation, concerning a gigantic Hawaian species of *Gunnera*, that "the rhizome is very soft, and can be severed by a single machete stroke[2]."

One special point of interest connected with the hypothesis of the origin of polystely through an aquatic ancestry, lies in the fact that, if it be accepted, it forms a particularly salient instance of the working of a certain principle of evolution which the present writer proposes to call "the Law of Loss[3]"; this law will be discussed in Chapter XXVIII.

[1] Gwynne-Vaughan, D. T. (1897); see also Chapter III, p. 37.
[2] MacCaughey, V. (1917).
[3] Arber, A. (1919²).

CHAPTER XIV

THE AERATING SYSTEM IN THE TISSUES OF HYDROPHYTES

THE existence of a highly-developed system of intercellular spaces, is one of the most marked anatomical characters of water plants. It is generally assumed that this lacunar system serves for the storage of the oxygen evolved in assimilation, and its conveyance to the parts of the body that stand in especial need of it, more particularly the roots and rhizomes buried in the asphyxiating mud. The mesophyll of the lamina, the ground tissue of the petiole, and the cortex of the stem and root, are the regions in which the air spaces reach their greatest development.

In the stem, the cortex, which is generally broad in proportion to the stele, is penetrated by lacunae, which may be so numerous as to render the whole organ extremely fragile in texture. Two features in the arrangement of the cortical cells, however, seem in some degree to obviate the dangers of this fragility. The air spaces are, in the main, confined to the middle cortex, while the outer cortex in many cases consists of elements which are more closely placed and thus form a firmer peripheral shell[1]; the septa, again, are radially arranged and thus are able to withstand pressures acting at right angles to the axis, which would otherwise be liable to crush the stem[2]. Support is also obtained by diaphragms[3], occurring chiefly at the nodes, which divide the air spaces into sections; these diaphragms are not air-tight, but are more or less water-tight, so that they form a safeguard against the flooding of the entire aerating system in the case of accidental injury. Fig. 118, p. 184, represents part of a transverse section of a stem of *Potamogeton natans*, in which

[1] Haberlandt, G. (1914). [2] Schenck, H. (1886).
[3] Duval-Jouve, J. (1872), Blanc, M. le (1912) and Snow, L. M. (1914).

the cortical lamellae are connected by a diaphragm (D) with small intercellular spaces (m) at the angles of the cells. Fig. 119,

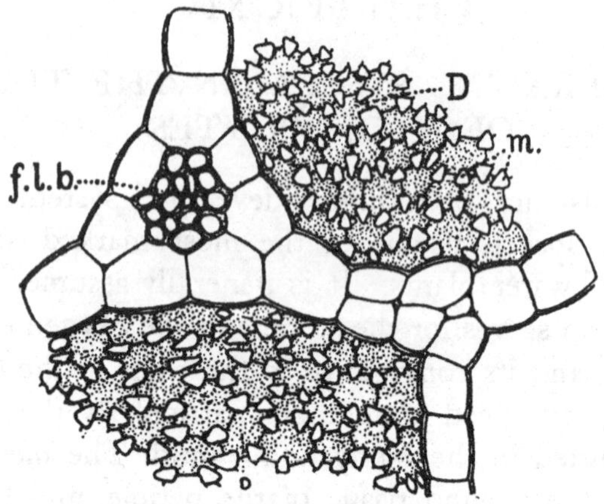

FIG. 118. *Potamogeton natans*, L. Part of T.S. of stem with diaphragm (D) penetrated by intercellular spaces (m). *f.l.b.* = vascular bundle. [Blanc, M. le (1912).]

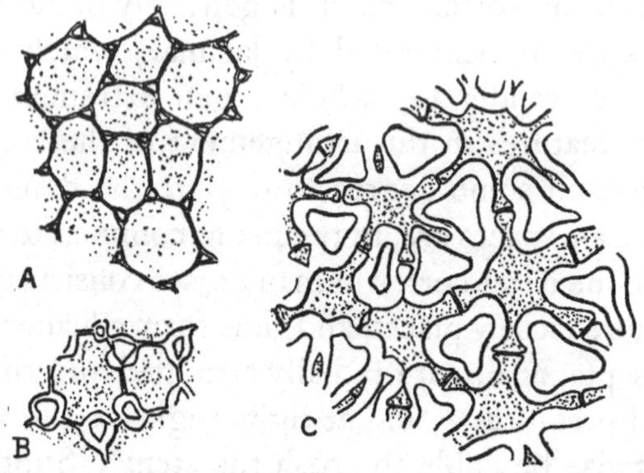

FIG. 119. *Hippuris vulgaris*, L. Three stages in the development of the nodal diaphragms of the stem, seen in T.S. (all × 318). *A*, young stem, intercellular spaces small and walls scarcely thickened. *C*, old stem, 7 mm. in diameter; intercellular spaces so much enlarged that the cells are stellate, walls much thickened. *B*, same stem as *C*, but from a region 3·5 mm. across, which shows intermediate characters. [A. A.]

A, *B*, *C*, shows the development of the nodal diaphragm-tissue in the case of *Hippuris*.

The air spaces may be either formed by the separation of cells (schizogenous) or by their destruction (lysigenous). When the air spaces are schizogenous, they may be arranged in the form of a single ring (e.g. *Myriophyllum*, Fig. 116, p. 179), or a number of rings may occur, giving a lace-like appearance to the stem, when seen in transverse section (e.g. *Hippuris*). The development of the air spaces in the cortex of *Hippuris vulgaris*[1] is illustrated by Fig. 120 *A* and *B*. The Water Crowfoot forms a transition to those plants in which the air spaces are lysigenous, for, in the young stem, irregularly placed schizogenous air spaces occur, especially in the pith (Fig. 113, p. 176),

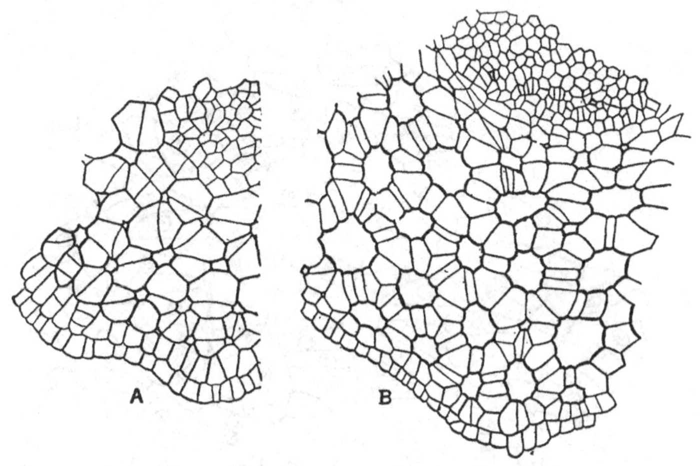

FIG. 120. *Hippuris vulgaris*, L. Parts of transverse sections through a younger stem (*A*) and an older stem (*B*) showing the origin of the cortical lacunae. [Barratt, K. (1916).]

while, in the older stem, the whole of the central parenchyma becomes torn and destroyed, leaving a large axial lacuna. *Peplis Portula*[2] is an example of a plant whose air spaces are mainly lysigenous. In transverse sections of the internodes, four such spaces are visible, each containing the torn remains of cells.

The aerating system of the roots of aquatics is to be found in the cortex. In some cases, e.g. *Vallisneria*, the intercellular spaces may be small, but more frequently they are of conspicuous size, and arranged with a regularity that gives a notable

[1] Barratt, K. (1916). [2] Schenck, H. (1886).

symmetry of pattern to the transverse section. The process of development of the intercellular spaces has been followed by the present writer in the case of *Stratiotes aloides*[1] (Fig. 121). The whole inner region of the cortex in the root of this plant must be visualised as consisting of radially arranged plates, one cell wide, which in the early stages are so placed as to leave no spaces between. The cells composing the plates divide very rapidly, and a number of new cell-walls are formed, almost all in planes

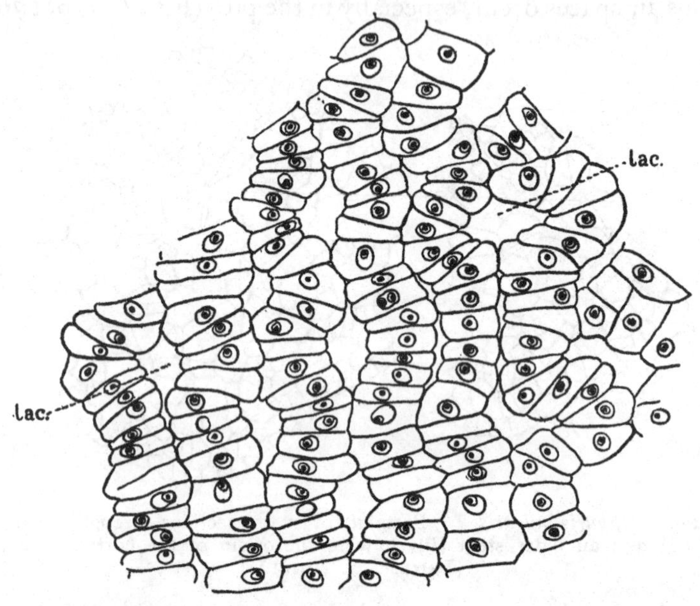

FIG. 121. *Stratiotes aloides*, L. Tangential section through middle cortex of a young root to show the origin of the lacunae (*lac.*). (× 318.) [Arber, A. (1914).]

at right angles to the long axis of the root. The result is that the plates elongate in the direction of growth of the root, but, owing to the rapidity of their cell-divisions, the plates grow in length faster than the rest of the root, and are thus forced into undulations, since they become too long to retain their normal vertical position. The possibility of their taking up this sinuous form is due to the fact that the root enlarges in diameter and thus allows room for the separation of the plates. It will readily

[1] Arber, A. (1914).

be seen that a series of plates, side by side, elongating independently, and at the same time prevented from stretching to their full length, will naturally become detached from one another at certain points, leaving spaces between. The result of these processes is that the middle cortex, as seen in transverse section, consists of radial plates of cells, like the spokes of a wheel, in contact or separated by lacunae, whereas in tangential section the plates are found to meet their neighbours at intervals, so as to form a network.

In some plants, e.g. *Myriophyllum* and *Callitriche verna*[1], the air spaces in the root cortex may be increased by the replacement of small schizogenous air spaces by large cavities of a partially lysigenous nature, due to the disruption of the septa.

Remarkable as is the aerating system developed in the primary tissues, that formed in the course of secondary growth is often even more conspicuous. This secondary aerating system, or aerenchyma, arises in some cases from a phellogen, in others from a typical cambium. We will first consider that which is produced by a phellogen, and may be regarded as a special modification of an ordinary periderm. It is well known that in land plants the impervious corky mantle, which so often covers the older parts, is interrupted at intervals by lenticels, or patches of powdery cork, in which the cells are slow in becoming suberised, and are separated by intercellular spaces, instead of being closely applied to one another as in normal periderm. These lenticels form a channel by which gaseous exchange takes place between the atmosphere and the interior of the plant. We have thus, in the lenticel tissue, an example of an aerenchyma formed on a small scale by ordinary terrestrial plants, and, moreover, this aerenchyma has a tendency to become hypertrophied when the plant is submerged. The case has been described, for instance, of a Poplar branch which had been a long time under water, and in which masses of whitish tissue protruded from the surface in many places. On examination these protrusions proved to be due to a great development of the

[1] Schenck, H. (1886).

aerenchyma of the lenticels[1]. *Salix viminalis* and *Eupatorium cannabinum*, again, have been shown to develop spongy tissue beneath the lenticels when grown in water or on marshy soil[2]. In the course of evolution, this tendency to hypertrophy of the lenticel tissue under the influence of water, may have formed the starting point for the development of the special air-containing phelloderm which is so marked a feature of a number of plants to which we must now refer.

It was recorded more than forty years ago, by a Russian observer, that the stems and roots of *Epilobium hirsutum*, *Lycopus europaeus*, and two species of *Lythrum* produced aerenchyma, when grown in water[3]. In *Lythrum Salicaria* the aerenchyma, which appears on the submerged parts when grown in shallow water, enlarges the stem to as much as four times its normal thickness[2]. It can be induced to occur in this and other cases (e.g. *Lycopus europaeus*) by merely keeping the cut branches in water for a few weeks[1]. The list of our native waterside plants, in which aerenchyma occurs under suitable conditions, includes *Lysimachia*, *Lotus*, *Oenanthe*, and *Scutellaria*, in addition to the genera already named[4]. Schenck[2], to whom our knowledge of aerenchyma is largely due, showed that this tissue was particularly characteristic of Onagraceae, where it occurred in twelve species belonging to three genera; Leguminosae, where it was found in six species representing five genera; and Lythraceae, where it appeared in six species belonging to three genera. It was Schenck who proposed the useful term 'aerenchyma' for this non-suberised ventilating tissue produced by a phellogen. The cells are not dead and empty, as in normal cork, but are lined with a delicate protoplasmic pellicle and generally contain clear cell-sap; they are separated by extensive lacunae. That they are homologous with cork-cells is indicated by the fact that, in the roots of *Jussiaea*, the cork, formed when the plant grows on land, is replaced by aerenchyma when it grows in

[1] Goebel, K. (1891–1893). [2] Schenck, H. (1889).
[3] Lewakoffski, N. (1873[1]); on *Epilobium* see also Batten, L. (1918).
[4] Glück, H. (1911).

water[1]. It has been suggested[1] that the stimulus that causes the phellogen to develop aerenchyma in lieu of cork, is the lack of oxygen in the inner tissues. The present writer would prefer, however, to express the same idea somewhat differently, and to say that the presence of some minimum of oxygen is possibly a necessary condition for the process of suberisation, which is inhibited when the oxygen-content of the cell-sap falls below a certain point.

Some remarkable cases of aerenchyma development are found in the tropical Onagraceous genus *Jussiaea*[2]; in *J. peruviana* (Fig. 122, p. 190), the submerged parts of the shoots are clothed with this tissue, which is also developed on the normal roots which enter the mud (*m.r.*), and in certain erect roots which seem to serve entirely for aeration (*a.r.*). Fig. 122 *B* exhibits the origin of the stem aerenchyma (*a*) from a phellogen (*pg*). Special breathing roots also occur in the case of *Jussiaea repens*. They show, in transverse section, a tiny stele, surrounded by a voluminous aerenchyma. That the modification of these roots is directly related to the aquatic environment, is indicated by the fact that *Jussiaea grandiflora*, when cultivated for some years in the botanical garden at Marburg as a land plant, produced only normal adventitious roots, but when it was transferred to water it developed roots with aerenchyma[3].

The aerenchyma of certain members of the Leguminosae has been recognised for many years. Humboldt and Bonpland[4], for instance, more than a hundred years ago, recorded that in "*Mimosa lacustris*" (*Neptunia oleracea*, Lour.), the Floating Sensitive Plant (Fig. 123, p. 191), the stems and branches were covered by "une substance spongieuse, blanchâtre." They made the mistake, however, of supposing that this tissue was a foreign body, and not an integral part of the plant. More recent observations[5]

[1] Schenck, H. (1889). [2] Martins, C. (1866).

[3] Goebel, K. (1891–1893).

[4] Humboldt, A. de, and Bonpland, A. (1808).

[5] Rosanoff, S. (1871). This author uses the name "*Desmanthus natans*" for the plant now called *Neptunia oleracea*.

have made it clear that the spongy mass (*f* in Fig. 123) is an aerenchyma developed from a phellogen. That it also acts

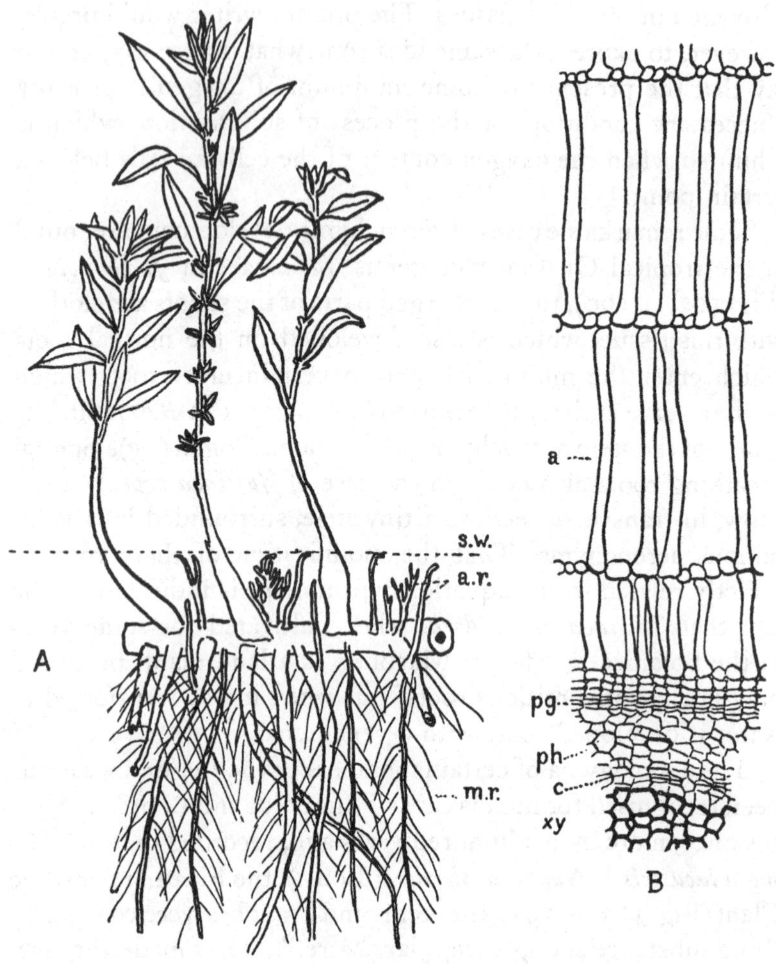

FIG. 122. *Jussiaea peruviana*, L. *A*, habit drawing. The shoots are clothed with aerenchyma up to the water level (*s.w.*). *m.r.*, mud roots; *a.r.*, air roots. Aerenchyma occurs in both types of root. (Reduced.) *B*, Transverse section of submerged part of a stem to show aerenchyma (*a*) developed from phellogen (*pg*). The phloem (*ph*), normal cambium (*c*) and xylem (*xy*) are also shown. [Adapted from Schenck, H. (1889).]

as a float is indicated by Spruce's[1] account of the plant as he saw it growing in South America. He describes the buoyant

[1] Spruce, R. (1908).

"cottony felt" as serving to hold the delicate bipinnate leaves and the heads of pale yellow flowers above the surface of the water. In *Sesbania*[1], again, another Leguminous genus not at all closely related to *Neptunia*, a similar air-tissue occurs, arising from a cork-cambium in the inner cortex, just outside the endodermis.

It is a curious fact that among the Leguminosae we not only meet with the case just described, in which an aerenchyma arises externally from a phellogen, but we also find instances in which a tissue of somewhat similar nature is produced internally from

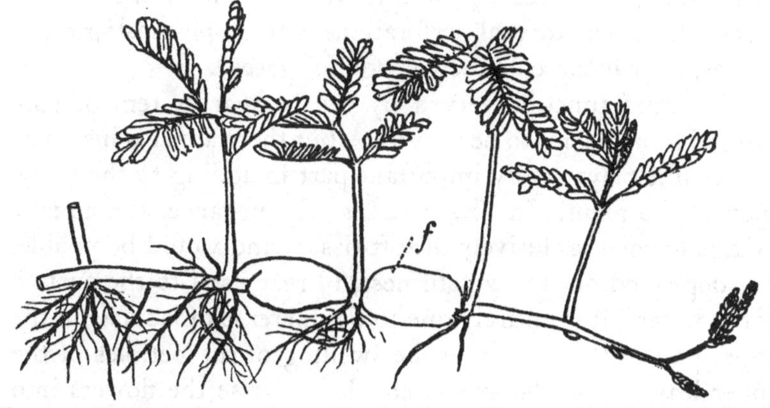

FIG. 123. *Neptunia oleracea*, Lour. Floating shoot. The two oldest internodes have lost their floating tissue, *f*, while the three youngest have not yet developed it. (Reduced.) [Adapted from Rosanoff, S. (1871).]

a normal cambium, and is thus of the nature of secondary wood. In these cases, the air is contained within the xylem elements. *Aeschynomene aspera*, Willd.[2] is a Leguminous shrub, frequent in India on the margins of fresh waters, in which a pith-like tissue, white and homogeneous, occupies the greater part of the stem. This substance is, in fact, secondary xylem. It is so extremely light in weight that it is collected to make toys, floats for fishermen's nets, and 'pith' helmets. Another member of the same genus which grows in Venezuela, *Ae. hispidula*, H. B. K.[3], has remarkable swellings on the submerged

[1] Scott, D. H. and Wager, H. (1888). [2] Moeller, J. (1879).
[3] Ernst, A. (1872[2]).

parts of its stem, said to be due to aerenchyma. A third Leguminous plant, which has been described under the name of *Herminiera elaphroxylon*, G. and P.[1] but which is perhaps better regarded as another member of the genus *Aeschynomene*, also has aerenchyma[2]. The floating wood of this plant, which is known as the "Ambatsch," is employed on the Blue Nile to make rafts. The pieces used are as thick as a man's arm, and show under the bark a shining white woody mass, penetrated by numerous rays. The wood is exceedingly light; a segment of stem 2½ feet long and about 4 inches in diameter, is described as weighing less than 1½ ounces. It has been shown that the pits of the xylem are real perforations with no pit-closing membranes, so that there is free passage for gases[3].

The chief function served by the lacunar system of submerged stems seems to be aeration[4], but there are also instances in which it plays a very important part in adding to the buoyancy of the plant. In *Trapa natans*, for instance, the aquatic stem is formed exclusively of soft tissue, and would be unable, if it depended on its own stiffness, to rear itself to the surface of the water. It is entirely due to the increase of lacunae in the upper part of the stem, and the swelling of the petioles of the upper leaves, that the axis is enabled to raise the flowers into the air. In the deeper regions, the pith is a compact tissue, and there are only two circles of lacunae in the cortex, but in the upper part of the stem the pith is lacunate and the number of circles of air spaces increases to four or five[5].

The secondary lacunar tissues were always assumed by the earlier writers to serve for flotation alone; in certain cases (e.g. some of the Leguminosae already mentioned) it is quite possible that they were correct. Martins[6], who long ago described and figured the air roots of *Jussiaea*, regarded them merely

[1] Also called *Aedemone mirabilis*, Kotschy.

[2] Kotschy, T. (1858), Hallier, E. (1859), Jaensch, T. (1884[1]) and (1884[2]), Klebahn, H. (1891). See also Hope, C. W. (1902).

[3] Goebel, K. (1891–1893).

[4] Schenck, H. (1889).

[5] Costantin, J. (1884).

[6] Martins, C. (1866).

as floating organs. For this particular case, this view can scarcely be maintained, since Goebel[1] has shown that *Jussiaea repens* floats quite well, even if the roots be all removed. A good case has been made out, however, for regarding the aerenchyma of *Nesaea verticillata*[2], one of the Lythraceae, as a true floating tissue. Many of the wand-like stems of the plant, growing on the borders of ponds in America, are described as reaching a length of six to eight feet. In July and August they bend with their own weight until the stem apex touches the water, when it curves upwards again. In the region of contact between the stem and the water a swelling occurs, and roots also arise from

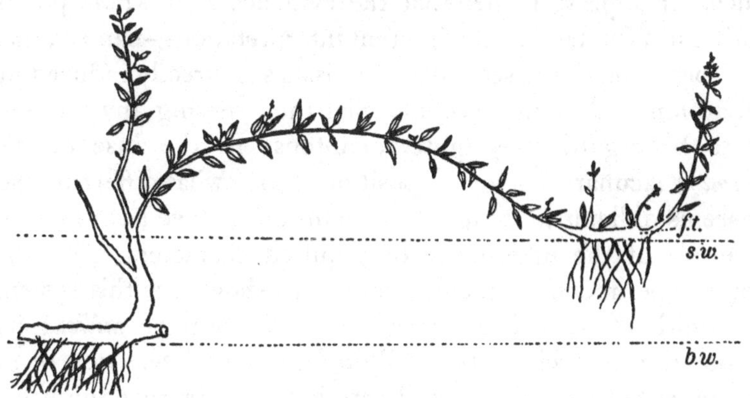

FIG. 124. *Nesaea verticillata*, H. B. and K. Plant at beginning of August; *s.w.*, surface of water; *b.w.*, bottom of water; *f.t.*, floating tissue. [Adapted from Schrenk, J. (1889).]

this region, anchoring the floating part of the stem to the ground (Fig. 124). The epidermis of the swollen region becomes fissured, disclosing a snowy white, soft, elastic, spongy tissue, which arises from a pericyclic phellogen. Contraction of the roots draws the swollen part down into the water, and the spongy layer gradually extends over the submerged regions. In the autumn the long slender stems die, except those portions that have produced floating tissue around themselves, and have rooted in the mud. A new root-stock is thus developed, sometimes at a considerable distance from the mother-plant. As

[1] Goebel, K. (1891–1893). [2] Schrenk, J. (1889).

evidence for the view that the aerenchyma in this plant is not respiratory in function, Schrenk, who described it, points out that in old stems the surface of this tissue is covered by a layer which is air-tight and suberised, and that a similar layer is also sometimes found separating it from the interior of the stem. He accounts for its occurrence in regions where it cannot serve for flotation, by supposing that the meristem spreads there automatically from the floating parts.

To the present writer, however, the question whether the *secondary* air-containing tissues of water plants serve mainly for aeration or for flotation, seems to be a matter of minor importance. It appears to her that the evidence as a whole points rather to a fundamentally different interpretation—namely, that the formation of the secondary air-tissues is directly induced by environmental conditions, and that their serving any purpose is to be regarded as quite fortuitous. In the case of the *primary* lacunar system, the position is somewhat different, and it seems difficult to escape the conclusion that we have here an example of the inheritance of acquired characters. There is some experimental evidence tending to show that this system was initiated as a direct response to the aquatic *milieu*; its elaboration may either be attributed to natural selection or to the inherited effects of use. There is no doubt that the habit of developing an elaborate aerating system has now become in many cases an inherited character, for though it can be modified and reduced by terrestrial conditions, it cannot be altogether eliminated.

CHAPTER XV

LAND FORMS OF WATER PLANTS, AND THE EFFECT OF WATER UPON LAND PLANTS

THE majority of water plants, with the exception of those most highly specialised for aquatic life, are capable of giving rise to land forms. Those plants which, when mature, produce floating or air leaves, can obviously develop a land form with less change in their structure and mode of life than those which normally live entirely submerged. *Limnanthemum nymphoides*[1], for instance, has been found growing on damp ground with abbreviated internodes and petioles, and with reduced laminae. Land forms of *Hydrocharis*[2], and many Nymphaeaceae[3] and Alismaceae[4] are known, either in nature or in cultivation. Successful terrestrial forms can also be produced by those Potamogetons which possess coriaceous, floating leaves, or have the power to develop such leaves on occasion. The land form of *Potamogeton natans* is shown in Fig. 125, p. 196. *P. varians*, a form allied to *P. heterophyllus*, Schreb., can exist for season after season without being under water at all, tiding over the winter by means of its bead-like tubers[5]. Even *P. perfoliatus* has also been recently stated to produce a land form[6], though it is generally regarded as a typically submerged type, which is incapable of terrestrial life[7].

Myriophyllum, *Callitriche*[8] and the Batrachian Ranunculi (Fig. 126, p. 196) agree in producing land forms which are close-growing and tufted. When *Myriophyllum spicatum*[1], for

[1] Schenck, H. (1885).
[2] Mer, É. (1882[1]).
[3] Bachmann, H. (1896), and Mer, É. (1882[1]). See also p. 32, Ch. III.
[4] See Chapter II and Glück, H. (1905).
[5] Fryer, A. (1887).
[6] Uspenskij, E. E. (1913).
[7] Fryer, A., Bennett, A., and Evans, A. H. (1898–1915).
[8] Lebel, E. (1863).

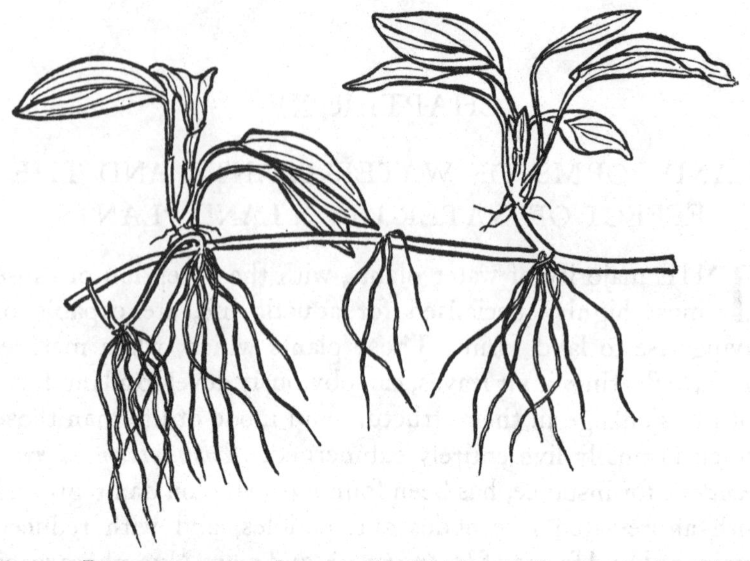

FIG. 125. *Potamogeton natans*, L. or possibly *P. polygonifolius*, Pourr. Land form from a dried-up swamp, New Forest, September 2, 1911, after a very dry summer. Only the blades of the leaves, and sometimes not even the whole of these, were visible above ground. (Reduced.) [A. A.]

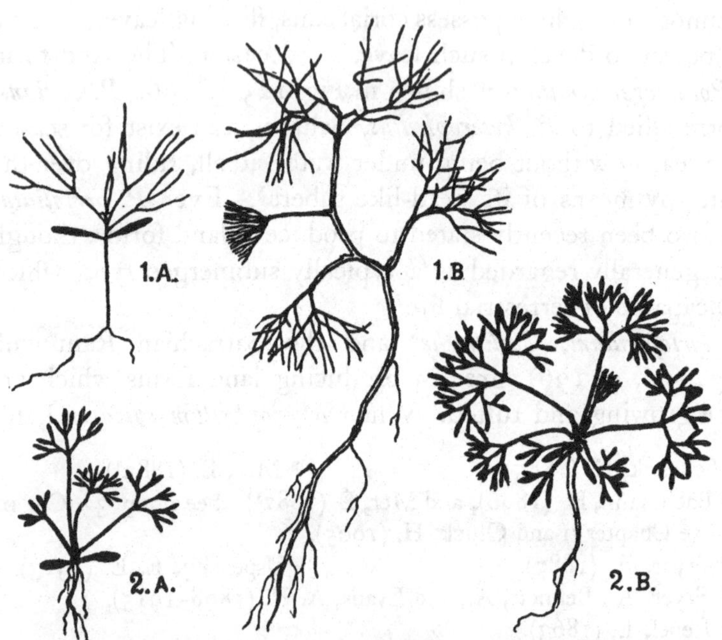

FIG. 126. *Ranunculus aquatilis*, L. 1*A*, seedling which germinated in water, and which is shown in 1*B* at a somewhat older stage. 2*A*, seedling which germinated on land, and which is shown in 2*B* at a somewhat older stage. (Nat. size.) [Askenasy, E. (1870).]

instance, is left stranded, the water leaves are apt to dry up, but the ends of the shoots grow into a land form entirely different in habit from the water form. It develops as a minute turf, an inch high; the stems are frequently branched, the internodes are short instead of being elongated as in the water form, and many adventitious roots are produced from the nodes. The leaves are smaller than in the submerged form, and the segments are fewer, broader and thicker.

A close connexion between submerged and aerial 'forms' has in recent years been demonstrated in the case of *Hottonia*, the Water Violet. In this plant, which previous observers had

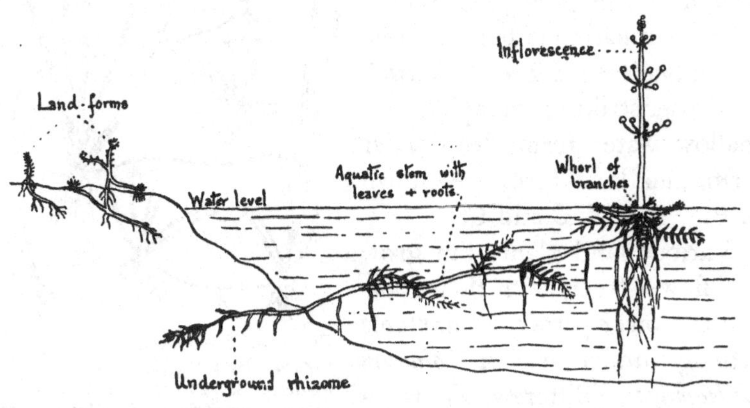

FIG. 127. *Hottonia palustris*, L. Diagrammatic sketch of typical land and water forms. [Prankerd, T. L. (1911).]

erroneously described as free-floating, it is now known[1] that the oldest part of the rhizome is generally embedded in mud, and that from it arise vertical aerial branches, which may become detached by the dying off of the older part of the stem, thus giving rise to so-called 'land forms,' which are similar in anatomical structure to the submerged parts of the aquatic plant, rather than to the aerial inflorescence region (Fig. 127).

The differences between the land and water leaves of *Polygonum amphibium*, have already been mentioned, and are illustrated in Figs. 99 and 100, p. 152. It is notable that in this case

[1] Prankerd, T. L. (1911).

the plant reaches its optimum development as an aquatic, and flowers freely in water. As a land plant it rarely blossoms and, indeed, under xerophilous conditions, flowering seems to be entirely inhibited[1].

In the case of amphibious plants, which can produce land or water forms according to circumstances, the difference in external appearance is often very marked. *Limosella aquatica*, for instance, produces a land form with leaf-stalks half-an-inch to one inch long, while the water form may have petioles six inches long, terminating in tender translucent blades[2]. *Littorella lacustris* is another striking example. The shallow water form, deep water form, and land form are shown in Fig. 128 *A*, *B* and *C*.

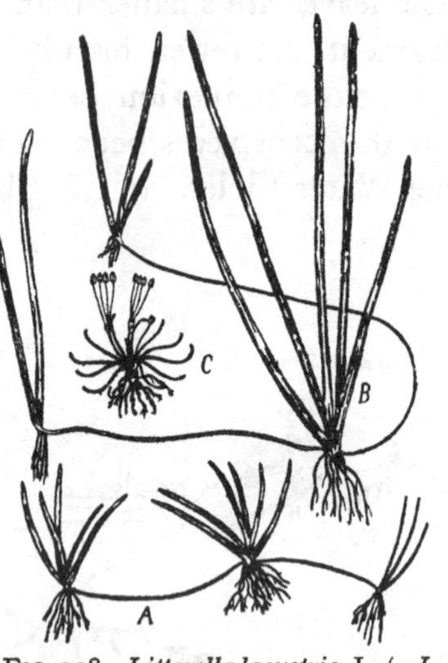

FIG. 128. *Littorella lacustris*, L. (=*L. juncea*, Berg.). *A* and *B*, water forms; *C*, land form. *A* is from water 30 to 40 cms. deep; *B* is from water 100 cms. deep; *C* shows three male flowers one of which has lost its stamens. (Reduced.) [After Glück, H. (1911), *Wasser- und Sumpfgewächse*, Bd. III, Fig. 34, p. 346.]

Various land plants can grow and flower freely with their roots and the lower parts of their stems actually under water; *Solanum Dulcamara* (Bittersweet) is a species to which these conditions seem especially favourable. Such plants form a transition to those which frequent the margins of fresh waters, and are capable of responding to changes in the water level by producing, at need, actual aquatic forms. Glück[3], who has given great attention to this subject, has shown that, in nature, submerged forms, often with reduced vegetative organs, are produced not only by plants which normally inhabit damp or marshy situations, such as *Ranunculus Flammula* (Figs. 134 and 135, p. 203), *Caltha palustris* (Fig. 129), *Cnicus pratensis*[4] (Fig. 130

[1] Massart, J. (1910). [2] Schenck, H. (1885).
[3] Glück, H. (1911). [4] Glück uses the name *Cirsium anglicum*, D.C.

A and *B*) and *Menyanthes trifoliata*, but also by typically terrestrial plants such as *Achillea ptarmica*, *Trifolium resupinatum* (Fig. 131 *B*) and *Cuscuta alba* (Fig. 131 *A*). Glück[1] has also produced experimentally a submerged form of *Iris Pseudacorus*. Seeds of terrestrial plants may sometimes germinate and reach a considerable development while entirely submerged. The

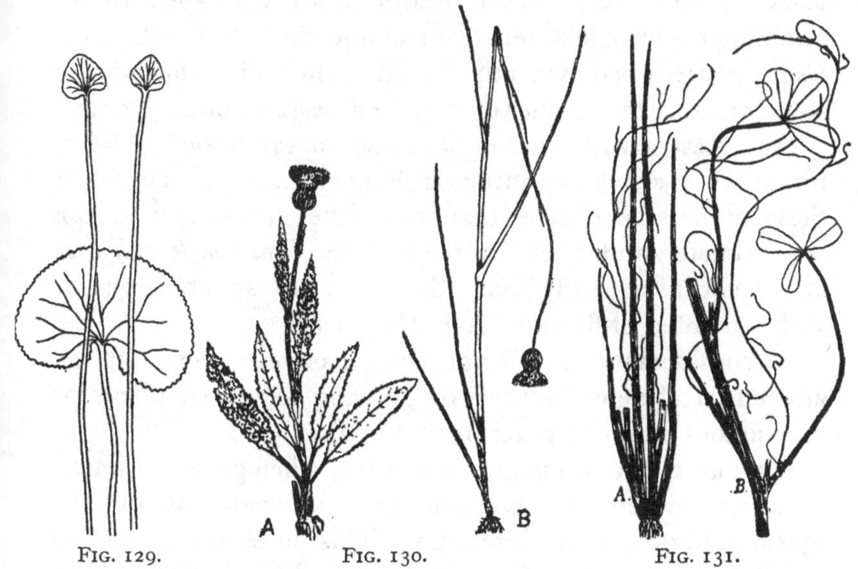

FIG. 129. FIG. 130. FIG. 131.

FIG. 129. *Caltha palustris*, L. The two leaves with long petioles belong to the submerged form: the middle leaf is a corresponding air leaf of the land plant. (Reduced.) [After Glück, H. (1911), *Wasser- und Sumpfgewächse*, Bd. III, Fig. 3, p. 65.]

FIG. 130. *Cirsium anglicum*, D.C. (=*Cnicus pratensis*, Willd.). *A*, land form, *B*, water form. [After Glück, H. (1911), *Wasser- und Sumpfgewächse*, Bd. III, Figs. 1 *a* and *b*, p. 16.]

FIG. 131. *Cuscuta alba*, J. and C. Presl, forma *submersa*. *A*, parasitic on water form of *Echinodorus ranunculoides*, (L.) Engelm. *B*, parasitic on the form of *Trifolium resupinatum*, L. with floating leaves. (Reduced.) [After Glück, H. (1911), *Wasserund Sumpfgewächse*, Bd. III, p. 114, Figs. 7*A* and *B*.]

present writer has noticed Horse Chestnuts sprouting freely in the mud at the bottom of a stream: one which was measured had a plumular axis more than 1 inch in length, and a primary root of 3½ inches.

In connexion with Glück's record of a submerged form of

[1] Glück, H. (1911).

Cnicus pratensis, it is interesting to note that a somewhat different water form has been described in the case of *C. arvensis*[1]. The plant in question had suffered nine months' inundation in a fenland flood; when observed in November, at first sight "the leaf-rosette appeared normal;...but on lifting it, it was found to be attached to the ground by about 2 or 3 feet of slender leafless stem of very soft and flexible consistency—exactly resembling the woodless stem of a true aquatic. During the flood *Cnicus arvensis* had evidently floated at the end of this aquatic stem, much in the manner of, say, a *Potamogeton* or *Callitriche*."

The present writer has noticed *Ranunculus repens*[2] growing by the water-side and putting out long runners into the water; these runners bore leaves that were either submerged or rose approximately to the level of the surface. *Hydrocotyle vulgaris*[2] is also not infrequently seen either more or less submerged or with a number of floating leaves (Fig. 132).

A considerable amount of work has been done on the anatomical changes induced by growing terrestrial plants or amphibious plants in water instead of air.

Among terrestrial plants, *Vicia sativa*, when grown in water, does not develop aquatic characters in its epidermis, but the xylem suffers marked diminution. This enfeeblement of the xylem is characteristic of various other land plants when grown in water, and, in the case of *Ricinus* and *Lupinus*, there is a similar reduction in the thickening of the bast fibres[3]. *Rubus fruticosus*, when grown in water, showed no change in the microscopic structure of its sub-aquatic leaves and stem, except that, in both organs, the chlorophyll was developed nearer the surface than in the normal condition in air, while the hairs on the stem tended to be unicellular instead of multicellular[4]; in the shoots of *Salix*, also, little anatomical change was induced by submergence[5].

[1] Compton, R. H. (1916).

[2] The existence of these forms was noted by Glück, H. (1911). On *Hydrocotyle* see West, G. (1910).

[3] Costantin, J. (1884).

[4] Lewakoffski, N. (1873[2]).

[5] Lewakoffski, N. (1877).

In the case of amphibious plants, the comparison of air and water shoots gives results of greater interest. Costantin[1] described the anatomy of a plant of *Mentha aquatica* growing on dry land, which happened to have the apex of one of its shoots plunged into water. The young part of the stem, which had thus grown in an aquatic *milieu*, when compared with the older part growing in air, was found to be glabrous and to have a greater diameter and larger air spaces. The same increase in the air

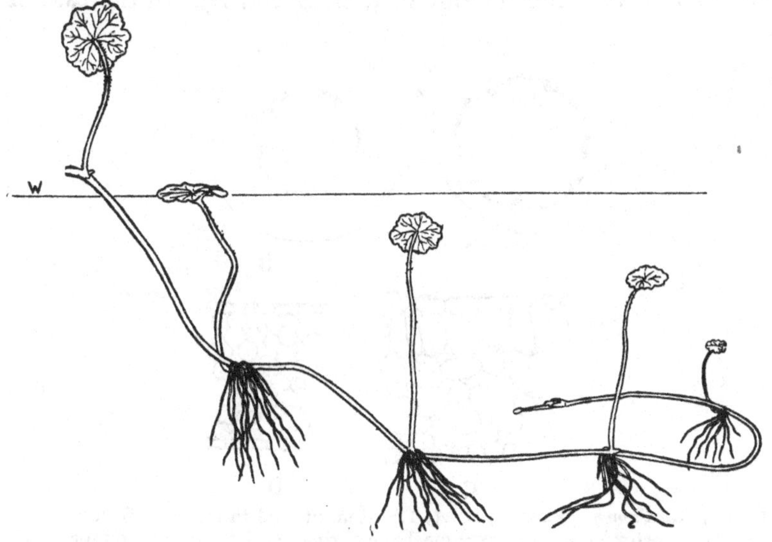

FIG. 132. *Hydrocotyle vulgaris*, L. A branch sent out into water from a plant growing on the bank; *w*, water level. The under surfaces of the five expanded leaves were examined for stomates, which were present on all. The petiole of the air leaf was more hairy than that of the succeeding leaves. July 14, 1910. ($\frac{1}{3}$ nat. size.) [A. A.]

spaces and of the diameter of the stem, was observed in submerged shoots of *Veronica Anagallis* and *Nasturtium amphibium*. Costantin notes that, in general, when submerged plants are grown in deep water, the fibrous and tracheal elements diminish markedly.

Cardamine pratensis is an example of an amphibious plant which seems to pass with remarkable ease from the water to the air condition. The present writer has found, on more than one

[1] Costantin, J. (1884).

occasion, that an entirely submerged plant, when placed in soil under ordinary aerial conditions, rapidly developed into a typical land plant. Schenck[1] has described the comparative anatomy of submerged and aerial plants of this species. The anatomy of the submerged stem showed several points of interest. The intercellular spaces and the diameter of the cortex were increased; the vascular cylinder had approached nearer the centre of the stem; all mechanical elements were absent, and the xylem was reduced (Fig. 133, cf. *A* and *B*). In the case of

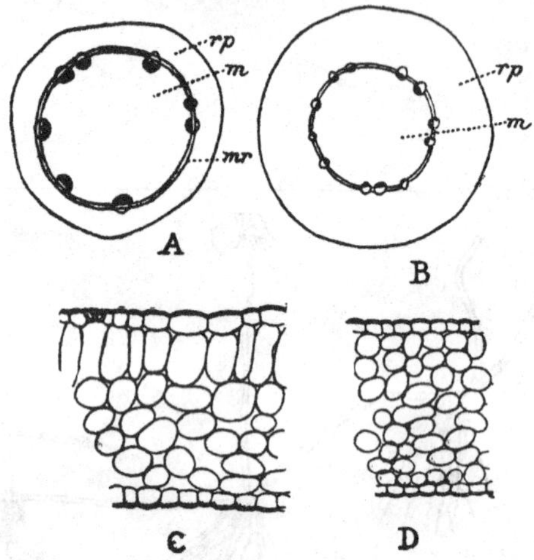

FIG. 133. *Cardamine pratensis*, L. *A*, T.S. stem of land form. *B*, T.S. submerged stem; *rp* = cortex, *m* = pith, *mr* = mechanical ring. *C*, T.S. leaf of land form. *D*, T.S submerged leaf. [Schenck, H. (1884).]

the leaves, those that were submerged had developed no palisade tissue (Fig. 133, cf. *C* and *D*).

Such anatomical work as that briefly outlined above, leads to the general conclusion that when *amphibious* plants are grown in water they readily acquire the characters which we regard as typical of aquatic plants, but that, when *terrestrial* plants are grown under similar conditions, the changes which occur, though trending in the same direction, are very much less marked. There seem to be two possible, alternative explanations of this difference of behaviour. On the one hand it may be that

[1] Schenck, H. (1884).

amphibious plants were not originally gifted with any special aptitude for aquatic life, but that they have gradually acquired, and passed on to their descendants, the capacity for reacting in an advantageous way to the stimuli of an aquatic environment, and that we are thus dealing with a case of the inheritance of acquired characteristics. But the second alternative, which appears to the present writer to have most in its favour, is that, in general, those species which are capable of a suitable response to aquatic conditions have already been sifted out by nature, and now inhabit situations where such conditions, at least occasionally, arise; or, in other words, that the various species of flowering plants were all endowed, from the first moment of their appearance, with different constitutions which gave them varying degrees of capacity for the adoption of water life; and that their habitats have been determined by this capacity and not *vice versâ*[1].

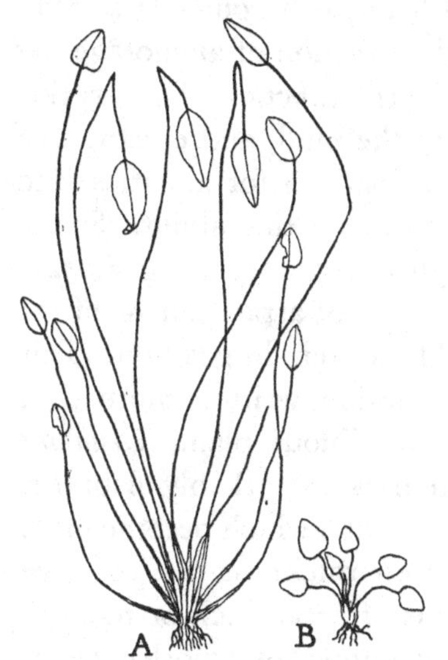

FIG. 134. *Ranunculus Flammula*, L. *A*, form with floating leaves. *B*, land form. (Reduced.) [After Glück, H. (1911), *Wasser- und Sumpfgewächse*, Bd. III, Figs. 84 and 85, p. 494.]

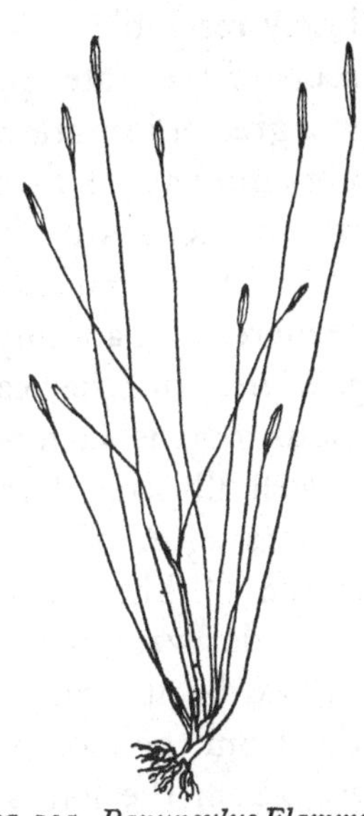

FIG. 135. *Ranunculus Flammula*, L. Submerged form. The short upright stem replaces the inflorescence. (Reduced.) [After Glück, H. (1911), *Wasser- und Sumpfgewächse*, Bd. III, Fig. 86, p. 496.]

[1] See Footnote 1, p. 162.

CHAPTER XVI

THE ROOTS OF WATER PLANTS

THE roots of certain of the more specialised water plants, are extremely reduced or even in some cases entirely absent, e.g. *Ceratophyllum*, *Aldrovandia* and *Utricularia*. In other instances, such as *Nymphaea*, although the primary root is very short-lived, a considerable system of adventitious roots may be developed. As we shall show in Chapter XXI, among aquatics, absorption by the roots is by no means of such negligible importance as some writers have suggested; but at the same time, when plants rooted at the bottom of water are compared with those terrestrial herbaceous plants which they most closely resemble in size and habit, it becomes clear that, in the roots of the water plants, the function of anchorage has assumed a greater importance, while the function of absorption is less pre-eminent. A firm hold in the mud, and erectness of the flowering stem, are often a *sine quâ non* for aquatics, and their roots help in various ways to bring this about. Sometimes we merely get a richly ramifying root system, e.g. *Ranunculus aquatilis*[1]. In other cases the type of arrangement of the adventitious roots is such as to hold the stem in position. This point is well illustrated in a description written more than seventy years ago[2], of a certain amphibious plant, *Oenanthe Phellandrium*. "The flowering stem is remarkably fistulose, furnished under water with frequent joints, which become more distant upwards: it attains its greatest thickness two or three internodes from the base, where it is often an inch or more in diameter. From the joints proceed numerous whorled pectinated fibres [adventitious roots], of which the lower ones are as stout as the original fusiform root: these, descending in a conical manner to the bottom of the water, form a beautiful

[1] Hochreutiner, G. (1896). [2] Coleman, W. H. (1844).

system of shrouds and stays to support the stem like a mast in an erect position, while the pressure on the soft mud is lessened by the buoyancy of the hollow internodes."

There are other cases, again, in which anchorage depends on some modification of the adventitious roots. *Brasenia Schreberi* (*peltata*)[1], for instance, is fixed by its well-developed root-caps, which are of the nature of anchors, and prevent dislodgment of the buoyant plant, when it is swayed about by the agitation of the water surface. A still more remarkable method is the production of spirally twisted roots, which in some cases fully deserve the name of tendrils. Most of the known examples occur in the Potamogetonaceae, but they have also been recorded in the Hydrocharitaceae (*Hydrilla*)[2], Fig. 136, and

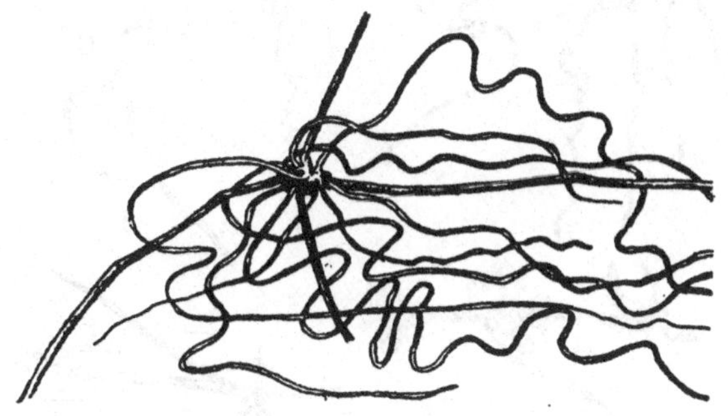

FIG. 136. *Hydrilla verticillata*, Presl. Tendril roots. [Kirchner, O. von, Loew, E. and Schröter, C. (1908, etc.).]

Gentianaceae (*Menyanthes*)[3], while the present writer has noticed them in *Myriophyllum verticillatum* (Haloragaceae). The first case among the Potamogetonaceae in which spirally twisted roots were observed, seems to have been *Cymodocea antarctica*[4]. At a later date the corkscrew roots of *Zannichellia palustris* were fully discussed by Hochreutiner[5] (Fig. 137 *A–F*, p. 206). He describes these roots as long, unbranched, and twining about other objects like tendrils—to use his own ex-

[1] Schrenk, J. (1888).

[2] Graebner, P., in Kirchner, O. von, Loew, E., and Schröter, C. (1908, etc.).

[3] Irmisch, T. (1861).

[4] Tepper, J. G. O. (1882).

[5] Hochreutiner, G. (1896).

pression, "elles grimpent en bas." He adds that *Potamogeton densus* (Fig. 137 *G* and *H*) shows the same peculiarity. A more recent writer[1] has recorded that, when the turions of *Potamogeton obtusifolius* germinate, they produce spirally coiled roots, which apparently serve to anchor the plantlets in the mud.

Twining roots are not confined to water plants; a case is recorded by Darwin[2], on the authority of Fritz Müller, in

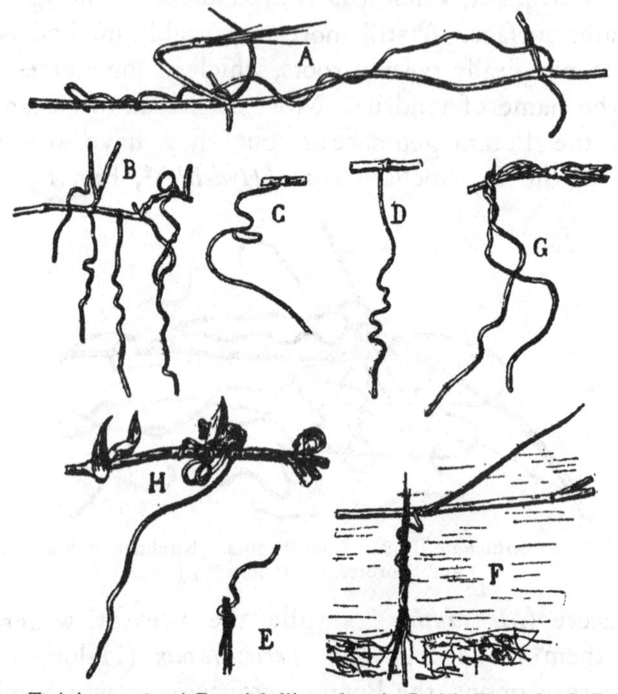

FIG. 137. Twining roots of *Zannichellia palustris*, L. (*A–F*) and of *Potamogeton densus*, L. (*G, H*). [Hochreutiner, G (1896).]

which the aerial roots of an epiphytic *Philodendron* in the forests of S. Brazil, twined spirally downwards round the trunks of gigantic trees. That root tendrils merely represent a further development of the general tendency to nutation common to stems and roots, is indicated by C. and F. Darwin's[3] record

[1] Graebner, P., in Kirchner, O. von, Loew, E., and Schröter, C. (1908, etc.). [2] Darwin, C. (1891). [3] Darwin, C. and F. (1880).

of a slight and tentative circumnutation in the seedling roots of several ordinary terrestrial plants. When the radicles of *Phaseolus*, *Vicia* and *Quercus* "were compelled to grow and slide down highly inclined surfaces of smoked glass, they left distinctly serpentine tracks."

Hildebrand[1] has described a differentiation between absorbing and anchoring roots in the case of *Heteranthera zosteraefolia*. He states that from each leaf-base two roots arise, one of which remains short and branches freely, while the other grows rapidly in length and serves for anchorage. Plants cultivated in England do not, however, so far as the present writer has been able to observe, show this distinction; it would be interesting to know whether other botanists, who have seen this species growing in Brazil, can confirm Hildebrand's description. In the case of *Phragmites communis*[2], there is a similar differentiation between long, thick, unbranched mud-roots, and thin water-roots, branched to the third degree.

The roots of free-floating plants obviously do not serve for anchorage, but they seem sometimes to perform a corresponding rôle in preserving equilibrium; this is particularly obvious in the cases of *Lemna* and *Stratiotes*. Aquatic roots often exercise another function, which is more remote from those generally assumed in the case of terrestrial plants—namely, that of assimilation; their colour is sometimes quite conspicuously green. In the Water Chestnut, *Trapa natans*[3], the later roots, developed adventitiously below the leaf-bases, are free-floating and branched. These feathery structures have been supposed by some authors to be of foliar nature; this is erroneous, although physiologically they correspond to the divided leaves of *Myriophyllum*[4]. It is an indication of the extraordinarily acute mind of Theophrastus, the Father of Botany (born B.C. 370), that he avoided the morphological pitfall which has been fatal to so many subsequent writers, for in describing *Trapa* he says, "quite peculiar to this plant is the hair-like

[1] Hildebrand, F. (1885).
[2] Pallis, M. (1916).
[3] Barnéoud, F. M. (1848).
[4] Goebel, K. (1891–1893).

character of the growths which spring from the stalk; for these are neither leaves nor stalk[1]." We have already alluded to the thalloid roots of the Podostemaceae, which also serve for assimilation.

Like the stems of aquatics, the roots show certain anatomical divergences from those of land plants[2]. Root hairs are occasionally absent, e.g. *Lemna trisulca*. The roots of *Elodea* bear no absorbent hairs so long as they are immersed in water, but they develop them freely on entering the soil[3]. In other hydrophytes, e.g. *Hydrocharis*, the root hairs are unusually long. It is rather curious that in the roots of water plants the piliferous layer, and the layer immediately below it, are often cuticularised. The aerating system, which occurs in the primary cortex, or as a secondary formation, has been dealt with in Chapter XIV.

As in the case of submerged stems, the vascular system of the roots tends to be very much reduced. The simplest root among Dicotyledonous water plants is that of *Callitriche stagnalis* (Fig. 138), which has two protoxylems—each consisting of a single tracheid—separated by a single median metaxylem element. This simple xylem group is flanked on either side by a single sieve-tube with companion-cells. In certain Monocotyledons, a still more extreme degree of simplification is reached. *Vallisneria spiralis* (Fig. 139), for instance, has merely a central channel, corresponding to the central vessel of other forms, surrounded by a ring of cells, three of which are apparently sieve-tubes, each accompanied by a companion-cell. *Naias*, again, has a root of a very simple type, in which the phloem is more conspicuously developed than the xylem[4] (Fig. 140). The reduction series in the roots of the Potamogetons is illustrated in Fig. 41, p. 65.

Plasticity is certainly a marked feature of the roots of water plants, for though they have to some extent given up the work of absorption, they have assumed and developed various other functions to which their terrestrial ancestors must have been comparative strangers.

[1] Theophrastus (Hort) (1916).
[2] Schenck, H. (1886).
[3] Snell, K. (1908).
[4] Sauvageau, C. (1889[1]).

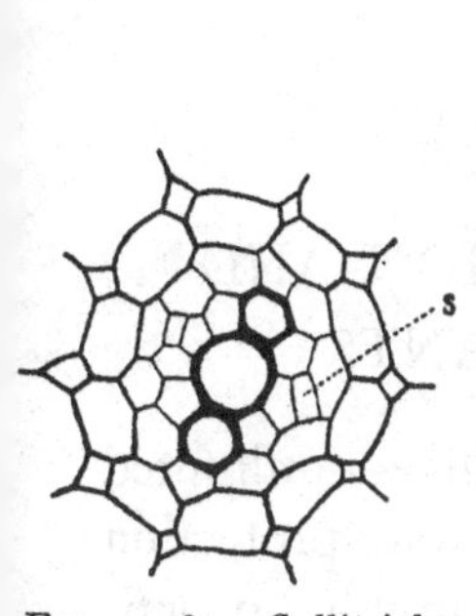

FIG. 138. *Callitriche stagnalis*, Scop. T.S. central cylinder of adventitious root of water form (×470); *s*, sieve tube. [Schenck, H. (1886).]

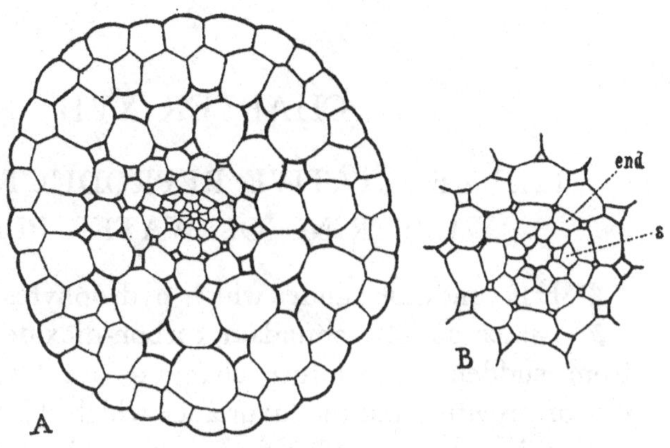

FIG. 139. *Vallisneria spiralis*, L. *A*, T.S. adventitious root (× 240). *B*, T.S. central cylinder (× 470); *end*, endodermis, *s*, sieve tubes of which three are present. The central vessel is unthickened. [Schenck, H. (1886).]

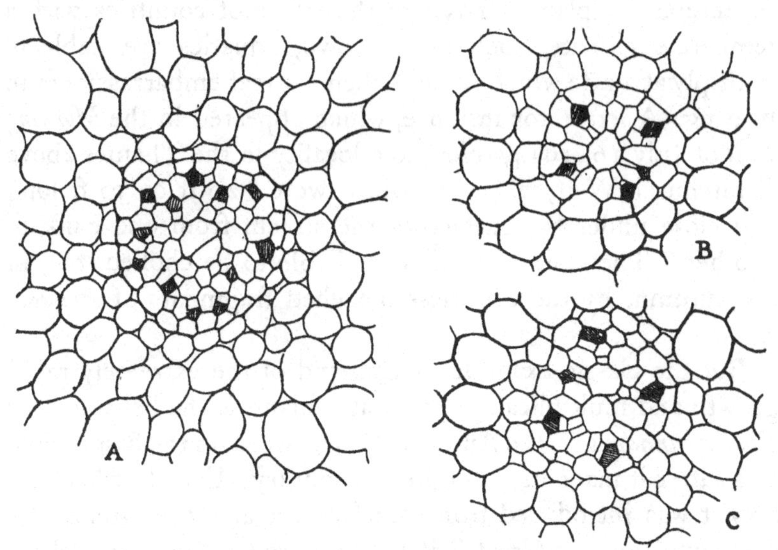

FIG. 140. T.S. central region of roots of *Naias*, sieve tubes shaded. *A*, *Naias major*, All., two central vessels. *B* and *C*, *N. minor*, All., one central vessel in *B* and two in *C*. [Sauvageau, C. (1889[1]).]

CHAPTER XVII

THE VEGETATIVE REPRODUCTION AND WINTERING OF WATER PLANTS

THE conditions under which hydrophytes live—unlimited water supply, abundant carbon-dioxide and protection from sudden temperature changes—are favourable to activity of growth[1], and the luxuriance which this vegetation often attains is a matter of common note; indeed it sometimes becomes such a hindrance to navigation as to compel the attention, not only of botanists, but also of those who normally take no interest in plants. Even in the rivers of countries with a temperate climate, such as our own, aquatics are liable to multiply at a rate which renders them a great embarrassment in boating. A letter, for instance, which appeared in the *Morning Post* of July 16, 1914, refers to a locality in the Thames above Whitchurch Weir, where the weeds were "20 ft to 30 ft long and close under the surface of the stream from one bank to another." The sluices of mills are liable to be choked, too, in the autumn, by the countless detached fragments of *Potamogeton*.

But the classic example in England of the extremely rapid growth and multiplication of a water plant, is the behaviour of *Elodea canadensis*[2], the American Waterweed, in the first decade or so after it made its début in this country. Exactly when and how it was introduced from America remains a mystery. Its first appearance in Great Britain is said to have been in Ireland in 1836, while its first recorded occurrence in England was in Berwickshire in 1842[3]. It travelled south, and by 1851 was so

[1] Schenck, H. (1885).

[2] Marshall, W. (1852) and (1857), Caspary, R. (1858[2]), and Siddall, J. D. (1885). For the continental history of the plant see Bolle, C. (1865) and (1867).

[3] Johnston, G. (1853).

luxuriant at Burton-on-Trent—where it had been recorded in 1849[1]—that it bid fair to block up one of the two streams into which the Trent there divides. Unfortunately the Curator of the Cambridge Botanic Garden, who had received the plant from Professor Babington in 1847, introduced it into a tributary of the Cam in 1848. By 1852 it had spread into the river, and so completely choked it as to raise the water level several inches, and to prevent fishing, swimming and rowing, and greatly to hinder the towing of barges. At this date it first invaded the fen district, and in a few years so choked the dykes as seriously to impede drainage. The difficulties caused by the presence of excessive quantities of the plant were so acute that an adviser was sent down by the Government to consider the best method of dealing with the pest. No successful plan for coping with it was discovered, but in a few years the luxuriance of the *Elodea* diminished without any apparent cause. Siddall[2], to whom we owe the most exhaustive treatment of the subject, concludes that, "The experience of those who have had most to do with it seems to indicate that if left alone, its habit is, upon first introduction into a new locality, to spread with alarming rapidity; so much so as literally to choke other water plants out of existence. But this active phase reaches a maximum in from five to seven years, and then gradually declines, until at last the *Anacharis* [*Elodea*] ceases to be a pest, and becomes an ordinary denizen of the pond, river, or canal, as the case may be."

As has been already stated in Chapter IV, *Elodea canadensis* never reproduces itself sexually in this country, and the history of the plant suggests that possibly the whole *Elodea* population of England may be regarded, in one sense, as a single individual, with an enormous vegetative output, mechanically sub-divided into vast numbers of apparently distinct plants; in other words, it is not improbable that it may represent the soma developed from a single fertilised ovum. It would thus be a "major plant unit," whose soma consists of a vast number of minor individuals. Pallis[3], in a most suggestive study of the problem of

[1] Caspary, R. (1858[2]). [2] Siddall, J. D. (1885). [3] Pallis, M. (1916).

individuality in the case of *Phragmites communis*, has brought forward evidence which strongly suggests that the "major plant unit," i.e. the total vegetative output which one fertilised egg is capable of initiating, is to be regarded as a constant for each species, its mass being the measure of specific vital energy. She has shown that, in the case of the reed swamps of the Danube, there are distinct indications of a definite life-cycle of vegetative growth, terminating in senescence and death, whose arrival is not fortuitous or due to external conditions, but is a necessity inherent in the very nature of the species itself. Such a thesis is obviously very difficult to substantiate, but the history of *Elodea*, outlined above, certainly seems to the present writer to lend itself more readily to some such interpretation, than to the accepted explanation, which sees in the aggressive phase of this introduced plant, merely the direct stimulating effect of change of environment. *Elodea* has passed through a period of great luxuriance, followed by a gradual diminution in vigour, occurring more or less contemporaneously in all the localities which have been colonised by its rapid vegetative multiplication. By 1883 its period of maximum abundance was apparently over. In 1909 an enquiry[1] was set on foot to determine the condition of the species at that date, i.e. sixty-seven years from its first recorded appearance in England. This enquiry resulted in reports from many localities indicating that *Elodea* had sunk everywhere into the condition of a mere denizen, displaying no greater luxuriance than the other water plants with which it was associated. Siddall, in this year, wrote that he had some difficulty in finding a specimen of *Elodea* in a locality where in 1873 all other vegetation was choked with it. He also made the extremely interesting statement that the circulation of the protoplasm was very feeble in 1909 as compared with its condition in 1873—a statement which the present writer feels must be accepted with some reserve, for it is a point on which a really critical comparison would be attended with obvious difficulties.

The general history of *Elodea* seems at least to point towards

[1] Walker, A. O. (1912).

the conclusion that the "individual," which was introduced into this country, has run its course, through an opulent maturity, to a point approaching senility, which may ultimately lead to complete extinction. Water plants certainly appear to offer a favourable field for the study of the "major individual," since, in this biological group, reproduction by sexual means is often deferred for long periods.

In warmer climates the rapidity of growth of water plants is even more remarkable than in temperate regions. The way in which *Eichhornia speciosa*, Kunth[1], the Water Hyacinth, may sometimes choke a wide river, forms a really startling example of excessive quickness of growth and multiplication. About the year 1890, this plant was accidentally introduced into the St John's River in Florida, which, being a sluggish stream, was particularly well-suited to serve as its home. After seven years, two hundred miles of the river bank had become fringed with a zone of *Eichhornia* from twenty-five to two hundred feet in width. In the summer of 1896, a strong north wind drove the plants up stream from Lake George, forming a solid mass entirely covering the river for nearly twenty-five miles. The growth was so dense that small boats with screw propellers could not get through the mass. Formerly, when the stream was clear, logs used to be rafted down the river, and it is estimated that, at the time when the Water Hyacinth was at its maximum, the lumber industry of the region suffered an approximate annual loss of $55,000 from the difficulty of rafting.

In Africa, the River Lettuce, *Pistia Stratiotes*, plays a similar part to the Water Hyacinth of America in hindering navigation. Miss Mary Kingsley[2] gives a characteristically racy description of its behaviour on the Ogowé and the neighbouring rivers in the French Congo. "It is," she writes, "very like a nicely grown cabbage lettuce, and it is very charming when you look down a creek full of it, for the beautiful tender green makes a perfect picture against the dark forest that rises from

[1] Webber, H. J. (1897). [2] Kingsley, M. H. (1897).

the banks of the creek. If you are in a canoe, it gives you little apprehension to know you have got to go through it, but if you are in a small steam launch, every atom of pleasure in its beauty goes, the moment you lay eye on the thing. You dash into it as fast as you can go, with a sort of geyser of lettuces flying up from the screw; but not for long, for this interesting vegetable grows after the manner of couch-grass. I used to watch its method of getting on in life. Take a typical instance: a bed of river-lettuces growing in a creek become bold, and grow out into the current, which tears the outside pioneer lettuce off from the mat. Down river that young thing goes, looking as innocent as a turtle-dove. If you pick it up as it comes by your canoe and look underneath, you see it has just got a stump. Roots? Oh dear no! What does a sweet green rose like that want roots for? It only wants to float about on the river and be happy; so you put the precious humbug back, and it drifts away with a smile and gets up some suitable quiet inlet and then sends out roots[1] galore longitudinally, and at every joint on them buds up another lettuce; and if you go up its creek eighteen months or so after, with a little launch, it goes and winds those roots round your propeller[2]."

The luxuriance of hydrophytes as compared with other herbaceous plants can be demonstrated not only by examples of their multiplication on a large scale, but also when the dimensions of individuals are considered. A striking instance is afforded by Caspary's[3] measurements of the leaves of a plant of *Victoria regia* cultivated in a hot-house; the maximum growth of the lamina recorded in 24 hours was as much as 30·8 cms. in length and 36·7 cms. in breadth. Even in our climate the growth of aquatics must be rapid, to produce the length of stem sometimes observed; in the case of *Ranunculus fluitans*, shoots twenty or more feet in length have been recorded[4], while floating

[1] Botanically these " roots " are of course lateral stems.

[2] For other cases of plant accumulations which are on a sufficient scale to form serious obstructions, see Hope, C. W. (1902).

[3] Caspary, R. (1856²). [4] Schenck, H. (1885).

branches of *Utricularia vulgaris* may be six feet long[1]. The shoot system, as a whole, sometimes attains a remarkable development. The present writer examined, for instance, a plant of *Polygonum amphibium* growing at Roslyn Pits, Ely, on June 30, 1913, which showed at the surface of the water only one flowering branch with seven foliage leaves. The plant was pulled up with a boat-hook and inevitably somewhat mutilated in the process, but, notwithstanding the breakages, the various axes forming the shoot system were found to measure altogether approximately forty-two feet. Besides the two visible leafy shoots, eight of the branches terminated in leaf buds, which looked as though they would probably have reached the surface in the course of that season. The longest internode in the horizontal part of the stem measured as much as sixteen inches.

The great development often reached by individual water plants is no doubt an expression of the same tendency as that which leads them so generally to perenniation. Annuals are quite rare among hydrophytes; only a few examples are known, such as *Naias minor*, *Naias flexilis* and certain species of *Elatine*[2]. There is of course no dry season to be spanned, and many aquatics can continue their vegetation all the year round, in some cases paying little regard to the passage from summer to winter. *Zannichellia palustris*, for instance, may be found in flower in November, while *Aponogeton distachyus*, cultivated out-of-doors in England, flowers sometimes in December and January. The strength of the tendency to perenniation may be illustrated by the fact that the following plants have at different times passed successfully through one or more winters in so unsympathetic a location as a rain-water tub in the present writer's garden—*Hydrocharis Morsus-ranae*, *Stratiotes aloides*, *Spirodela polyrrhiza*, *Lemna trisulca*, *Myriophyllum* sp., *Oenanthe Phellandrium* var. *fluviatilis*, *Ceratophyllum*, *Hippuris*, and two species of *Potamogeton*. That the perennial habit is directly related to the environment, seems to be indicated by the fact that, in the case of *Callitriche*[2], the land forms are annual while

[1] Burrell, W. H. and Clarke, W. G. (1911). [2] Schenck, H. (1885).

the water forms are perennial. In the aquatic Callitriches, rooted internodes bearing lateral buds may remain in the mud and tide over the winter[1]. *Montia fontana*, also, is biennial in places where the water is liable to dry up, but, in springs and permanent streams, it grows strongly and becomes perennial[2].

Those water plants which have not adopted special methods of perenniation, generally retain their leaves through the winter, e.g. *Peplis Portula*, *Ceratophyllum*, *Hottonia*, and the submerged species of *Callitriche*. In the case of such plants, any detached shoot will generally grow into a new individual with extreme readiness. In *Hottonia* the branches forming a whorl below the inflorescence become separated from the axis and give rise to new plants in the spring[3]. The present writer has noticed that, in the case of *Peplis Portula* and *Ceratophyllum*, the submerged stems are very brittle, and, in the early autumn, quantities of detached floating shoots may be observed. The behaviour of *Callitriche*[4] is particularly striking, for in this case new plants can be formed from a node with only a very small piece of internode attached. *Lawia zeylanica*, Tul.[5], one of the Podostemaceae of Ceylon, can recommence its growth from any portion of the thallus, however small, if it be submerged under favourable conditions, and other members of the family have a similar power. A very notable capacity for vegetative multiplication is exhibited by some Cruciferae. In the case of the North American *Nasturtium lacustre*[6], the pinnately dissected, submerged leaves become detached about the middle of August and float at the surface of the water; an adventitious bud arises at the base of each leaf and develops into a new plant. The same production of buds from foliar tissue has long been known in *Cardamine pratensis*, the Lady's Smock, where it can easily be observed at various times of year (Fig. 141). On May 21, 1919, the present writer saw countless plantlets growing from detached leaflets in a dyke in the fens near Lakenheath Lode.

[1] Vaucher, J. P. (1841) and Lebel, E. (1863). [2] Royer, C. (1881–1883).
[3] Prankerd, T. L. (1911). [4] Hegelmaier, F. (1864).
[5] Willis, J. C. (1902). [6] Foerste, A. F. (1889).

The caddice worms, which also abounded in this dyke, seemed to have a great fancy for using the leaflets in constructing their cases, and, in consequence, their armour was often elegantly crested with tiny adventitious plants of Lady's Smock.

In addition to those aquatics which retain their leaves through the winter, there are others which perenniate in or upon the substratum by means of rhizomes or tubers. Plants which adopt this habit, may be described as aquatic geophytes. *Limnanthemum* (Figs. 22 and 23, p. 41), *Castalia* (Fig. 11, p. 26) and *Nymphaea* (Figs. 10, p. 25 and 12, p. 27) are rhizomatous. In some cases—e.g. *Sagittaria*, certain Potamogetons and Nymphaeaceae—special tubers are formed which outlast the winter These afford a means of vegetative multiplication, since an individual plant may in some cases give rise to numerous tubers; a single plant of *Sagittaria sagittifolia*, for instance, may produce as many as ten tuber-bearing stolons. Another method of vegetative reproduction is illustrated by *Littorella lacustris*[1], which puts out runners in the spring, bearing at their apices young plants not easily distinguishable from seedlings; these plantlets become independent by the late summer or autumn. A plant of this species with a runner is shown in Fig. 142, p. 218.

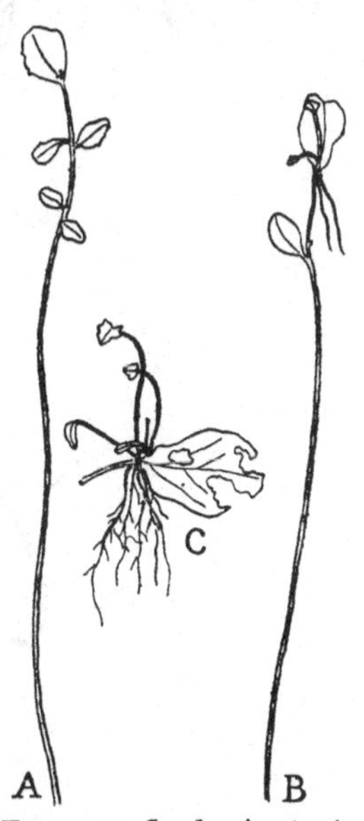

FIG. 141. *Cardamine pratensis*, L. *A* and *B*, leaves of submerged type growing among *Utricularia* in shallow pool, Commissioners' Pits, Upware, June 27, 1914. In each case the terminal leaflet bears an adventitious plantlet at the base. *C*, single, much-decayed pinnule bearing a well-developed plantlet; same locality, August 17, 1917. (½ nat. size.) [A. A.]

The most distinctive mode of wintering and of vegetative reproduction found among hydrophytes, is, however, by means of winter-buds or turions; these specialised shoots, which are

[1] Buchenau, F. (1859).

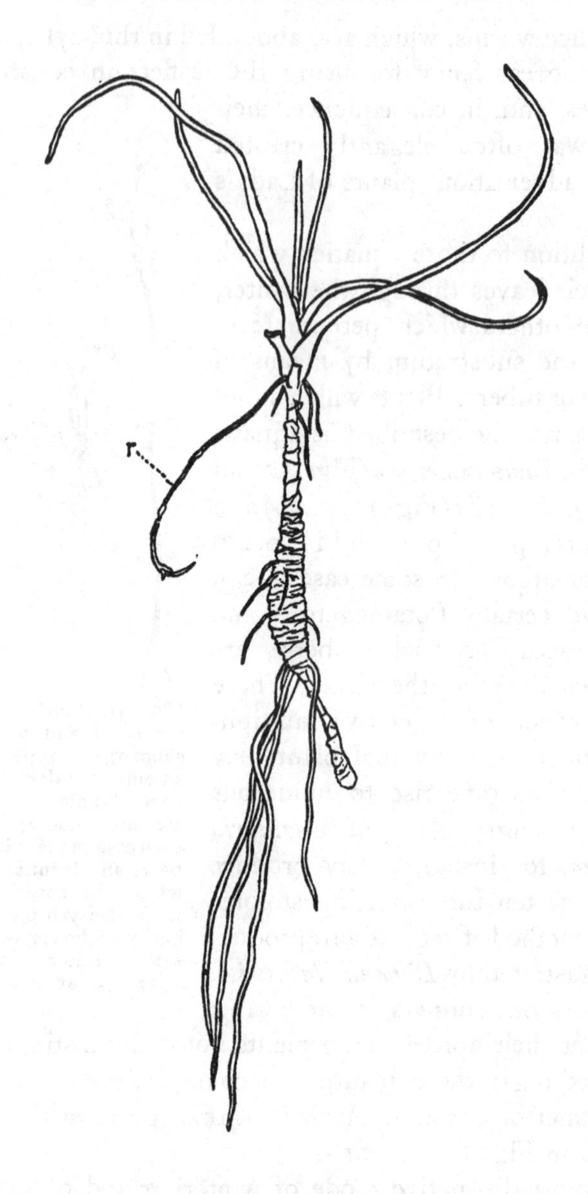

FIG. 142. *Littorella lacustris*, L. Plant drawn in February 1912. The collapsed region at base of stem probably represents the part formed in 1910; *r* = runner arising in a leaf axil. (Reduced.) [A. A.]

stored with food material, and protected externally in some way, become detached from the parent and pass the winter either floating, or resting at the bottom of the water. In the spring they expand, produce adventitious roots, and rapidly develop into full-fledged individuals. Certain plants, also, which do not actually produce independent turions of a specialised type, show transitions towards such a development. If shoots of the Greater Spearwort, *Ranunculus Lingua*[1], are left in water over the winter, they rise to the surface in the spring in a partly decomposed state, but bearing healthy buds in the axils of their leaves; these become detached to give rise to new plants. *Elodea canadensis* (Fig. 34, p. 55) and *Stratiotes aloides* (Fig. 32, p. 53), again, produce primitive reproductive buds, which do not immediately become free, but germinate while attached to the parent plant[2]. The apices of the shoots of *Ceratophyllum* are clothed in autumn with leaves which are more crowded and of a deeper green than those of the rest of the shoot, but, as we have already pointed out[3], they can scarcely be said to form definite winter-buds.

Certain turions showing a high degree of specialisation have already been mentioned, e.g. those of *Hydrocharis* (pp. 47–49), *Potamogeton* (pp. 66–69), the Lemnaceae (pp. 75–77), *Aldrovandia* (p. 110), and *Utricularia* (pp. 101–103). The difference between the normal foliage leaf and the protective outer leaf of the turion, in the case of *U. intermedia*, is shown in Fig. 143, p. 220. Among the British plants to whose wintering habits we have not yet referred, *Myriophyllum verticillatum*[4] affords a striking example of turion formation. In August the plant may be found simultaneously producing flowers and winter-buds (Fig. 144, p. 221). Early in October the ragged shoots may be seen floating, with here and there a compact turion (*T*), distinguished against the faded brownness of the parent plant by its vivid, dark-green hue. These winter-buds become detached during the cold season, and

[1] Belhomme, (1862). [2] Glück, H. (1906). [3] See p. 87.

[4] The winter-buds of *Myriophyllum* were noted by Vaucher, J. P. (1841).

in the spring they expand into graceful shoots (Fig. 145, p. 222). The germination normally occurs in March or April, but it can be induced at any time if the temperature is favourable; if brought indoors and kept warm, the turions will develop into new plants in October, November, December or January[1]. Cold is inimical to the winter-buds, and, if frozen for a few days, many of them are killed. The turions of different aquatics vary very widely in their capacity to withstand freezing[1]. Those of *Utricularia vulgaris* are uninjured by inclusion in ice for as long as twelve days, while *Hydrocharis Morsus-ranae*, according to Glück's experiments, is still more sensitive than *Myriophyllum*,

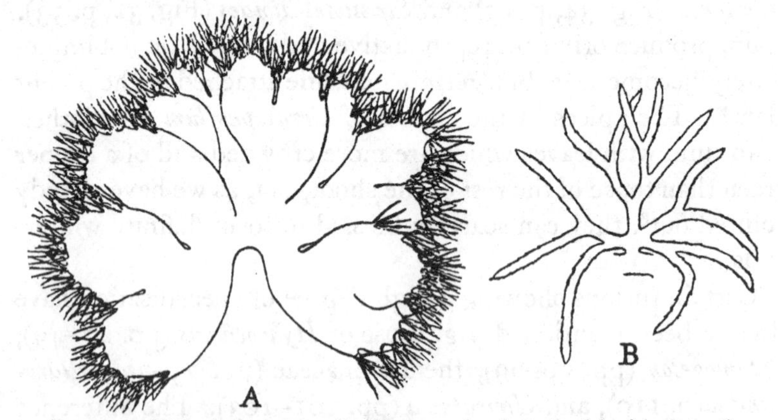

FIG. 143. *Utricularia intermedia*, Hayne. *A*, winter-bud leaf (enlarged). *B*, summer leaf (less enlarged). [Goebel, K. (1891–1893).]

for, after three to ten days in ice, nearly all the turions were killed. However, according to Guppy[2], they are able to withstand inclusion in ice for a period of some weeks; the discrepancy between these results requires some explanation, which may perhaps lie in the particular conditions of the experiments. The turions of many hydrophytes are saved from the risk of becoming frozen by their habit of wintering at the bottom of fairly deep water.

For many years botanists were inclined to interpret the development of 'winter-buds' on the simplest teleological

[1] Glück, H. (1906). [2] Guppy, H. B. (1893).

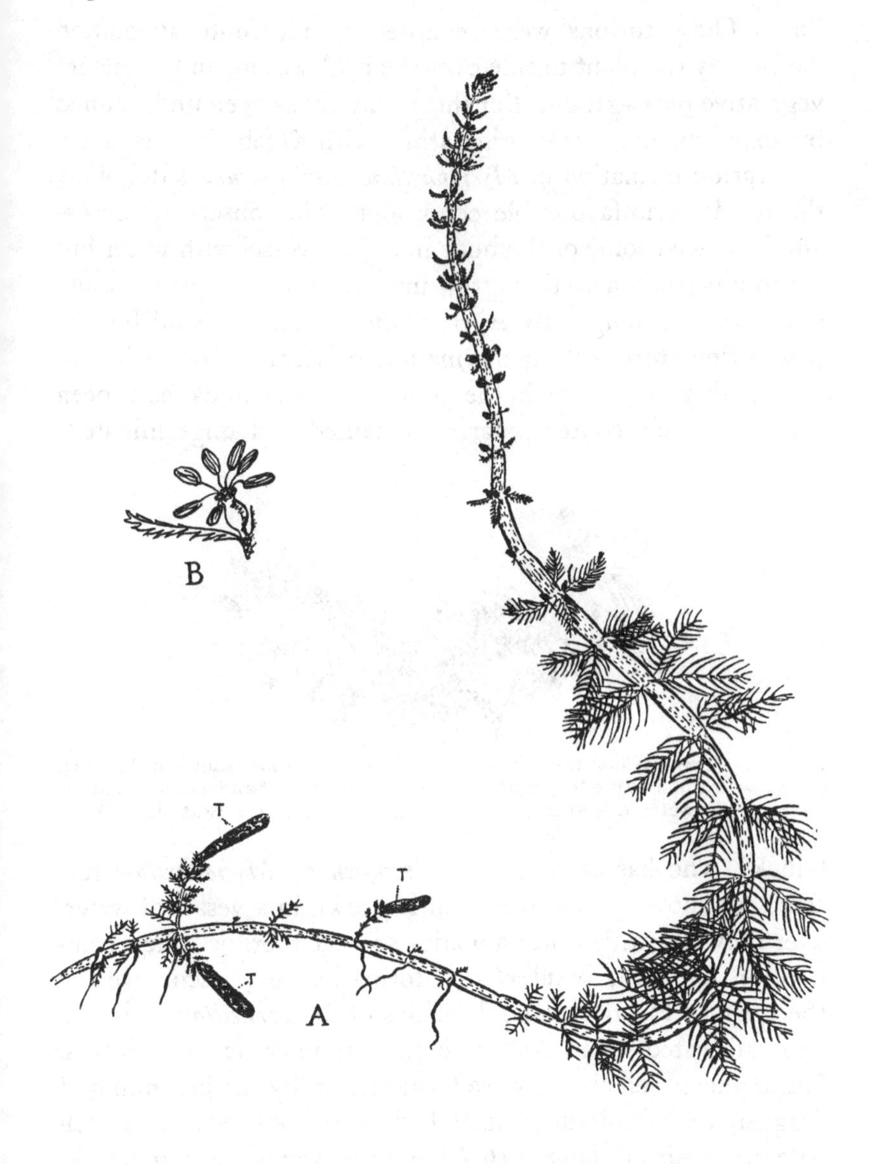

FIG. 144. *Myriophyllum verticillatum*, L. August 15, 1911. *A*, the inflorescence shows in succession female, hermaphrodite and female flowers. Three turions, *T*, occur on the lower part of the axis. (Reduced.) For the further development of one of these turions see Fig. 145, p. 222. *B* shows an hermaphrodite flower and its subtending leaf. (Enlarged.) [A. A.]

lines. These turions were regarded as a definite adaptation devised by the plant to tide over the cold season, and to ensure vegetative propagation. But this position has been undermined by experimental work originating with Goebel's[1] discovery that turion formation in *Myriophyllum verticillatum* is definitely the result of unfavourable conditions. This observer, for example, placed some of the buds in a glass vessel with water but without earth, where they grew into richly rooted plants, more than 30 cms. long. By April 1, these plants had all formed new turions terminating the main and lateral shoots, while in the locality from which the original winter-buds had been collected, their contemporaries remained still ungerminated!

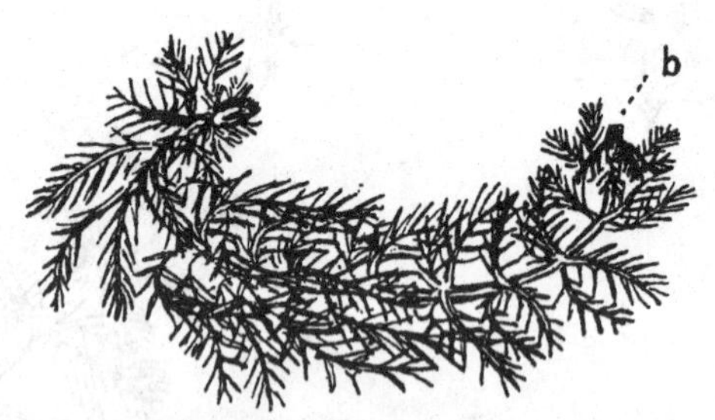

FIG. 145. *Myriophyllum verticillatum*, L. One of the turions shown in Fig. 144, p. 221, which had begun to germinate after the winter's rest and was found at the bottom of the water in this condition on March 16, 1912; *b*, base. (Nat. size.) [A. A.]

Glück[2], who has carried Goebel's work on *Myriophyllum* further, has shown that if the plant is grown in a vessel of water, over-crowded with other aquatics so that there is much competition for food, 'winter' bud formation may occur even in the spring. He also planted turions of *M. verticillatum* in soil, and cultivated them for an entire summer as land plants. Numerous green shoots were formed, but, by the beginning of August, each individual plant had also produced four to ten pale green turions (Fig. 146 *K*), most of which were under the soil. This early development of turions is attributed by Glück to the lack of water from which the plants suffered. On the other

[1] Goebel, K. (1891–1893). [2] Glück, H. (1906).

hand, luxuriant specimens growing in water in warm situations may vegetate throughout the winter without forming turions. It is most likely that, in normal life, it is the lowering of the temperature in the autumn which induces the formation of winter-buds.

That it is unfavourable conditions which bring about the development of turions, seems to be true not only of the Water Milfoil but of aquatics in general. Some remarkable experiments on the effect of starvation upon *Utricularia* have been quoted on pp. 102–103. Similar results have been obtained in the case of *Sagittaria sagittifolia*[1] in which, however, the vegetative multiplication is effected by tubers and not by turions. Tuber formation in the Arrowhead normally occurs when the plant has exhausted itself by the formation of inflorescences, and when cooler weather sets in. The land form, like that of *Myriophyllum*, produces tubers several weeks earlier than the form growing under the optimum aquatic conditions. Glück[1], one autumn, planted a tuber of the Arrowhead in a pot of earth and left it there, almost without water, until towards the end of the following July. The plant, which had failed to appear above the soil, was then examined, and it was found that the tuber had put out a few wretched-looking little ribbon-leaves, which had not possessed strength to penetrate the earth. It had also formed four tiny stolons, 1·5 to 2 cms. long, each terminating in a small tuber, 8 to 10 mm. in length. This tuber formation had apparently occurred as a

FIG. 146. *Myriophyllum verticillatum*, L. Land form with five subterranean turions, two of which are marked *K*. The turion of the previous generation, *W*, is still attached to the base of the plant and has formed a number of adventitious roots. The two lowest turions have grown out of the axis of the previous turion. (Reduced.) [After Glück, H. (1906), *Wasser- und Sumpfgewächse*, Bd. II, Pl. V, fig. 61.]

[1] Glück, H. (1905).

consequence of want of food, and at the cost of the reserve material in the parent tuber. At the same date no second generation of tubers had been developed by the normal control plants among Glück's cultures.

Myriophyllum verticillatum illustrates yet another character of winter-buds—namely the close relation which they sometimes bear to the flowers and inflorescences. Glück[1] records that in the case of an infructescence of this plant which had become submerged, the continued development of the axis produced an ordinary turion at the apex. The connexion between flowering and vegetative reproduction is also well shown in certain other hydrophytes, and notably in the Alismaceae.

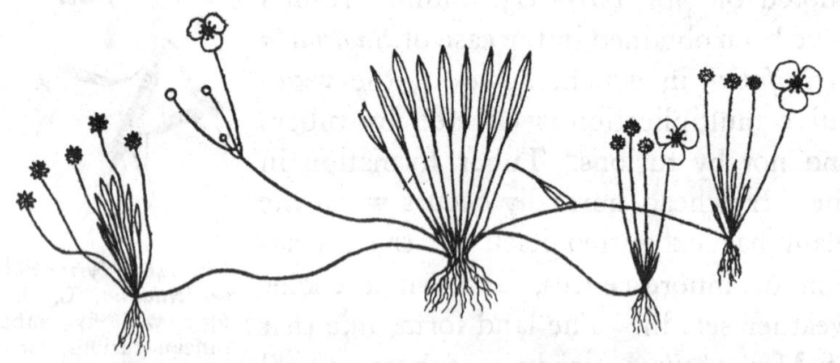

FIG. 147. ***Echinodorus ranunculoides,*** **(L.) Engelm. var.** ***repens*** **forma** ***terrestris.*** **Habit drawing to show transitions between inflorescences and entirely vegetative rosettes. (Reduced.) [After Glück, H. (1905),** ***Wasser- und Sumpfgewächse,*** **Bd. I, Pl. II, fig. 16.]**

Fig. 147 shows transitions between inflorescences and vegetative offshoots in the case of *Echinodorus ranunculoides*. *Caldesia parnassifolia* is an even more striking example, for here the inflorescences may be transformed into axes bearing turions (Fig. 148). Fig. 149 *A* shows part of an inflorescence of this plant, in which whorls of turions replace the flowers, while Fig. 149 *B* represents the germination of one of these turions. In *Caldesia*, and in other Alismaceae, the transformation of the inflorescences into vegetative shoots goes on step by step with the increase in depth of the water. In this connexion it may be

[1] Glück, H. (1906).

noted that certain Cryptogams (e.g. *Riccia*, *Fontinalis* and *Pilularia*[1]) behave like the Flowering Plants we have cited, in not producing sexual organs in deep or rapidly flowing water, while the replacement of sporangia by young plantlets has been recorded in the case of certain examples of *Isoetes lacustris* and *I. echinospora*[2], which grew at a considerable depth in one of the

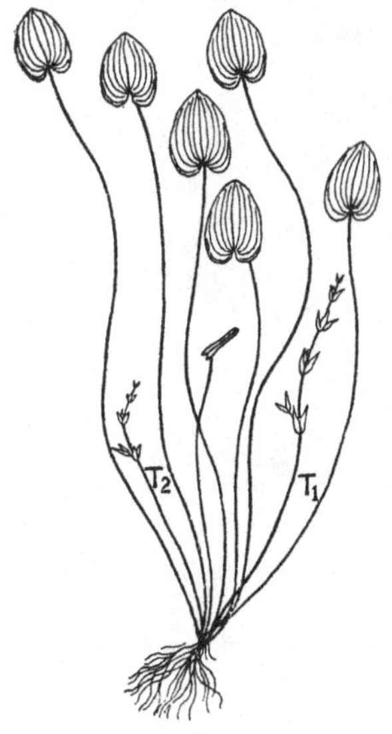

FIG. 148. *Caldesia parnassifolia*, (Bassi) Parl. forma *natans*. A deep water plant which has developed axes bearing turions (T_1 and T_2) in place of inflorescences. (Reduced.) [After Glück, H. (1905), *Wasser- und Sumpfgewächse*, Bd. 1, Pl. IV, fig. 25.]

FIG. 149. *Caldesia parnassifolia*, (Bassi) Parl. *A*, 'turion inflorescence' from a plant growing in water 50 to 60 cms. deep. (Reduced.) *B*, turion which germinated in water 50 cms. deep. (Slightly enlarged.) [After Glück, H. (1905), *Wasser- und Sumpfgewächse*, Bd. 1, Pl. IV, figs. 27 *a* and 28.]

lakes of the Vosges country. The occurrence of tubers in lieu of flowers in *Castalia Lotus*[3] can be paralleled among such terrestrial plants as *Polygonum viviparum*, but aquatic conditions seem particularly to favour a general tendency to replacement of sexual by vegetative reproduction.

[1] Schenck, H. (1885). [2] Goebel, K. (1879).
[3] Barber, C. A. (1889); see Fig. 19, p. 37.

There is good reason to suppose that, as Schenck[1] long ago suggested, the vegetative reproduction of water plants merely illustrates the general rule that vegetation and fructification stand in inverse ratio to one another. Orchards bear better when the trees are pruned, while in wet years when leafage is over-luxuriant, fruit formation diminishes. And thus the excessive vegetative activity of water plants acts, in all probability, as a deterrent to sexual reproduction.

[1] Schenck, H. (1885).

CHAPTER XVIII

THE FLOWERS OF WATER PLANTS AND THEIR RELATION TO THE ENVIRONMENT

THE most notable characteristic of the flowers of the majority of aquatic Angiosperms is that they make singularly little concession to the aquatic medium, but display the utmost conservatism in form and structure. The plants which have, in the course of evolution, adopted water life, have, as we have already shown, profoundly modified their vegetative organs in connexion with their new environment, but their methods of sexual reproduction in general depart little from those which had already become stereotyped in their terrestrial ancestors. This sharp distinction, between the degree of modification of the vegetative and reproductive parts, is particularly well shown in the case of so highly specialised a water plant as *Utricularia vulgaris*. Here the vegetative body is entirely submerged, but the aerial inflorescence axis and the flowers, which are adapted to entomophilous pollination, in no way differ from those of a terrestrial plant. The extreme divergence in mode of life, and even in internal structure, between the aerial reproductive region and the submerged vegetative region in this species, led an anatomist to speak of the plant as consisting of "an aquatic being, vegetating horizontally without roots," and "a vertical aerial being, producing flowers at its apex, and implanted in the first, which serves it as soil, or rather as roots[1]."

Those hydrophytes which still retain a type of flower adapted for aerial life, are under the absolute necessity of raising their inflorescence axis well above the water level, if cross-pollination is to be secured. This is sometimes very incompletely achieved,

[1] Tieghem, P. van (1868).

and even within the same genus we find differing degrees of success in the avoidance of submergence of the flower. *Ranunculus fluitans*, for instance, which does not hold its peduncles well erect and grows in rapidly flowing water, very often suffers from the inundation of its flowers, and, in consequence, fails to set seed[1]. Sometimes the attempt to rise above the water surface seems to have been entirely given up. *Ranunculus trichophyllus* is described as growing in the River Inn in enormous masses, and frequently blooming under water, opening its flowers at a depth of 1 to 1½ feet, but whether it can set seed under these conditions does not seem to have been observed[2]. Those Batrachian Ranunculi which flower successfully in rapidly flowing water, prove to be species such as *R. carinatus*, Schur. (*R. confusus*, Gen. et Godr.) which produce long flowering stalks rising erect above the water, and not readily submerged by slight changes in level[1]. In the case of the heterophyllous Water Buttercups, the leaves associated with the flower are often floating and relatively undivided; this must be an assistance in maintaining the equilibrium of the pedicel[3]. In *Heteranthera zosteraefolia*, also, the leaf next the inflorescence is described as always being of the floating type[4]. The association of floating leaf and flowers in *Limnanthemum nymphoides*, which is so close that the inflorescence appears at first sight to spring from the petiole, must also play a part in holding the flowers above water. If any locality in which *Limnanthemum* grows freely be visited in August, the way in which the fringed, yellow flowers are held clear above the water will be found to be one of their most striking characters.

The early development and whorled arrangement of the branches springing from the base of the inflorescence axis in *Hottonia palustris*[5], the Water Violet, serve to support it on all sides, and to keep it vertical, while the numerous adventitious roots arising from the base of the erect shoot probably have a

[1] Freyn, J. (1890). [2] Overton, E. (1899).
[3] Askenasy, E. (1870). [4] Hildebrand, F. (1885).
[5] Schenck, H. (1885) and Prankerd, T. L. (1911).

similar effect (Fig. 127, p. 197). The part played by the roots in holding the stem of *Oenanthe Phellandrium* in an upright position has already been mentioned[1], as well as the specialised branches which in some Bladderworts keep the inflorescence erect[2]. Fig. 150 shows the whorl of six branches surrounding

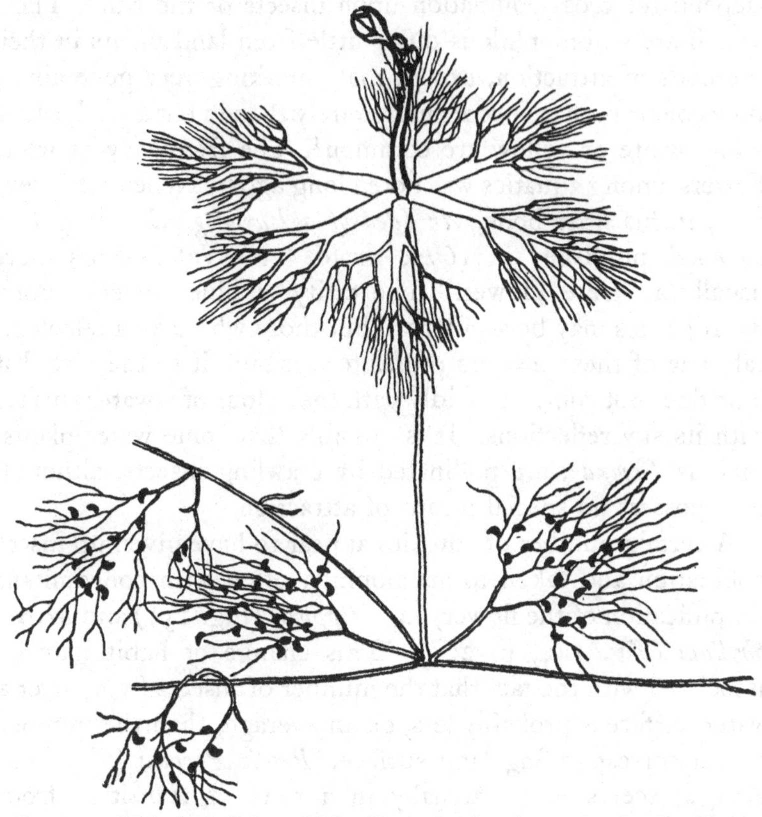

FIG. 150. *Utricularia inflata*, Walt. Part of swimming water shoot, with an inflorescence axis bearing six floating organs. [Goebel, K. (1891–1893).]

the flowering axis in *Utricularia inflata*. Spruce[3], in his account of his travels in the Amazon region, mentions, as a general observation, that those hydrophytes which rear themselves erect and thus raise the flowering part of their stem well out of the water, prove on examination to have the sub-aquatic leaves grouped

[1] See p. 204. [2] See p. 99. [3] Spruce, R. (1908).

in whorls, even when their terrestrial relatives have a different arrangement. He states that *Jussiaea amazonica* has the narrow submersed leaves so closely whorled as to resemble the Mare's-tail of our ponds, while the emersed leaves are solitary.

Those water plants whose inflorescences rise into the air, depend for cross-pollination upon insects or the wind. Those which are entomophilous differ little from land plants in their methods of attraction, except that, speaking very generally, a blue colour perhaps occurs more rarely than in terrestrial plants, while white or yellow are common[1]. The frequency of white flowers among aquatics was noted long ago by Nehemiah Grew, who, in his little book, *An Idea of a Phytological History Propounded*, published in 1673, wrote, "to Water-plants more usually a White Flower." The rarity of blue flowers among hydrophytes may be accidental, but those who take a teleological view of these matters prefer to attribute it to the fact that blue does not contrast vividly with the colour of a water surface with its sky reflections. It is possible that some water plants, such as *Lemna*[2], are pollinated by crawling insects, although they possess no special means of attraction.

A certain number of aquatics appear to have given up insect pollination and taken to anemophily, often with concomitant simplification of the flower, e.g. *Hippuris* (Fig. 151) and *Myriophyllum* (Fig. 144, p. 221). This change of habit may be associated with the fact that the number of insects flying over a water surface is probably less, on an average, than the number over a corresponding land surface. *Peplis Portula* (Fig. 152, p. 232) seems to be actually in a state of transition from entomophily to anemophily. There are six fugacious little white petals, and a small amount of honey is secreted[3]. But the flowers are very inconspicuous, and no insect visitors appear to be attracted. The stigma becomes ripe a little sooner than the stamens, but they bend inwards over it and pollinate it[4].

Myriophyllum is an example of a wind-pollinated genus, in

[1] Schenck, H. (1885). [2] See p. 80. [3] MacLeod, J. (1894).
[4] Willis, J. C. and Burkill, I. H. (1895).

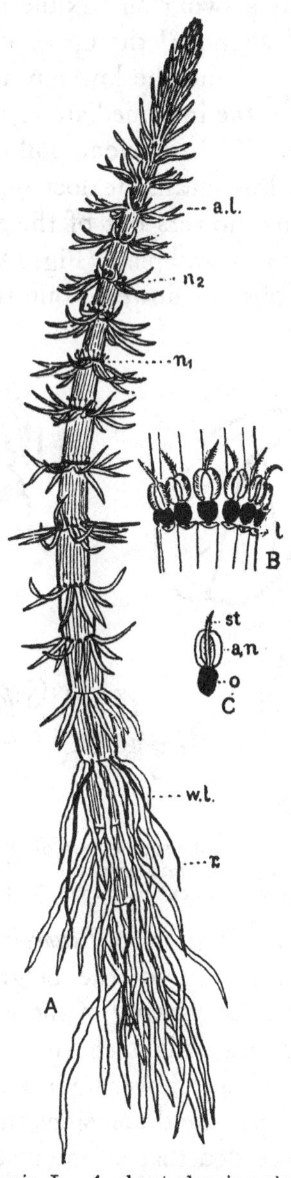

FIG. 151. *Hippuris vulgaris*, L. *A*, shoot showing air leaves (*a.l.*), water leaves (*w.l.*) and roots (*r.*). Whorls of flowers at the upper nodes; n_1, node with flowers whose anthers have dehisced; n_2, node with flowers whose anthers are still closed. *B*, whorl of flowers enlarged, leaves (*l*) cut away. *C*, a single flower seen from adaxial side; *st* = feathery style; *an* = anther; *o* = ovary. (Reduced.) [A. A.]

which the long anthers swing on flexible filaments (*B* in Fig. 144, p. 221). In *M. spicatum*[1] the upper flowers of the spikes are generally staminate, and the lower pistillate, while perfect flowers often occur in the intermediate region.

Littorella lacustris, which is anemophilous, sets a full complement of seeds by this means; it does not, like *Myriophyllum* and *Hippuris*, raise its flowers out of the water, but is sterile except when it grows as a land plant (Fig. 128 *C*, p. 198). When submerged it develops no flowers, but reproduces itself by

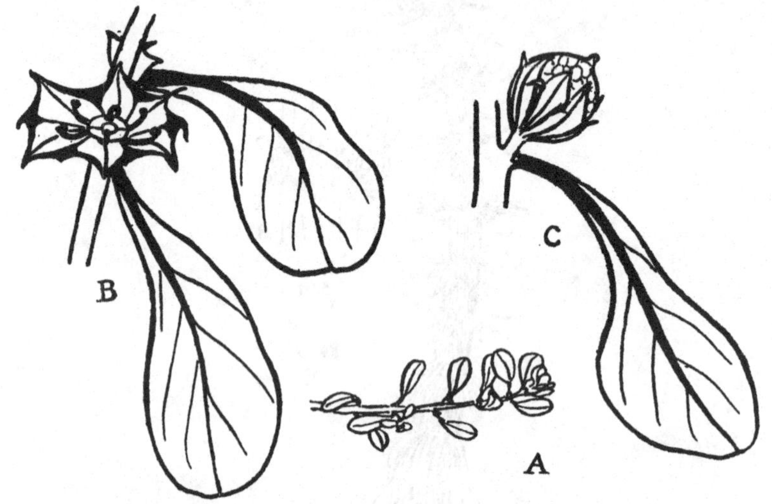

FIG. 152. *Peplis Portula*, L. Land form, Forest of Wyre, September 13, 1911. *A*, part of branch. (Nat. size.) *B*, flower and leaves. (Enlarged.) *C*, fruit with seeds showing through transparent fruit coat. (Enlarged.) [A. A.]

runners (Fig. 128 *A* and *B*, p. 198). *Littorella* has been described as flowering so luxuriantly, in the height of summer in a dried-up swamp, that the shaking of the white stamens in the wind gave the whole area a silken sheen[2], while another record relates to a case of this plant flowering in a dry year, when it had only attained to such minute dimensions that the length of the filaments actually exceeded that of the rest of the plant[3]! In this genus we are probably not dealing with a case of loss of

[1] Knupp, N. D. (1911). [2] Buchenau, F. (1859).
[3] Preston, T. A. (1895).

entomophily associated with the water habit, since the immediate ancestors of *Littorella* were most likely closely related to the typically wind-pollinated Plantagos.

The difficulty of keeping entomophilous or anemophilous flowers above water seems to have led, in the case of certain aquatics, to the formation of cleistogamic flowers which can set seed even when submerged. But Prankerd's[1] work has suggested that records of cases of cleistogamy among water plants ought to be received with some caution, unless they are based on evidence of a highly critical nature. Concerning the Water Violet, this author writes, "Cleistogamy has been attributed to *Hottonia*, but I have found no trace of it during three summers' field work. The idea is probably due to some small, closed flowers, which occur sometimes among those fully developed, but serial sections have shown that these are merely abortive." It is possible that similar detailed investigations of other water plants, which have the reputation of bearing cleistogamic flowers, might considerably reduce the list; *Subularia* for instance, which has been called cleistogamic, seems to open its flowers even if submerged[2]. There are however a certain number of cases in which the existence of cleistogamy is adequately established. Hooker[3], for example, described the phenomenon in detail in *Limosella aquatica*, L. This plant in Kerguelen's Land was, he writes, "found in the muddy bottom of a lake, and probably flowers all the year round. I gathered it in the month of July (mid-winter), beneath two feet of water, covered with two inches of ice; even then it had fully-formed flowers, whose closely imbricating petals retained a bubble of air, the anthers were full of pollen and the ovules apparently impregnated. The climate of Kerguelen's Land being such, that this lake is perhaps never dried, it follows that the plant has here the power of impregnation when cut off from a free communication with the atmosphere, and supplied with a very small portion of atmospheric air, generated by itself." *Ranunculus fluitans*, Lmk.,

[1] Prankerd, T. L. (1911). [2] Hiltner, L. (1886).
[3] Hooker, J. D. (1847).

R. aquatilis, L. and *R. divaricatus*, Schr. are also said to flower under water, pollination occurring in a bubble of air formed within the perianth[1].

An Indian species of *Podostemon*, *P. Barberi*, Willis[2], has cleistogamic flowers, with one stamen standing close up against the stigmas (Fig. 82, p. 121). *Alisma natans*[3] is described as being cleistogamic in deep water, while *Echinodorus ranunculoides*[4] has an entirely submerged form which flowers under water at a depth of three feet. Other recorded cases of cleistogamy are *Heteranthera dubia*[5] (Fig. 153) and *Hydrothrix Gardneri*[6] (Pontederiaceae), *Euryale ferox*[7] (Nymphaeaceae), *Illecebrum verticillatum*[3] (Caryophyllaceae), *Tillaea aquatica*[8] (Crassulaceae), *Trapella sinensis*[9] (Pedaliaceae), and a number of species of Lythraceae with apetalous or subapetalous flowers, belonging to the genera *Rotala*, *Peplis* and *Nesaea*[10].

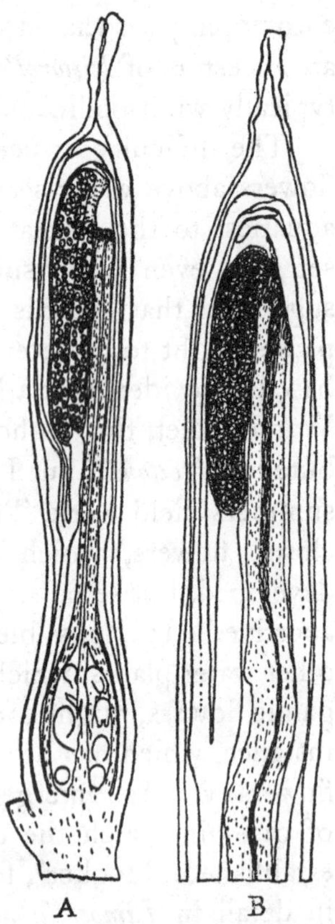

FIG. 153. *Heteranthera dubia*, (Jacq.) MacM. *A*, L.S. through an immature flower cut slightly obliquely in the adaxial-abaxial plane. The tip of the stigma lies below the upper ends of the anthers, and the style at this stage is straight. *B*, upper portion of an older flower cut in the same general direction as *A*. The stigma has been shoved up into the upper end of the flower in contact with the tips of the anthers where the stigmatic hairs touch the pollen grains through the breaks in the stamens. Pollen tubes are passing from the anther into the stylar chambers. The style is beginning to fold on account of its excessive elongation. [Wylie, R. B. (1917[1]).]

The pollination of cleistogamic flowers, though it may occur beneath

[1] Royer, C. (1881–1883).
[2] Willis, J. C. (1902).
[3] Schenck, H. (1885).
[4] West, G. (1910).
[5] Wylie, R. B. (1917[1]).
[6] Goebel, K. (1913).
[7] Goebel, K. (1891–1893).
[8] Caspary, R. (1860).
[9] Oliver, F. W. (1888).
[10] Gin, A. (1909).

the water surface, is no more truly aquatic than are the vital processes of a man in a diving bell, since, as Hooker points out in the case of *Limosella*, the transference of the pollen takes place within a bubble of gas. Certain plants, however, present transitional methods of pollination, which without being actually hydrophilous, show approaches to this state. The oft-quoted case of *Vallisneria spiralis* (Hydrocharitaceae) is perhaps the best instance of such a transitional method. The male and female plants are distinct. The female flowers are solitary within a spathe, and are carried up to the surface of the water by the elongation of the peduncle below the spathe. When mature they lie horizontally on the water surface[1]. The submerged male spathes contain over 2000[2] small flowers each with two stamens; the perianths are hermetically sealed, each enclosing a bubble of air. These male flowers become detached and rise to the surface of the water, where they open. The floating male flowers were figured early in the eighteenth century by Micheli[3], an Italian botanist. A later observer in India[4] speaks of "seeing under a noonday sun the innumerable florets freed from their spathes and ascending like tiny air-globules till they reach the surface of the water, where the calyx quickly bursts—the two larger and opposite sepals, reflex, forming tiny rudders, with the third and smaller recurved as a miniature sail, conjointly facilitating in an admirable manner the florets' mission to those of the emerging females." The male flowers are thus conveyed over the water surface by air currents, and some of them get carried into the neighbourhood of the female flowers, where the sticky pollen of the dehiscing anthers is likely to be rubbed off against the exposed stigmas. Each female flower, owing to its weight, is surrounded by a minute depression in the surface film of the water; the male flowers easily slide down the slope thus produced, and so approach the female[2]. After pollination the spiral peduncle contracts, carrying the maturing fruit deep down into the water; it is said that the

[1] Chatin, A. (1855[2]).
[2] Wylie, R. B. (1917[2]).
[3] Micheli, P. A. (1729).
[4] Scott, J. (1869).

contraction does not actually bring it to the bottom of the water, but that the last stages in the descent are accomplished by its own weight, when it is ripe[1]. Other Hydrocharitaceae, e.g. the marine genus *Enhalus*[2], possess a pollination mechanism resembling that of *Vallisneria*. Others again, e.g. *Elodea callitrichoides*[3], have, by a further modification, arrived at a type of pollination which is strictly hydrophilous, for the pollen, instead of being rubbed off against the stigmas, is shed explosively and falls on to the surface film, reaching the stigmas by flotation. The ultimate stage in the series of the Hydrocharitaceae is reached by the marine genus *Halophila*, in which neither male nor female flowers emerge from the water, and the process of pollination takes place in complete submergence[4]. The stigmas are thread-like and the pollen-grains, being united into strings, adhere readily to the stigmas, which present elongated receptive surfaces.

The family Hydrocharitaceae is, indeed, of unique interest from the standpoint of the evolution of submerged pollination, since it includes within itself all stages in the transition from entomophily to hydrophily[5]. It contains insect-pollinated flowers, such as *Hydrocharis Morsus-ranae* and *Elodea densa*, with attractive perianths, and, sometimes, nectaries; flowers in which the unwetted pollen is conveyed over the water by the 'boat mechanism,' e.g. *Vallisneria*; flowers in which the pollen floats on the surface of the water, e.g. *Elodea callitrichoides*; and, finally, flowers with entirely submerged pollination, such as *Halophila*.

Callitriche[6], among the Dicotyledons, provides another group of species in which the transition from aerial to aquatic pollination can be followed. The genus is subdivided into two sections: *Eu-callitriche*, to which the ordinary amphibious species of Water Starwort belong, and of which *C. verna* is the type, and

[1] Royer, C. (1881–1883). [2] Delpino, F. and Ascherson, P. (1871).
[3] Hauman-Merck, L. (1913[2]). See p. 55. [4] See p. 130.
[5] See pp. 55–57.
[6] Hegelmaier, F. (1864), Jönsson, B. (1883–1884), and Schenck, H. (1885).

Pseudo-callitriche, which consists of submerged plants grouped round the species *C. autumnalis*. *C. autumnalis* has no land form, but vegetates, flowers and fructifies below the level of the water surface. Throughout the genus the simple male and female flowers occur separately (Fig. 154); the female flowers are commonly found lower down the inflorescence than the male, but, in *C. autumnalis*, several male and female regions may alternate with one another. Insects, and possibly wind, carry the pollen of the Eu-callitriches, which is of the terrestrial type and is clothed with an exine insoluble in sulphuric acid. That of the Pseudo-callitriches, on the other hand, is of the aquatic type; it has no differentiated exine and contains oil globules which render it lighter than water. It is carried to the stigmas by water currents.

FIG. 154. *Callitriche verna*, L. July 19, 1910. Flowering shoot showing three female (♀) and one male (♂) flower. In the case of the male flower both bracts can be seen. (Enlarged.) [A. A.]

The aquatic pollination of *Ceratophyllum* (Hornwort) has already been considered[1], as well as that of three members of the Potamogetonaceae, *Cymodocea*[2], *Zostera*[3], and *Zannichellia*[4]. In connexion with the submerged pollination of *Naias graminea*—also belonging to the Pondweed family—a picturesque incident which has been placed on record by Bailey[5], suggests that aquatic animals may occasionally play a part in the pollination of submerged plants. He writes, "While...examining portions of a living plant on which were ripe anthers, I noticed a colony of *Vorticellidae* attached to one of the fascicles of leaves; the grace and activity of its movements led me to watch it for a considerable time, and whilst so watching it I witnessed grains of pollen whirled in all directions, or drawn into the vortex of the animal by its marginal cilia. The alternate contraction and

[1] See pp. 84–85. [2] See p. 126. [3] See pp. 127–129.
[4] See pp. 70–71. [5] Bailey, C. (1884).

elongation of the elastic and thread-like pedicles of the colony kept the pollen-grains in constant motion, which left me no doubt that at times the grains would be directly borne to the stigmatoid appendages of the pistilliferous flowers."

It seems to the present writer conceivable that, in future phases of evolution, if more Angiosperms reach the highly specialised stage of complete submergence, the water fauna may come to play an important part in their pollination. There may even arise a parallelism of development and an interdependence between aquatic animals and submerged plants comparable with that which has obtained in the case of aerial insects and the flowers which they pollinate!

In general, the consideration of the flowers of hydrophytes seems to lead to the conclusion that submerged pollination is a relatively modern development. It is, from some points of view, merely a further advance on lines similar to those already marked out in the case of anemophily. The great majority of hydrophilous plants have near relatives—sometimes even members of the same genus—which retain anemophilous or entomophilous habits; this may be regarded as a proof that plants with submerged pollination have arisen in comparatively recent times from ancestors with the aerial type of flower. *Ceratophyllum* forms an exception, since it is entirely hydrophilous, and has no intimate affinities with any other genus. It is probable, from its extreme adaptation to aquatic conditions and its isolated position in the relatively primitive Ranalean plexus, that it is a genus whose ancestors took to aquatic life at a very early stage in the race history of the Angiosperms.

CHAPTER XIX

THE FRUITS, SEEDS AND SEEDLINGS OF WATER PLANTS[1]

As we have shown in the preceding chapter, submerged pollination represents an advanced stage in acclimatisation to water life, to which only a small proportion of hydrophytes have attained. But it is by no means so rare to find the events subsequent to pollination taking place beneath the water surface. A great many aquatics—not only those which are hydrophilous, but also a number of those which raise their flowers into the air for pollination by wind and insects—after fertilisation draw down their gynaeceum into the water where the ripening processes take place. In fact, the water plants which retain an entirely aerial method of fruit-ripening are relatively few; examples of these exceptions are *Utricularia*, *Hottonia* and *Lobelia*, all of which lift their many-seeded capsules on long infructescence axes above the water level. Numerous examples of those aquatics which are pollinated in air but ripen their fruit in water, might be quoted, but it will suffice to recall *Aldrovandia*[2], the Aponogetonaceae[3], *Limnanthemum Humboldtianum*[4], *Victoria regia*[5], the Batrachian Ranunculi (Fig. 93, p. 145), *Pontederia rotundifolia*[6] (Fig. 155, p. 240) and other members of the Pontederiaceae[4]. Among the Hydrocharitaceae[7], the ripening ovary is conveyed down into the water by several different methods; in *Limnobium* and *Ottelia* the flower-stalk bends down, in *Vallisneria* it contracts spirally, while in *Stratiotes*

[1] For a good general account to that date, see Schenck, H. (1885).
[2] Caspary, R. (1859 and 1862).
[3] Krause, K. and Engler, A. (1906).
[4] Müller, F. (1883).
[5] See p. 34.
[6] Hauman-Merck, L. (1913[1]).
[7] Montesantos, N. (1913).

the fruit is carried down by the sinking of the entire plant. The lowering of the fruit must not, however, be regarded as a special innovation due to aquatic conditions, since countless examples

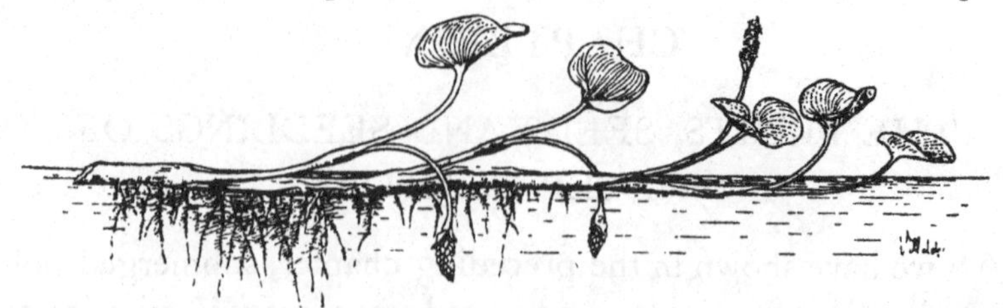

FIG. 155. *Pontederia rotundifolia*, L. Branch bearing inflorescence (negatively geotropic) and infructescences (positively geotropic). (Reduced.) [Hauman-Merck, L. (1913[1]).]

occur among terrestrial plants, e.g. the spiral contraction of the fruit stalk of *Cyclamen* and the downward curve of the peduncle of *Linaria Cymbalaria*.

In those submerged fruits which are many-seeded, the method of dehiscence is necessarily different from that obtaining among terrestrial plants, since desiccation can play no part. The irregular opening of the fruit of *Nymphaea lutea* has already been described[1]. In the case of *Limnanthemum nymphoides*,

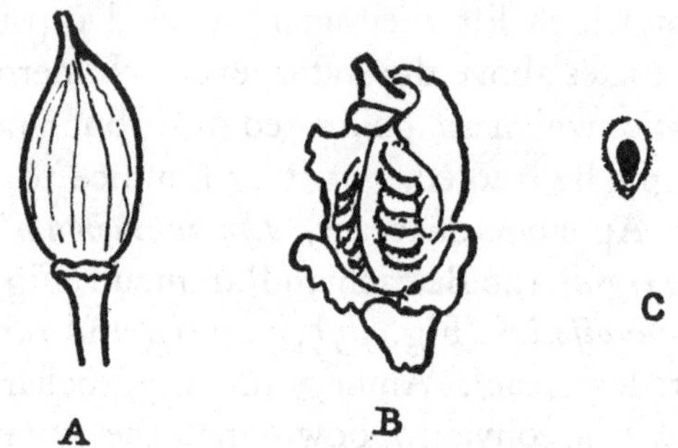

FIG. 156. *Limnanthemum nymphoides*, Hoffmgg. and Link. *A*, fruit from surface of water, October 1, 1914 (nat. size). *B*, fruit kept in water in unheated greenhouse since October 1, which had dehisced by November 23 (nat. size). *C*, seed, November 24, 1914 (× 2). [A. A.]

dehiscence takes place in a somewhat similar fashion. The present writer found a number of infructescences of this plant with green fruits (Fig. 156 *A*) floating on the surface of the

[1] See p. 35.

water at Roslyn Pits, Ely, on October 1, 1914. At this stage the seeds were unripe and white. The fruits were brought to the laboratory and kept in water. After a considerable time the pericarp split irregularly, after a fashion closely recalling *Nymphaea*; by November 24, the fruits were in this bursting condition and the seeds, which had darkened in colour, had all the appearance of being ripe. The embryos are said to be protected by the cuticularised epidermis of the testa[1]. The seeds are flat and ciliated at the edge (Fig. 156 *C*). That these hairs serve for flotation is indicated by the fact that if they are cut off the least touch makes the seeds sink[1]. It has also been ascertained that the seeds may become firmly attached to the downy plumage of a bird's breast, by means of this fringe of hairs[2]. The splitting of the ovary wall takes place mostly near the base—the lobes that are thus produced curving up until the outer epidermis of the pericarp, which was originally convex, becomes concave. This curvature is due to decay and loss of tissue on the inner surface of the fruit-wall, followed by swelling of the rest of the tissues, with the exception of the outer epidermis and adjacent layers (Fig. 157 *A* and *B*, p. 242).

The fruits of *Stratiotes aloides* and *Hydrocharis Morsus-ranae* are said to be burst open by the swelling of mucilage produced from the testa of the enclosed seeds.

A remarkably large proportion of aquatics, on the other hand, have fruits which are either one-seeded and indehiscent, or else take the form of schizocarps or heads of achenes, separating into one-seeded segments. The seeds are thus protected both by pericarp and testa, which is possibly of value in enabling them to resist the rotting effect of prolonged submergence[3]. It is interesting in this connexion to compare, for instance, the fruits of *Plantago major* and of the closely related aquatic, *Littorella lacustris*[1]. The Plantain has a pyxidium capsule, with a thin elastic wall, opening by means of a lid and containing a number

[1] Fauth, A. (1903). [2] Guppy, H. B. (1906).

[3] The protection of the embryo in certain aquatics is considered by Marloth, R. (1883).

of seeds. The fruit of *Littorella* on the other hand is reduced to a nut developed from the two-celled gynaeceum. Only one chamber is fertile and the embryo is protected by means of the sclerised fruit wall, with its aperture closed by a stopper formed from the funicular region of the seed. A protective endocarp, with an opening closed by a plug, is also found in the four one-seeded segments of the schizocarp of *Myriophyllum spicatum*,

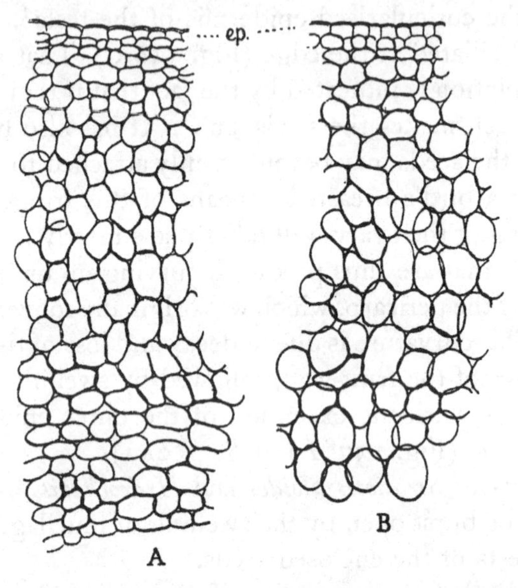

FIG. 157. ***Limnanthemum nymphoides***, **Hoffmgg. and Link.** ***A***, **T.S. wall of fruit represented Fig. 156** ***A***, **p. 240, October 1, 1914.** ***B***, **T.S. wall of fruit represented Fig. 156** ***B***, **November 23, 1914. (Both × 78** ***circa.***) ***ep.*** **= epidermis. [A. A.]**

and in the drupe-like one-seeded nutlet of *Hippuris vulgaris*[1]. The seeds of the latter species winter in mud at the bottom of the water, protected by the stony endocarp. At germination the radicle emerges from the stone through a foramen which was previously filled by a cuticularised stopper, formed from part of the funicle and integument. In *Alisma Plantago* the embryo is protected by a chaffy carpel wall and a testa described by different authors as suberised[1] or as composed of pectic substances[2]. In the case of the four nutlets into which the

[1] Fauth, A. (1903). [2] Crocker, W. and Davis, W. E. (1914).

schizocarp of *Callitriche* divides, the same function is performed by the pericarp, which is thin, but tough and elastic. The well-protected seeds of hydrophytes can in many cases withstand inclusion for a considerable length of time in ice or frozen mud. The fruits of *Sagittaria sagittifolia*, *Alisma Plantago* and *Myriophyllum spicatum*, and the seeds of *Castalia alba* and *Nymphaea lutea* can tolerate freezing for a week or two, or, in some cases, much longer[1].

With the particularly effective protection of the embryo in hydrophytes, their characteristic habit of delayed germination is probably to be associated. The sprouting of the seed may in some cases be deferred until the third, fourth, or fifth year[2], the embryo remaining uninjured by this prolonged period of dormancy.

Several investigators have studied the subject of delayed germination, and the rather curious fact has emerged that this delay only occurs if the seeds are continuously immersed in water; if they are subjected to a period of drying, they germinate promptly. It has been noted, for example, that the seeds of *Mayaca fluviatilis*, a Brazilian water plant, which were dried for six weeks after gathering, germinated at once, while seeds harvested at the same time, but put immediately into water, showed no sign of sprouting at the end of three months[3]. The seeds of some water plants can tolerate drying for a very long period, e.g. thirty months in the case of *Limnanthemum nymphoides*[4]. The result of experimental work on the subject seems to be to show that the delayed germination of undried seeds is due to the mechanical pressure exerted by the seed coats[5]; if these are artificially ruptured, the development of the embryo presents no further difficulties. It has been found, for instance, that in

[1] Guppy, H. B. (1893) and (1897).

[2] Guppy, H. B. (1897); on delayed germination in Potamogetons see pp. 71, 72, and in *Nymphaea*, p. 36.

[3] Ludwig, F. (1886).

[4] Guppy, H. B. (1897).

[5] Sauvageau, C. (1894), Crocker, W. (1907), and Crocker, W. and Davis, W. E. (1914). For a somewhat different view see Fischer, A. (1907).

comparative cultures of the achenes of *Alisma Plantago*, examined at the end of ten days, those in which the protective coats were intact, had not germinated at all, while 98 per cent. of those whose walls had been ruptured, had begun to sprout. The reason why preliminary drying favours germination, may possibly be that it gives rise to some cracking of the seed coats; a speeding-up of germination also occurs, in some cases, if the seed passes through the alimentary canal of a bird[1], a result which again may be due to some disintegrating chemical or mechanical action exerted on the wall. Freezing may also assist germination by means of its effect on the outer covering of the seed[2].

It should be noted, that delayed germination, though specially characteristic of water plants, is by no means peculiar to them. That the causes which bring it about are of a similar nature in aquatics and terrestrial plants, is indicated by the fact, well known to gardeners, that a large proportion of such seeds as those of *Canna*, fail to germinate unless the shell is filed through. The phenomena of delayed germination suggest that Nature, in her solicitude for the protection of the embryo, is liable to defeat her own ends by enclosing it in a prison from which it can only escape with difficulty.

The germination and development of the seedling in aquatics vary according to the natural affinities of the plants in question, and are characterised by few peculiarities related to the environment, except a very frequent reduction of the primary root. In *Utricularia* (Fig. 67, p. 100), *Stratiotes aloides*[3], *Hydrocharis*, *Ruppia*, *Ceratophyllum* (Fig. 55, p. 86), the Podostemaceae, *Nymphaea lutea*, *Aldrovandia*[4], *Hippuris*, *Naias*, *Trapa*[5], etc., the radicle is either quite undeveloped or very short-lived. In *Aponogeton distachyus*[6] the primary root does not attain to more than 0·5 cms. in length, and eventually it disarticulates by

[1] Guppy, H. B. (1897).
[2] Guppy, H. B. (1893).
[3] Irmisch, T. (1865).
[4] Korzchinsky, S. (1886).
[5] Queva, C. (1910); see also Fig. 160, p. 247.
[6] Serguéeff, M. (1907).

means of an absciss layer. There are exceptions, however, to the general rule that the radicle of water plants is poorly developed: in *Lobelia Dortmanna*, for example, it attains fair dimensions[1].

In the case of those water plants which grow rooted in the soil, the poor development of the radicle is often compensated, at an early seedling stage, by the production of a garland of very long root-hairs, which grow out from the 'collet,' or junction of hypocotyl and root, e.g. *Hippuris*[2], *Elatine hexandra*[3] (Fig. 158) and many Helobieae[3,4], such as *Zannichellia* (Fig. 159 *C*, p. 246). This type of seedling is, however, by no means confined to hydrophytes, but is also found in a number of land plants.

The weight of the large seed of *Nelumbo*[3], and of the achene wall in the case of the small seedling of *Zannichellia*[5] (Fig. 159), are sufficient to keep the seedling steady at the bottom of the water until the epicotyl and first leaves are produced. Other

FIG. 158. *Elatine hexandra*, D.C. Germination of seed; *s*, seed-coat; *h*, wreath of hairs growing from collet and surrounding the primary root which forms a minute conical structure. [Klebs, G. (1884).]

seedlings are anchored for some time by the fruit wall and associated structures: the grappling apparatus of *Cymodocea antarctica*, for instance, has been already described[6]. In *Trapa natans*[7] (Fig. 160, p. 247) the fixation of the seedling is accomplished in an unusual way, for here the heavy nut sinks to the bottom of the water, where it is held by hooks derived from the calyx. Two structures of very unequal size (Co_1 and Co_2) are generally interpreted as the two cotyledons, though possibly this view is open to revision. The hypocotyl, including even its extreme apex, which presumably is of root nature, is negatively geotropic. The first lateral roots, borne by the hypotocyl, curve downwards and anchor the plant in the soil, while many of the

[1] Buchenau, F. (1866).
[2] Irmisch, T. (1859[1]).
[3] Klebs, G. (1884).
[4] Warming, E. (1883[1]).
[5] Hochreutiner, G. (1896).
[6] See p. 127.
[7] Goebel, K. (1891–1893).

later roots borne on the hypocotyl and plumule are negatively geotropic.

An exceptional case is that of *Littorella lacustris*[1], in which the seeds remain *in situ*. The gynaecea are borne close to the axis, between the leaves, near the base of the little plant. On

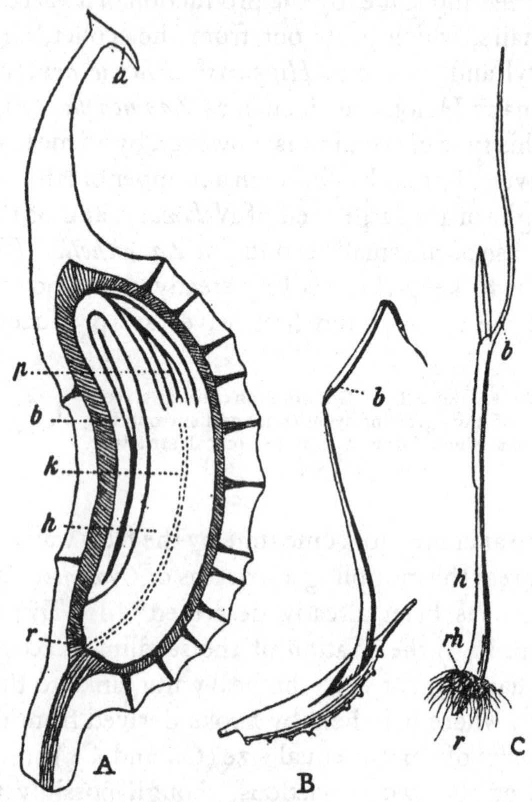

FIG. 159. *Zannichellia polycarpa*, Nolte. *A*, L.S. fruit (× 15); *a* = stigma; *b* = cotyledon; *h* = hypocotyl; *k* = vascular tissue; *r* = primary root; *p* = plumule. *B*, cotyledon emerging from fruit (× 6). *C*, seedling (× 4); *rh* = root hairs. [Raunkiaer, C. (1896).]

the death of the parent, the fruits are left surrounded by the decaying remains; they germinate where they were produced, only being dislodged in rare instances. The somewhat similar behaviour of *Cymodocea aequorea* has been discussed on p. 127.

[1] Fauth, A. (1903).

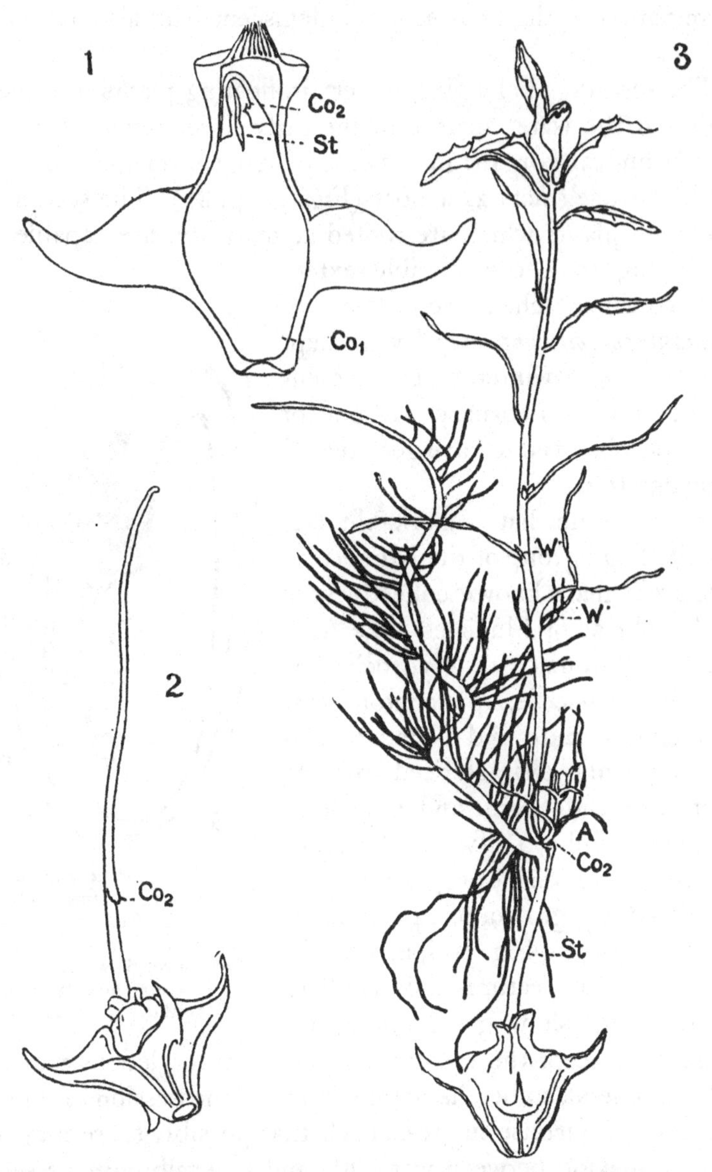

FIG. 160. *Trapa natans*, L. 1, L.S. through seed (Enlarged); Co_1, the larger and Co_2 the smaller cotyledon; *St*, stalk of larger cotyledon. 2 and 3, seedlings (Reduced); *A*, shoot arising in axil of smaller cotyledon; *W*, roots arising in the region of the leaf insertions. [Goebel, K. (1891–1893).]

In the case of *Farmeria metzgerioides*, one of the Podostemaceae, germination of the two-seeded, indehiscent fruit also occurs *in situ*[1].

The seedlings belonging to certain floating plants owe their station at the water surface to the early development of some type of buoyant organ: in the case of *Lemna*, for instance, the cotyledon itself acts as a float (Fig. 52, p. 81). The seedlings of certain plants which are rooted at maturity, are capable of developing to a considerable extent while still unattached. Some seeds of *Limnanthemum nymphoides*[2] were kept in water over a winter by the present writer, and on February 11, one of them was observed to have germinated while floating.

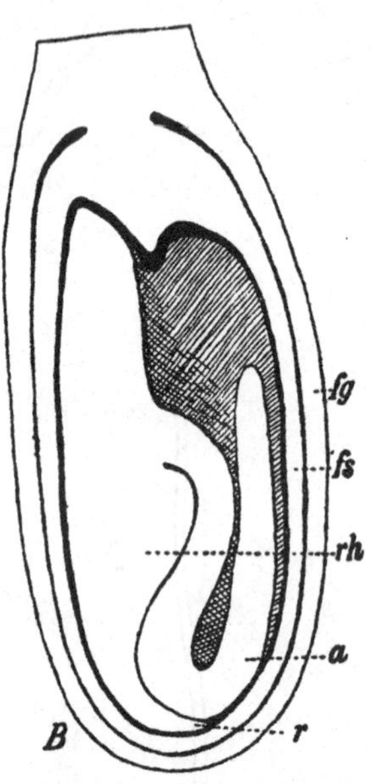

FIG. 161. *Zostera marina*, L. Fruit in longitudinal section. (×15.) *fg*=fruit coat; *fs*=seed coat; *r*=primary root; *rh*=hypocotyl which at its base enwraps the cotyledon *a*. [Raunkiaer, C. (1896).]

It is a somewhat remarkable fact that the large group of the Monocotyledons which are known collectively as the Helobieae or Fluviales—the Alismaceae, Butomaceae, Hydrocharitaceae, Juncaginaceae, Aponogetonaceae, Potamogetonaceae and Naiadaceae—are uniformly characterised by the absence of endosperm and by a 'macropodous' embryo, in which the hypocotyl reaches excessive proportions (e.g. *Zannichellia*, Fig. 159, p. 246, *Zostera*, Fig. 161, *Ruppia*, Fig. 166, p. 319); in almost all other features the members of the group show great range and diversity. Except the Helobieae, the Monocotyledons may be said, in general, to be characterised by the possession of endosperm. The surmise suggests itself that possibly there may be some connexion between water life and an exalbuminous seed with an enlarged hypocotyl. The predominance among aquatics

[1] Willis, J. C. (1902). [2] Fauth, A. (1903).

of seeds with elaborate and impervious coats, seems to indicate that plants with imperfectly protected embryos have been unable to enter upon aquatic life. Possibly there is a danger of rotting if the contents of the seed are at all freely exposed to the water. If this is so, it may be that an embryo which keeps its reserves inside its own tissues is better adapted for water life than one whose storehouse is outside its own body, even if it is enclosed in a resistant coat; the food is probably more secure from the depredations of Bacteria and from other harmful external influences, if it is incorporated within the cells of the embryo instead of being merely surrounded by the testa. In the opinion of the present writer, Monocotyledons have, in general, reduced their seed-leaves to a single cylindrical or tubular structure by means of the fusion of the petiolar or sheathing regions and the loss of the blades. They are thus not in a position to store food in the laminae of the cotyledons, as is done, for instance, in the case of such Dicotyledons as the Pea or the Bean. The radicles of aquatic seedlings are, as we have already shown, markedly reduced, so a second possible location for food storage is thus eliminated. In this connexion we may recall the fact that, whereas mature Dicotyledons often store food in their tap roots (e.g. Carrot, etc.) this method is unsuitable for Monocotyledons, owing to the ephemeral nature of the primary radicle, and they are hence almost wholly restricted to storage in leaf structures, stem structures, or adventitious roots. We are thus left with the fact that if a Monocotyledonous embryo is to store its food in its own body, the only region where this can be conveniently accomplished is the hypocotyl, since both cotyledon and primary root have suffered reduction. From these considerations we may perhaps conclude that the non-endospermic type of seed with a macropodous embryo, whose hypocotyl has become enlarged for food storage, represents a form of Monocotyledonous seed which is particularly well fitted for aquatic life.

PART III

THE PHYSIOLOGICAL CONDITIONS OF PLANT LIFE IN WATER

" For the student of the conditions of aquatic life, the real inquiry has yet to be begun."

H. B. Guppy, 1896.

CHAPTER XX

GASEOUS EXCHANGE IN WATER PLANTS

THE problems which a water plant has to solve, in connexion with its assimilation and respiration, differ widely from those which confront a terrestrial plant, since, instead of being surrounded by atmospheric air, it passes its life in water holding only a certain amount of air in solution. Owing to the varying solubility of the atmospheric gases, the dissolved air differs from free air in composition. At 15° C., the proportions in which the constituents should occur have been calculated to be as follows[1]:

	FREE AIR	DISSOLVED AIR
Carbon dioxide	0·04%	2·19%
Oxygen	20·80%	33·98%
Nitrogen	79·16%	63·82%

In practice, however, the air dissolved in the surface layers of the water of lakes and streams, under natural conditions, yields varying figures when analysed, but all observers appear to agree that, as regards carbon dioxide it is supersaturated, sometimes highly so[2]. It seems clear that the excess cannot be obtained by diffusion from the air, for an American writer[3], who has experimented with *Elodea canadensis*, has shown that sufficient carbon dioxide to keep this plant growing, or even alive, does not diffuse into water exposed to atmospheric air at Baltimore during the winter months. He demonstrated that all the carbon

[1] Devaux, H. (1889). The proportion of nitrogen given in this table naturally includes the other inert gases which were not distinguished in Devaux's time; the amount would be more correctly stated as including approximately 78 per cent. of Nitrogen and 1 per cent. of Argon.

[2] Forel, F. A. (1892–1904); Regnard, P. (1891).

[3] Brown, W. H. (1913).

dioxide which a 3-litre jar of water would absorb from the air at ordinary temperatures, could be used up by ten shoots of *Elodea* in two minutes. His view is that the substratum serves as the chief source of carbon dioxide for submerged plants, the amount of this gas given off into the water from soil containing organic matter being greater than that obtained by diffusion from the air.

Whether the excess of carbon dioxide is, in general, derived from the substratum, or whether it is due to the oxidation of carbonaceous substances in the water or to other causes, the fact remains that hydrophytes growing under natural conditions live in an environment particularly rich in carbon dioxide. This advantage tends to be neutralised, however, by the slow diffusion of gases in water. There is also the further drawback that the absorption capacity of water sinks as the temperature rises, so that, in warm weather, when the life processes of the plant are proceeding most vigorously, the supply of carbon dioxide is reduced[1]. Assimilation is nevertheless remarkably active among water plants, several features which they commonly show being well suited to the prevailing conditions; one of these is the development of chlorophyll in the epidermal cells, so that the epidermis forms part of the assimilating system, which is thus not shut off from the surrounding medium by a layer whose function is purely protective, as in the case of terrestrial plants. Cuticle is relatively little developed, and the cell-walls seem to offer no more hindrance to the direct passage of dissolved gases than if they were merely thin plates of water[2]. That the waxy cuticle of such leaves as those of the submerged Potamogetons is no obstacle to the entry of liquids, has been proved by plasmolysis experiments in which the whole leaf was used[3].

Submerged plants show various characteristics which have the effect of increasing the surface relatively to the volume of the leaf, and thus bringing a large proportion of the assimilating cells into direct contact with the dissolved carbon dioxide. The

[1] Goebel, K. (1891–1893). [2] Devaux, H. (1889).
[3] Sauvageau, C. (1891[1]).

leaves may, for instance, be very thin, but extensive in area, as in the case of the submerged leaves of the Waterlilies, or they may be sub-divided into hair-like segments, as in *Myriophyllum*, etc.[1] In certain Podostemaceae belonging to the genus *Oenone*[2], curious hair-like outgrowths, rich in chlorophyll, are developed on the leaves (Fig. 81, p. 119). These outgrowths, from their presumed analogy with the breathing organs of water animals, have been called gill-tufts (Kiemenbüschel), though it has not been proved that they possess a respiratory function. The negatively geotropic roots of *Trapa natans*[3], the Bull Nut (Fig. 160, p. 247), provide another example of a finely divided submerged organ, by means of which gaseous exchange can readily take place. The intimate contact achieved between these organs and the water, probably assists not only assimilation but also respiration.

It is true that dissolved air is richer in oxygen than atmospheric air, about one-third of its volume consisting of this element, but the essential point to bear in mind is that the *total volume* of air held in solution in water at ordinary temperatures is so extremely small that in a litre of water the maximum amount of oxygen present is 10 cubic cms., as compared with more than 200 cubic cms. in a litre of atmospheric air[4]. The result is that water plants have considerably less oxygen at their disposal in each unit volume of the surrounding medium than is the case with land plants[3]; as far as hydrophytes are concerned, oxygen is a rare and precious commodity.

Thus, on account of the poverty of the medium in this element, no plant can be a successful aquatic unless it has a special capacity either for obtaining an adequate oxygen supply, or for husbanding it when obtained.

Every green plant forms oxygen as a by-product of carbon

[1] See Chapters XI and XII.
[2] Goebel, K. (1891–1893) and Matthiesen, F. (1908).
[3] Goebel, K. (1891–1893).
[4] Regnard, P. (1891); see also Forel, F. A. (1901).

assimilation, through the disintegration of carbon dioxide[1]. The greater part of the oxygen, in the case of terrestrial plants, is at once returned, by means of the stomates, to the atmosphere whence it came. But water plants show a marked tendency to retain this element, and we find that their tissues are generally penetrated by an elaborate system of intercellular lacunae, by means of which the oxygen evolved in the assimilating cells presumably finds its way to other parts of the plant, where it may be used for respiration[2]. The aerating system arises very early, as it also does, indeed, in many terrestrial plants; we find, for instance, that, in the growing apex of the stem of *Elodea*, there is a network of intercellular spaces reaching to within two or three cells of the summit, while, in the winter-buds of *Myriophyllum*, a complete ring of large air canals occurs only 1 mm. from the stem apex[3]. In many water plants the air system is so elaborately developed that almost all the cells are in contact with the internal atmosphere by means of some part of their surface. Unger[4], who has measured the quantity of air contained in various plant tissues, finds that 71·3 per cent. of the volume of the leaves of *Pistia Stratiotes*, L., the floating River Lettuce, is occupied by air, while in land plants, especially xerophytes, the percentage is much lower; for instance, the leathery leaves of *Eucalyptus Preissiana*, Schauer, contain 9·6 per cent. of air, and the succulent leaves of *Begonia hydrocotylifolia*, Hook., only 3·5 per cent.[5]

The exception that proves the rule that the tissues of water plants are characterised by the unusual development of inter-

[1] Cloëz, S. and Gratiolet, P. (1850); Cloëz, S. (1863); and later literature.

[2] The gases bubbling from wounds in the green shoots of submerged plants in sunlight have been described as containing about 90 per cent. of oxygen. Tieghem, P. van (1866).

[3] Devaux, H. (1889).

[4] Unger, F. (1854[2]).

[5] Unger, F. (1854[2]). The leaf of *Pistia* has no elongated stalk, while, in the case of *Eucalyptus* and *Begonia*, Unger includes the petioles in the calculation. This might tend slightly to exaggerate the difference in the percentages, but, even if corrected for this detail, the figures would doubtless remain sufficiently striking.

cellular spaces, is provided by the Podostemaceae, which form in other respects a highly anomalous group. The members of this family, which we have discussed in Chapter IX, flourish in rapidly moving water, even "at the sides of the waterfalls, with the furious current rushing right over them[1]." The tissues are found to include no large lacunae (Fig. 80, p. 118) and it is probably for this reason that these plants are confined to water which, on account of its movement, is necessarily well aerated. That the constitution of the Podostemaceae does actually render them dependent on high aeration of the water, is shown by the fact that, if, owing to a fall in the level of the stream, they are left behind in a stagnant pot-hole, death quickly ensues[1].

In ordinary hydrophytes, living in still or slowly moving waters, there must be a liability to asphyxiation in the case of the roots or rhizomes more or less buried in the saturated mud. The elaborate air-system, developed in the long petioles of such plants as the Waterlilies, probably plays some part in obviating this danger. These petioles form the connecting link between the submerged rhizome and the floating leaves, which not only themselves produce oxygen in the process of assimilation, but also have free access to the oxygen of the atmosphere. In many cases, the air-canals traversing elongated organs, such as stems and petioles, are crossed at intervals by diaphragms, which are not, however, air-tight. Their structure is illustrated in Fig. 119, p. 184, which shows phases in the development of the cells forming the partitions that, at every node, cross the stem of *Hippuris vulgaris*, the Mare's-tail. From these drawings it will be recognised that intercellular spaces occur at the angles of the cells, both in youth and age, so that gases can pass freely.

Although it seems to be generally agreed that oxygen is conveyed by means of the internal air-passages from the assimilating organs to other parts of the plant, there is still much obscurity with regard to the nature and causes of the movements of gases in water plants. These movements have been studied more particularly in the Nymphaeaceae. In *Nelumbo*, for in-

[1] Willis, J. C. (1902).

stance, a remarkable bubbling of gas from the leaves of an intact plant may sometimes be observed[1], but there seems little agreement among different observers as to the reasons for this curious phenomenon, or even as to the actual facts of its occurrence[2]. The whole subject needs to be reinvestigated by a botanist who is also a competent physicist. The only point about which there is some degree of certainty, seems to be that, at least, while assimilation is actually proceeding, high gas pressures occur in the air passages. This can be demonstrated by various direct means, for instance by cutting into the plant beneath the water-surface, when a stream of bubbles arises from the wound. A curious piece of indirect evidence, bearing on the same point, is perhaps worth recalling. It has been shown that, when a Waterlily petiole suffers from a wound which involves any of the air-canals, the cells bounding these cavities grow out in the form of hairs, until they choke the channel[3]. The suggestion has been made that this growth is induced by the temporary diminution of the high pressure in the air-canals, due to their sudden connexion with the external atmosphere[4].

The cause of the high pressure in the canals during assimilation is doubtless to be sought in the continual production of oxygen, which accumulates in these intercellular spaces. In the dark, when respiration is the only form of gaseous exchange that persists, the high pressure is often replaced by a negative pressure, since the relatively small quantity of carbon dioxide, produced partly at the expense of the oxygen in the internal atmosphere, diffuses away with considerable rapidity, in contrast to the oxygen, which diffuses slowly. The high pressure of the oxygen, in the lacunae adjoining the assimilating cells, may have an effect in inducing movement towards regions of lower pressure, such as the roots and rhizomes, where oxygen is presumably in great request. Differences of temperature, between the sun-warmed upper parts of the plant and those in the relatively cold lower layers of the water, may also have their effect in causing currents in the internal atmosphere.

[1] Raffeneau-Delile, A. (1841), Ohno, N. (1910). [2] Ursprung, A. (1912).
[3] Mellink, J. F. A. (1886). [4] Schrenk, J. (1888).

It has been suggested by Goebel[1] that the origin of the development of intercellular spaces in water plants may be attributed to the direct action of the medium—an enlargement of the air spaces resulting mechanically from the pressure of the gases evolved, which are prevented by the surrounding water from escaping freely. But he points out that the lacunar system, thus initiated, has ultimately become hereditary. Some support is given to Goebel's view by experimental work on amphibious plants, and by the study of the comparative anatomy of specimens growing under different conditions. It is found, for instance, that if such a plant as the Water Speedwell, *Veronica Anagallis*, grows with one of its shoots submerged, while the others develop in the air, the submerged shoot shows an increase in intercellular spaces, as compared with the air shoots[2]. But the presence of lacunae is something more than a mere direct effect of environment, since they persist, even if in a diminished form, when aquatics are grown on land. For example, stems of *Peplis Portula*, when grown in water, are characterised by four large lacunae in the cortex. On examination of plants growing terrestrially, it has been found that they also show four lacunae; the only difference between the aquatic and aerial plant is that, in the former, the bands of tissue separating the main lacunae are riddled by intercellular spaces, while, in the latter, they are relatively solid[2].

Whatever its origin may be, the aerating system in the stems, leaves and roots of water plants belonging to the most divergent cycles of affinity, is developed with a uniformity and an elaboration which undoubtedly indicate that it is definitely related to the *milieu*[3]. It is perhaps scarcely too much to say that the difficulty of breathing is the principal drawback to life in water, and that only those plants which have an inherent capacity for coping with this difficulty, can make their home permanently in an aquatic environment.

[1] Goebel, K. (1891–1893). [2] Costantin, J. (1884).

[3] For a consideration of the aerating system from the anatomical standpoint see Chapter XIV, p. 183.

CHAPTER XXI

ABSORPTION OF WATER AND TRANSPIRATION CURRENT IN HYDROPHYTES

ONE of the unfortunate results, which followed the publication of *The Origin of Species*, was the acutely teleological turn thus given to the thoughts of biologists. On the theory that every existing organ and structure either has, or has had in the past, a special adaptive purpose and "survival value," it readily becomes a recognised habit to draw deductions as to function from structure, without checking such deductions experimentally. Many points in connexion with the study of aquatics, and, notably, the whole subject of the absorption of water by such plants, have suffered profoundly from this tendency. Two of the most conspicuous anatomical characters of hydrophytes, as compared with land plants, are the relatively small amount of cuticle[1] on the surface of the epidermis, and the poor development and lack of lignification of the xylem. From these facts it has been lightly concluded that submerged plants, being able to absorb water over their entire surface, have simply dispensed with the transpiration current from root to leaf which is universal among land plants, and that their roots have lost all function except as attachment organs. These ideas have become text-book platitudes, and may still be found even in the writings of professed physiologists[2], despite the fact that they have been, to a large extent, refuted by a series of experimental investigations by different observers, the first[3] of which

[1] Cuticle, though small in amount, is invariably present on the epidermal walls of aquatics. See Géneau de Lamarlière, L. (1906).

[2] See for example Hannig, E. (1912), where the author speaks of submerged plants "bei denen kein Transpirationsstrom existiert."

[3] Unger, F. (1862). For a recent discussion of the subject see Snell, K. (1908).

appeared more than half a century ago. It may further be recalled that, as early as 1858, a French botanist[1] concluded, from certain experiments, that the transpiration of a terrestrial plant can continue when it is grown in a saturated atmosphere, and even when the leafy portion is entirely immersed in water. It is also known that emersed water plants transpire very freely[2]. We shall only find it necessary here to refer to a few of the more outstanding of the researches which bear directly upon the transpiration of submerged plants.

The more modern work on the subject may be said to begin with Sauvageau[3], to whom we owe so much of our knowledge of aquatics. He used for his experiments detached branches of submerged plants, in which the cut end of the stem had been sealed with cocoa butter, and all the roots had been removed. He found that, even under these circumstances, the shoots could live and prosper and develop fresh buds—thus, up to a certain point, justifying the current view that water could be absorbed through the surface of the stem and foliage. He also performed a converse experiment, by means of which he attempted to prove that, under normal conditions, a definite transpiration current, passing upwards to the leaves, occurs in submerged plants. The apparatus used is shown in Fig. 162, p. 262. It was essentially a form of potometer, modified for use with a submerged shoot. This experiment, however, as has been pointed out by a more recent worker[4], is open to the criticism that water may have been passively forced through the plant, owing to the pressure exerted on the cut surface of the stem by the column of water in the small tube. It seems as if some slight modification of the apparatus might readily be contrived to obviate this difficulty.

A number of further experiments were devised by Hochreutiner[5], of which the following example may be taken as

[1] Duchartre, P. (1858).
[2] Bokorny, T. (1890) and Otis, C. H. (1914).
[3] Sauvageau, C. (1891[1]).
[4] Weinrowsky, P. (1899).
[5] Hochreutiner, G. (1896).

typical. He employed two branches of *Potamogeton pectinatus*, L., arranging one of these branches so that its base, to a depth of 2 cms., was immersed in eosin solution, while its summit was in pure water; the second, he placed with its summit in eosin and its base in pure water. After a couple of days, sections of these two shoots were cut at various levels, and it was found that, in the case of the first branch, the eosin had mounted to a height of 15 cms. in the main axis, which was itself 20 cms. long, and to 13–16 cms. in the lateral branches. *P. pectinatus* possesses no vessels, but the xylem lacunae had

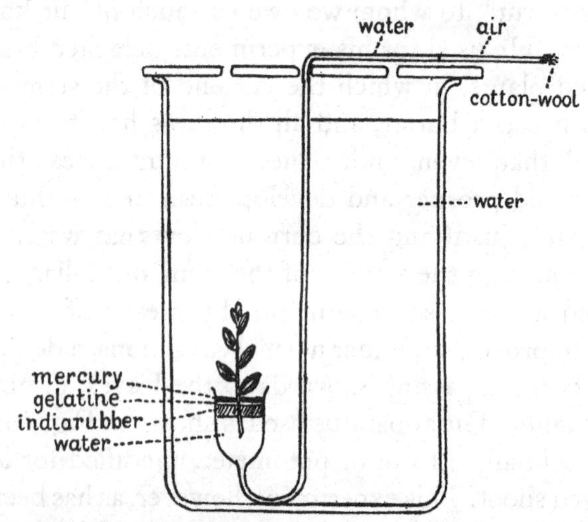

FIG. 162. Diagram illustrating experiment to show existence of 'transpiration' current in a submerged plant. [Sauvageau, C. (1891[1]).]

evidently formed the path for the transpiration current, the cell walls bounding them being alone coloured bright red. In the case of the second branch, only the epidermis was stained, the vascular tissue of the leaves and stem being unaffected.

Some experiments, similar in principle to those of Hochreutiner, but more striking in result, were made some years ago by two Cambridge botanists[1]. Their work had the advantage of being carried on *in situ*, so that the natural environment of the plant was, as far as possible, retained. *Potamogeton lucens* was

[1] Thoday, D. and Sykes, M. G. (1909).

chosen as the subject of the experiments, which were made in the River Cam during July and August. The method adopted was to attach a small glass bulb of aqueous eosin solution to the cut end of a submerged branch. A flourishing, leafy stem was selected, cut under water and left submerged for a short time. A little cotton-wool was then wrapped round the stem near the cut end, the small bulb of eosin brought down to the surface of the water, and the cut end lifted for a moment above the surface and inserted in the bulb. The plant was held beneath the water for a recorded time, and, at the end of the experiment, the bulb was removed and the stem at once examined. The rate of transmission of the eosin solution was found to be surprisingly rapid—the eosin travelling, in one case, at the rate of nearly 10 cms. per minute. In these cut shoots, root-pressure is obviously eliminated, and the upward stream was found to be due to the action of the leaves; the entire removal of the leaves rendered the current almost negligible, while, when some were cut off, the diminution in the rapidity of flow was roughly proportional to the number removed.

Such experiments as these seem to leave little room for doubt that an active water-current from base to apex, corresponding to the 'transpiration' current of land plants, occurs even in entirely submerged aquatics, or, in other words, that the absorption of water is polarised in the plant. Those who have denied the existence of the transpiration stream, have been led to do so rather on the *a priori* ground that such a current would be a superfluous feature in the economy of a plant surrounded by "water, water everywhere." This would in any case be a dangerous method of argument, and it is based moreover upon a misconception of the value of the transpiration current. Its use is not merely to supply the tissues with water, but also to convey to the assimilating and growing regions certain important elements of their food supply. Even the soil-water contains salts in solution in quantities that are relatively minute, and the only method whereby an adequate salt supply can be ensured is by the passage of a proportionately large volume of water through the plant. Further, in the case of submerged aquatics, the transpiration stream is, for two reasons, of even greater importance than in the case of terrestrial plants. Firstly, it has been shown that the water, in which submerged plants live, is generally still poorer in saline matter than that which percolates through the soil[1], and, secondly, there seems some reason to suppose that submerged plants depend upon their transpiration stream, not only for their salts, but also, possibly, for some part of their carbon dioxide supply. We have noted the possible importance of the substratum as a source of carbon dioxide[2] and, since this gas diffuses slowly, it is reasonable to suppose that the water absorbed by the roots from the soil may be richer in carbon dioxide than that in which the leaves are immersed. Hence it is not impossible that the transpiration stream in submerged plants may have its value in connexion with carbon assimilation[3].

The existence of a transpiration current throws light upon the otherwise inexplicable fact that many submerged plants have an elaborate system of roots, often bearing well-developed root-hairs. In the case of some Potamogetons, for instance, the root-hairs are said to survive and play their part after the death of the other cells of the piliferous layer[4]. Such a root system could scarcely be needed merely for purposes of anchorage, and, fortunately, we now have direct experimental proof that it serves also for absorption. An American observer, Raymond H. Pond[5], by means of an ingenious piece of apparatus, succeeded in actually measuring the water taken up by an individual root of one of the submerged Water Buttercups. The root in question, which was 14 cms. long and clothed with root-hairs, was found to absorb 5 cubic cms. of water in 24 hours.

Pond also carried out a number of indirect experiments on

[1] Sauvageau, C. (1891[1]). [2] See pp. 253, 254.

[3] The work of Brown, W. H. (1913), appears to support this view, though the author does not himself draw these conclusions, but regards the roots as mere organs of anchorage.

[4] Sauvageau, C. (1891[1]). [5] Pond, R. H. (1905).

the same subject, of which the interpretation is a less simple matter. He made comparative cultures of certain submerged species (*Vallisneria*, *Elodea*, etc.) rooted in soil, rooted in washed gravel, or anchored above the soil in such a way that the roots were unable to penetrate it. He found, throughout, that the rooted plants grew much better than those that were merely anchored. Very similar results have been obtained more recently by a German botanist[1], whose experiments may be illustrated by means of a single example. A number of shoots of *Elodea canadensis* were planted under water in soil in which they were allowed to take root. Another set of shoots, equal in number and approximately equal in size, were placed in the same glass receptacle, but were supported above the bottom in such a way that their roots were unable to penetrate the soil. After 28 days the experiment was interrupted, and the two sets of shoots were measured. It was found that the rooted shoots had grown much more rapidly, their total length amounting to 308·0 cms., as compared with 177·5 cms. in the case of those which had been prevented from taking root in the soil. The interpretation of these and similar results has been the subject of some controversy. Pond deduced that the primary cause of the retarded growth of the non-rooted plants was their inability to secure enough phosphorus and potassium and possibly other elements. He found that such plants, in the case of *Vallisneria*, were not only stunted in growth, but had their tissues loaded with an abnormal amount of starch; he came to the conclusion that lack of certain salts inhibited proteid synthesis and growth, though the conditions were favourable to photosynthesis. Another American author[2] has recently published results, bearing on this question, which it seems impossible to reconcile with the views of Pond. He finds that the difference in growth between rooted and unattached plants can be altogether eliminated by passing carbon dioxide through the water several times a day. He considers that the non-rooted plants do not suffer at all from lack of salts, but chiefly from lack of

[1] Snell, K. (1908). [2] Brown, W. H. (1913).

the supply of this gas which is given off from soil containing organic matter. The divergence of these workers' views indicates a direction in which further experimental work of a critical nature is markedly needed.

A piece of indirect evidence, which confirms, though it does not actually prove, the existence of a transpiration current in submerged vegetation, has recently been obtained in connexion with certain studies on the relative osmotic strength of the cell-sap in the leaves and roots of the same plant. In terrestrial species, the osmotic pressure in the leaves has been shown to be, as a general rule, less than that in the root, a result which is entirely in harmony with the known facts relating to root-pressure. In submerged plants (*Elodea*, etc.), the same osmotic relation has also been found to exist, a difference of as much as four atmospheres being recorded, in one case, between the pressures in leaf and root[1]. It seems impossible to explain these results on the hypothesis that the transpiration current in such plants is non-existent[2].

If it be granted that a transpiration[3] current occurs, even in plants which are entirely submerged, and that this current is, at least to some extent, dependent on the leaves[4], we are at once confronted with the problem of how the leaves eliminate the water, since the discharge of water-vapour obviously cannot occur in the manner characteristic of land plants. For a large number of submerged plants, though by no means all, the question has now been elucidated by the work of Sauvageau, von Minden and other observers[5]. In many cases the mecha-

[1] Hannig, E. (1912). [2] Snell, K. (1912).

[3] The word "transpiration" is deliberately used throughout this chapter, in lieu of "guttation," suggested by Burgerstein, A. (1904) as more appropriate for submerged plants. The expression, "transpiration," is not likely to cause any confusion, and the word "guttation," though perhaps more strictly accurate in many cases, is too awkward and ugly to be readily admitted into our language.

[4] Thoday, D. and Sykes, M. G. (1909).

[5] Oliver, F. W. (1888), Schrenk, J. (1888), Sauvageau, C. (1891[1]), Wächter, W. (1897[1]), Minden, M. von (1899), Weinrowsky, P. (1899), etc.

nism employed is one which is already very general in terrestrial plants, namely the development on the leaves of "water pores" which are able to extrude water in the liquid state[1]. These water pores, which occur singly or in groups in the neighbourhood of the nerve-endings, both in submerged leaves and on the under side of the floating leaves[2], resemble large stomates which remain permanently open. Beneath them, there is a marked expansion of the tracheal termination of the bundle, which is only separated from the epidermis by some layers of thin-walled turgid cells, known as the epithem. The epithem tissue is considerably developed in Dicotyledons, but less so in Monocotyledons. The intercellular spaces between the cells of this tissue are filled, normally, with water[3]; the epithem is believed to act as a regulator, preventing the expulsion of the drop until a certain root-pressure is reached[4]. Fig. 53, p. 82, illustrates the relations of the water pores and associated structures in the case of a floating leaf—that of *Pistia Stratiotes*. In this plant the vigorous excretion of drops of water may be readily seen, and we can scarcely doubt that, in the case of submerged leaves furnished with the same mechanism, the expulsion of drops also occurs, though it cannot be directly observed.

It is a curious fact—as yet unexplained—that the water pores of aquatics are often highly ephemeral, being resorbed and destroyed while the leaf is still quite young. This occurs, for instance, in *Callitriche*, in which the very young leaf bears two groups of water stomates at the apex (Fig. 163 *A*, p. 268). At an early stage the epidermis in the neighbourhood of the water pores becomes laden with a brownish, gummy or granular material, and the cells eventually die. Similar substances are apt to choke up the intercellular spaces of the epithem, and the mouths of the

[1] Burgerstein, A. (1904), enumerates more than 200 genera of flowering plants, belonging to nearly 100 families, in which the extrusion of liquid water from the leaves, either by means of water pores or apical openings, has actually been observed. The great majority of these are land plants.

[2] Schrenk, J. (1888).

[3] Volkens, G. (1883).

[4] Gardiner, W. (1883).

water pores, in other aquatics[1]. Possibly useless or poisonous substances, carried by the ascending sap, which, in the case of plants that get rid of their superfluous water through innumerable stomates, are too much diffused to do damage, may accumulate to a deleterious degree when they are localised by the elimination of the water through a relatively small number of pores. But, whatever its cause, the loss of the water pores of *Callitriche* seems more than compensated by the resulting development of "apical openings" (Fig. 163 *B*). The

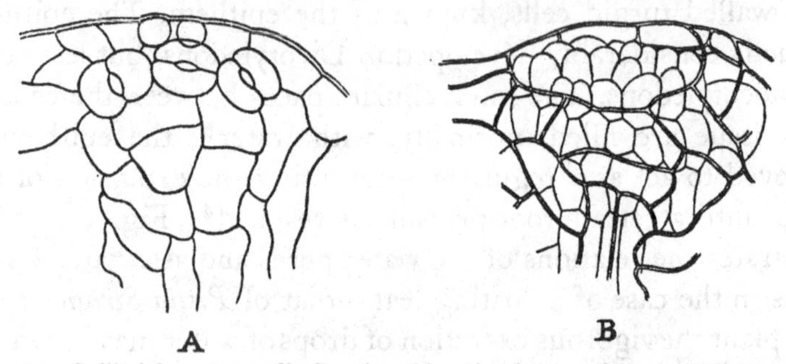

FIG. 163. *Callitriche autumnalis*, L. *A*, epidermis of apex of young leaf seen from below with a group of stomates. *B*, apex of an older leaf seen from below. The large opening in the epidermis is due to the resorption of five stomates; below the opening the small-celled parenchyma is exposed. [Borodin, J. (1870).]

destruction of the two groups of stomates exposes the substomatal chambers, which communicate directly with the apex of the vascular bundle, and apparently, through these two cavities, water is directly extruded.

In other cases the apical openings are said to have no connexion with the destruction of water stomates. The entire tissue clothing a bundle-end, including the epidermis, disappears, leaving the tracheids actually emerging at the surface. In *Heteranthera zosteraefolia*, water pores and an apical opening exist side by side, while Fig. 108, p. 167, represents a longitudinal section of the leaf-tip of *Potamogeton densus*, in which, by the death of the apical cells, the median nerve is brought into direct contact with the water.

That the elimination of water does actually take place

[1] Minden, M. von (1899).

through the apical openings of submerged leaves, is indicated by certain observations made independently by two different workers at the end of the last century[1]. In the natural situation of the leaves, it is not easy to devise a means of rendering this elimination visible, but it is found that, if the level of the water surface be lowered until the leaf apices emerge into the air, drops of water appear in the region of the apical opening; if wiped away they speedily re-form. This phenomenon has been witnessed in a considerable number of cases—as, for instance, the submerged leaves of *Littorella* and *Potamogeton crispus*—and we shall probably not be guilty of too great an assumption in supposing that the same thing goes on when the leaves are beneath the water surface. The exudation of water from water pores has been shown, in the case of land plants, to be dependent upon root-pressure, and the existence of identical pores in submerged species lends colour to the view that the roots of such plants are not mere holdfasts, but have to some extent retained their function as organs of absorption.

Notwithstanding the advances that have been made, many problems connected with the absorption and elimination of water by submerged plants remain to be solved. In *Hydrocleis nymphoides*, for instance, by the disappearance of a special transitory tissue at the leaf apex, the tracheids are left communicating freely with an empty space, but this space remains separated from the water by a persistent roof of cuticle, and can therefore play no part in the elimination of water (Fig. 164, p. 270)[2]. Again, side by side with *Zostera*, whose leaves are provided with apical openings, we have two other marine genera of the Potamogetonaceae, *Cymodocea* and *Posidonia*, in which no such openings occur. It seems that we must either suppose that the elimination of water from the apical openings is of relatively little importance, or that, in related genera, the main physiological activities of the plant may be differently performed.

In the case of such submerged, rootless plants as *Cerato-*

[1] Minden, M. von (1899) and Weinrowsky, P. (1899).

[2] Sauvageau, C. (1893).

phyllum and *Utricularia*, we are still far from understanding the mechanism of absorption and elimination. Here the liquid exchange presumably takes place entirely by means of osmosis and diffusion. But it should be noted that in both these unrelated genera, which are characterised by the total absence of true roots, there is a tendency to the production of subterranean shoots, which perform the function of roots[1]. This modification of other organs for subterranean work, appears to suggest that, in the course of evolution, some disadvantage has followed the reduction and ultimate loss of the root system, and that an attempt has been made to replace it.

The insectivorous habit of *Utricularia* may also perhaps be correlated with the reduction of the transpiration stream, and

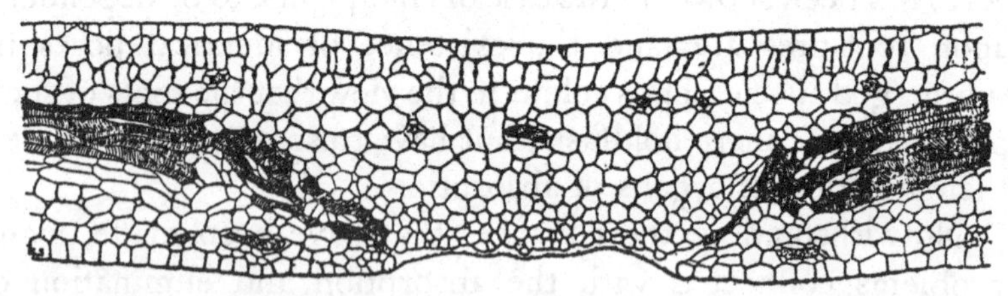

FIG. 164. *Hydrocleis nymphoides*, Buchen. T.S. leaf passing through the middle of the apical cavity which remains roofed in with cuticle. [Sauvageau, C. (1893).]

the consequent limitation of the food supply[2]. This is rendered more probable when it is remembered that the only other carnivorous genus among water plants, *Aldrovandia*, is also submerged and rootless. The resemblance of the two genera, in these respects, is the more remarkable since they belong to widely separated cycles of affinity. Their common insectivorous habit seems to indicate that a plant, which has dispensed with an active transpiration stream, needs some compensation for the loss of food materials involved.

In the present chapter, stress has been laid upon the difficulties besetting a submerged plant in connexion with the maintenance of a transpiration stream. In conclusion we must glance for a moment at an embarrassment incurred by such

[1] See pp. 88, 89, 96, 97. [2] Cohn, F. (1875).

plants, which is the very antithesis of the problem of keeping up the water supply—namely, the danger that the osmotic attraction of the cell-sap may draw an excess of water into the young tissues. A certain feature, occurring widely among water plants belonging to unrelated families, may possibly play some part in obviating this risk; this is the development of an outer layer of mucilage, clothing the young organs, whose epidermis has not yet matured to a resistant coat[1]. This slime is secreted by hairs or scale-like bodies, such as the "squamulae intravaginales[2]" occurring so frequently in the leaf axils of aquatic Monocotyledons. A similar secretion exists in many land plants: the young leaves of the Dock, for instance, are completely invested by it. Here, again, its power of delaying the passage of water, may be of some value to the plant, but, in acting as a protection against excessive transpiration, it has exactly the opposite influence to that exerted by the slimy coating of water plants. It has also been suggested that in the aquatics the mucilage may serve to prevent the soluble products of assimilation diffusing into the water, or that it may form a protection against animals and discourage parasitic and epiphytic growths. These theories, regarding the possible function of the slimy coating, are not easy to prove or to disprove, but there seems to be some experimental evidence that, in the case of submerged plants, the mucilage actually hinders the entry of water, while the distribution and mode of occurrence of the slime in different hydrophytes, furnish certain indications indirectly confirming this view. In some cases the development of mucilage begins very early, and it is thus present on the surface of the delicate organs of the seedling: the hypocotyl of *Callitriche stagnalis*, for instance, has scarcely emerged from the fruit before the epidermis shows the first rudiments of the secretory trichomes[3]. It has also been observed[4] that plants of tender structure, such as *Limnanthemum nymphoides* and *Polygonum amphibium*, retain their slimy coat to a much later stage than plants of tougher

[1] Goebel, K. (1891–1893).
[2] Irmisch, T. (1858[2]).
[3] Fauth, A. (1903).
[4] Schilling, A. J. (1894).

habit, such as *Potamogeton natans*. Again, it is found that all the Nymphaeaceae have their young leaves clothed with mucilage, with the one exception of *Nelumbo*, the Sacred Lotus. In this plant, on the other hand, the epidermal cells become cuticularised relatively early, and thus are able to exert a protective function. The mucilage of the Waterlilies may reach extraordinary proportions. In *Brasenia Schreberi* (*peltata*)[1], for instance, the thickness of the layer of slime coating the petioles and flower-stalks may exceed the diameter of the organ itself. Such an abnormal development can scarcely be regarded as a useful adaptation, and it is probably safest to look upon the production of mucilage, both in this and other aquatics, as a mere by-product of the plant's metabolism, any useful purpose that is served being purely secondary. There are certain cases which are particularly difficult to explain on the adaptational view. In *Ceratophyllum*[2], for instance, in which the growing point and young leaves are cuticularised, curious mucilage hairs occur, but do not seem to give rise to any protective layer. Again, the trichome-diaphragms, formed across the intercellular spaces in the petiole of *Nymphaea lutea*, are clothed with mucilage, although they are not in contact with water but with the internal atmosphere[3].

The problems in relation to water which confront a terrestrial plant, all hinge upon the difficulty of obtaining a constant and adequate supply. In the case of submerged aquatics, on the other hand, the supply is permanently excessive, and the plant can only live successfully in this *milieu* if it possesses the knack of controlling and regulating its absorption and elimination in such a fashion that a steady upward stream is ensured, while the tissues, especially those that are young and delicate, are preserved from supersaturation.

[1] For a detailed account of the mucilage of this plant see Schrenk, J. (1888), and Keller, I. A. (1893). See Fig. 20, p. 38, for the structure of the mucilage-secreting hairs.

[2] Strasburger, E. (1902).

[3] Raciborski, M. (1894[2]).

CHAPTER XXII

THE INFLUENCE OF CERTAIN PHYSICAL FACTORS IN THE LIFE OF WATER PLANTS

THE physical conditions, under which water plants have their being, differ widely from those which affect land plants. We have already considered the special features of the gaseous exchange and the water supply due to life in a liquid medium instead of in the atmosphere; it now remains to discuss the influence of certain other factors—especially temperature, illumination and gravity—upon plants growing in water.

When the thermal conditions of land and water plants are compared, the chief difference is found to be the smaller range of temperature variation—both diurnal and seasonal—which aquatics are called upon to endure. Though the truth of this statement is universally recognised, it is based upon relatively few exact observations, and further detailed field work is much needed upon the temperature variation in different types of waters, and the relation of this variation to vegetable life. A notable beginning in this direction has been made by Dr Guppy[1], to whom we owe many original observations on the bionomics of aquatics. He has shown that during a summer day and night, when the range of shade temperature in the air may be about 11° C., the range in the water of a river, such as the Thames at Kingston, may be as little as about 0·8° C. The smaller the stream, the greater the range of variation; a little brook, two or three feet across and only three or four inches deep, may show a variation in 24 hours of about 8° C., that is to say, about three-quarters of the range in the air, but ten times the range in the river. Irrespective of the size of the body of

[1] Guppy, H. B. (1894[1]); the results in this paper are given on Fahrenheit's scale, but in the present chapter they are quoted in Centigrade terms for the sake of uniformity.

water, depth and velocity are important factors in determining the extent of the variation; the more rapid the current and the shallower the stream, the greater is the daily range.

Besides the changes from hour to hour, the different temperatures, which occur simultaneously at different depths in the same body of water, must be noted. The heat received by a water surface is said to be absorbed almost completely (94 per cent.) by the topmost millimetre of liquid, warmth being conveyed to lower layers by means of currents only[1]. This explains a curious fact, to which attention is drawn by Guppy[2]. He points out that, in a river about 10 feet deep, the temperature at the surface and bottom are much the same, but that ponds and ditches differ from rivers in their liability to surface heating; this becomes especially marked where the water is crowded with plants, so that even the slight currents, that occur in stagnant pools, are checked by the mass of vegetation. A ditch full of plants, on a sultry afternoon, may exhibit a difference in temperature of 5° C. in nine inches, while a large pond, 4 or 5 feet deep, may be 6° C. to 7° C. warmer at the surface than the bottom. The result is that, on sunny days, the temperature of the ponds in the neighbourhood of a river generally stands some degrees above that of the river itself, and, in the height of summer, the variation may be nearly 7° C. As Guppy[3] remarks, "Everything in plant-life is behindhand in a river in comparison with a pond." This difference may possibly explain certain apparent anomalies in the distribution of aquatic plants in a single neighbourhood.

Guppy's observations relate only to comparatively shallow waters; in deep water the currents appear to be, as a rule, unable to convey the daily heat of the sun to a greater depth than about 10 metres. Beneath this level[1] the temperature sinks, until, at about 100 metres, it becomes constant at 4° or 5° C. Temperature is undoubtedly one of the principal factors

[1] Magnin, A. (1893).
[2] Guppy, H. B. (1894[1]), (1894[3]) and (1896).
[3] Guppy, H. B. (1896).

regulating the depth at which plants can grow. In deep lakes, in which the thermometer at 10 metres below the surface registers about 12° C. in summer, the higher plants are not found at a greater depth than 6 metres. In peat-bog lakes, however, the temperature of the lower layers is unusually high (17° C. to 21° C. at 10 metres) and, in these lakes, plants may be found even at a distance of 13 metres from the surface.

When we compare the aquatics of hot and cold countries, we do not find structural differences corresponding to the differences of temperature; there is, in fact, a remarkable uniformity in the general organisation of water plants, whether they live in tropical or temperate climates. On the other hand, they differ markedly in their life-cycles, since those in warm surroundings vegetate continuously, while those which have to pass through a cold season show the special features associated with hibernation, which we have discussed in Chapter XVII.

We owe to Guppy the discovery that the rarity, in this country, of the flowering and fruiting stages in the life-history of certain hydrophytes, is due to thermal conditions. He has shown, for instance, that *Ceratophyllum*[1] requires almost tropical temperatures for the maturation of its fruit, and that *Lemna gibba*[2] does not flower except in water which is heated, during the summer, to a degree unusual in this country. For many water plants, however, the temperature of optimum vegetative growth is decidedly low[3]. It has been recorded, for instance, that, in the case of a certain canal near Manchester, which is kept tepid by the entry of hot water from various mills, the vegetation does not develop with any luxuriance. A Pondweed, *Potamogeton crispus*, grows in this canal as a dwarfed variety, especially near spots where warm water enters[4]; critical experimental work would, however, be required before we could feel certain of the fact that this result is due to temperature alone.

To some aquatics, the fact that lakes and rivers remain in summer cooler than the surrounding atmosphere, may be a

[1] Guppy, H. B. (1894[1]).
[2] Guppy, H. B (1894[2]).
[3] Goebel, K. (1891–1893).
[4] Bailey, C. (1884).

drawback, and it has been suggested in this connexion that the development of anthocyanin, which is so frequent in hydrophytes, may be an adaptation for heat absorption[1]. In considering the general question of the pigmentation of water plants, however, it must be remembered that some of the most striking examples may possibly represent pigmented races derived from the normal specific form by the loss of an inhibiting factor; on this view, they are comparable with certain coloured varieties well known among terrestrial plants, and there is thus little reason to suppose that their pigmentation bears any relation to the aquatic *milieu*. *Nymphaea lutea*, var. *rubropetala*[2] for instance may perhaps be compared with the chestnut-red variety of the Sunflower, while a form of *Castalia alba*[3], which has been described as bearing rose-purple flowers, may be analogous to the red variety of the white Hawthorn. But, apart from such cases, there are certainly indications that anthocyanin is formed by water plants with special facility. The leaves of the Lemnaceae, *Hydrocharis*, *Limnanthemum*, and certain Nymphaeaceae, are often more or less pigmented. The Podostemaceae[4] also, are apt to develop anthocyanin in their surface cells.

There is, indeed, little room for doubt about the liability of water plants to produce red and violet pigment, but the attempt to explain this fact is fraught with difficulty and confusion. The simple teleological explanation which assumes that the development of anthocyanin is an adaptation for the absorption of heat rays, is probably far too facile; the fact that the Podostemads, growing in the tropics, in water which maintains a constant high temperature, very frequently produce these pigments, seems to tell against such a view. The few observations which the present writer has been able to make, do not seem to harmonise with any general statement about the adaptational distribution of red and violet pigments in water plants. For instance, in the Forest of Dean (September, 1910) *Peplis*

[1] Ludwig, F. in Kirchner, O. von, Loew, E. and Schröter, C. (1908, etc.).
[2] Caspary, R. (1861). [3] Fries, E. (1858). [4] See p. 113.

Portula was found growing at the bottom of a deep pool, and entirely free from anthocyanin; but a number of shoots had broken off, by the snapping of the brittle stems, and were floating at the surface, and putting out adventitious roots. In the case of these detached shoots, there was considerable pigmentation, and some of the leaves were quite red. Again, in an extremely hot sunny summer (August, 1911) in the dykes at Wicken Fen, many young Waterlily leaves of the floating type, which were still rolled and had not reached the surface, were noticed to be brilliantly red.

The whole subject of anthocyanin has recently been dealt with comprehensively by Miss Wheldale (the Hon. Mrs Huia Onslow)[1]. She puts forward the hypothesis that the pigment arises from a chromogen formed from sugars in the leaf, and that increase in the amount of carbohydrates leads to increased formation of chromogen with the resultant production of anthocyanin, unless the chromogen be removed. If translocation be slowed down for any reason, such as low temperature, production of pigment tends to occur. This seems entirely consistent with the facts so far as they relate to water plants. For instance, in the case of the detached shoots of *Peplis* mentioned above, there would be little possibility of material being rapidly translocated from the leaves, because there is nowhere for it to go to; Miss Wheldale's theory thus explains the relatively high pigmentation of these shoots. In the case also of the Lemnas and the Podostemads, practically the whole vegetative body consists of assimilating organs. The excess sugar cannot, therefore, be removed from those organs, and the theory thus fully explains their liability to coloration. It is also confirmed by the known fact that the Podostemaceae store large quantities of carbohydrate, which is used up in their rapid flowering period. In such cases as the Waterlilies, again, the relative coolness of river or lake water may be a hindrance to rapid translocation from leaves to rhizome. As regards the supposed functions of anthocyanin, Miss Wheldale concludes that "For the time

[1] Wheldale, M. (1916).

being we may safely say that it has not been satisfactorily determined in any one case whether its development is either an advantage or a disadvantage to the plant." It is therefore clear that the attractive theory that red coloration is developed by water plants as an adaptation to their mode of life, must be definitely abandoned, unless further evidence for its validity can be produced.

Although water plants live, on the whole, in a more equable and temperate climate than land plants, yet they are liable in winter to one very severe ordeal—the freezing of the water in which they occur. Some escape this trial by their habit of sinking to the bottom in the cold season, while others are able to withstand a temperature below freezing point for a long period, especially when they are in the turion or seed phase[1].

The illumination, to which submerged plants are exposed, is as much affected by the medium as are the thermal conditions. Free-swimming water plants and those with floating or aerial leaves, on the other hand, receive light in much the same way as land plants; as a result of their situation, the leaves are often exposed to all the available sunshine, mitigated by no shade whatever. Such plants thus present no problems of special interest in connexion with their light conditions, and they may be disregarded in the present discussion, which will be confined to those that are more or less completely submerged.

The light which reaches a submerged shoot has been reduced by four factors—reflexion from the water surface, absorption by the water, and darkening due to certain substances in solution or to solid particles in suspension[2]. The absorption and darkening may be very considerable in the less limpid waters. It has been shown by experiments with a recording galvanometer that 60 per cent. of the light may be absorbed by the first two metres[3]. Some observations made in the Lake of Geneva[4], with regard to the limit of visibility, show that a

[1] See pp. 220, 243.
[2] Goebel, K. (1891–1893).
[3] Regnard, P. (1891).
[4] Forel, F. A. (1892–1904).

white disc lowered into the water remains visible to a depth varying between 6·8 metres in summer and 14·6 metres in winter. The annual mean was found to be 10·2 metres. This method is a rough one, but it gives some idea of the penetrating power of the luminous radiations. The results obtained harmonise with the observation that chlorophyll may be developed without loss of intensity by plants living at a depth of 10 metres. In the Jura lakes, however, which are not very transparent, some etiolation is produced even at 4 to 5 metres, in the case of *Naias* and the submerged leaves of *Nymphaea lutea*[1].

Some hydrophytes are dependent upon direct sunlight; the Podostemaceae, for instance, are rarely to be found in shady places where the water does not receive at least some hours of sunshine during the day[2]. Certain water plants, on the other hand, such as species of *Utricularia* and *Ceratophyllum*, perish when exposed to strong illumination[3]; and, of submerged plants in general, it is undoubtedly true that the conditions, under which they live, approximate to those of 'shade plants' upon land[4]. Their response to these conditions is also similar, and they share the characteristics of delicacy of lamina, absence of a well-differentiated palisade-tissue and presence of chlorophyll in the epidermis[5]. An attempt has been made to trace the peculiarities of submerged plants to the direct etiolating action of the obscurity in which they live[6], just as it has been suggested that the aerating system in their tissues was originally due to the direct effect of the medium[7]. We may accept this view so far as to acknowledge that the influences in question may, in both cases, have played a part in the first initiation of the aberrant structure of submerged plants, but such direct effects are scarcely adequate to explain the structure of the most highly modified forms which have lost the power to live on dry land.

In certain water plants showing heterophylly, the intensity of the light is one of the factors concerned in determining which

[1] Magnin, A. (1893). [2] Willis, J. C. (1902).
[3] Goebel, K. (1891–1893). [4] Schenck, H. (1885).
[5] Stöhr, A. (1879). [6] Mer, É. (1880[1]). [7] See p. 259.

type of leaf shall be produced. For example, the submerged band-shaped leaves of *Alisma graminifolium*, Ehrh.[1] require a moderate illumination, while the air-leaves flourish in bright light. In shallow water, in which the plants would, under ordinary conditions, form air-leaves, the band-shaped leaves continue to be produced, if the surface of the water happens to be covered with a layer of Algae which reduces the light. The influence of sunshine in this case is perhaps only indirect, the activity of assimilation being probably the critical factor.

The effect of light upon the germination of the winter-buds of *Hydrocharis Morsus-ranae*, the Frogbit, has been studied experimentally[2], and it has been shown that it is impossible for these turions to develop into plantlets, unless they are exposed to a minimum degree of illumination, which is far removed from total darkness. The yellow and orange rays prove to be the most active in promoting germination. But, marked as is the effect of light on the vegetative growth of the Frogbit, its influence in connexion with flowering is far more striking. It has been shown[3] that a set of plants exposed daily from the spring onwards to nine hours of direct sunlight, produced more than a thousand flowers between the end of June and the end of August, while a corresponding set of plants, which were insolated daily for three hours only, produced no flowers at all. Individual plants from this second set, removed and placed in bright sunshine at the end of June, began to flower in four weeks. By artificially cooling the water in which the insolated plants grew, it was shown that these effects were produced by differences of illumination, and not by the heating influence of the sun's rays.

Darkness seems to inhibit the germination of certain water plants; this has been shown in the case of the achenes of *Ranunculus aquatilis* and the nutlets of *Callitriche*. The seeds of *Nymphaea lutea*, also, though they are able to germinate in the dark, do so in far greater numbers in diffuse light. In other cases, e.g. *Potamogeton natans*, darkness favours germination[4].

[1] Glück, H. (1905). [2] Terras, J. A. (1900).
[3] Overton, E. (1899). [4] Guppy, H. B. (1897).

On the subject of heliotropism, we do not appear, in the case of water plants, to possess much experimental evidence. The work of one observer seems to suggest that the heliotropism of stems is less intense in the case of submerged than of terrestrial plants[1]. Positive heliotropism has been recorded for the leaves of *Aponogeton distachyus* and *A. fenestralis*[2]; the floating leaves of *Trapa natans*[3], on the other hand, are described as transversely heliotropic and as owing their horizontal position on the surface of the water to their response to light. It was shown, in certain experiments, that, after a week in darkness, the new leaves, which had unfolded, stood upright out of the water. In this connexion it has been recalled that, among the near relations of *Trapa*, there are land plants with transversely heliotropic leaves.

The leaves of the water form of *Myriophyllum proserpinacoides* exhibit 'sleep' movements when living submerged. The young leaves, which, normally, are spreading, rise up at night and cover the growing point, thus returning more or less to the position they occupied in the bud. Sleep movements also occur in *Limnophila heterophylla*[4]. The leaves of *Myriophyllum* and *Ceratophyllum*—excluding those of the apical bud—are said to have the peculiarity of bending downwards on darkening[5].

As regards geotropism, aquatic plants seem to be generally comparable with land plants. In *Aponogeton*, for instance, it has been observed that the leaves are negatively, and the adventitious roots positively, geotropic[2]. The present writer has, however, noticed in the case of the seedlings of *Nymphaea lutea*, that the short-lived primary root, after the earliest stages are past, shows little response to gravity, sometimes pointing vertically upwards. But this is probably merely a sign of its early degeneration and decay. There are also instances of the stems of water plants, in certain specialised cases, responding to gravity in the reverse of the usual way. For instance, the lateral

[1] Hochreutiner, G. (1896).
[2] Serguéeff, M. (1907).
[3] Frank, A. B. (1872).
[4] Goebel, K. (1908).
[5] Möbius, M. (1895).

branches of *Potamogeton pectinatus*, when swelling up to form tubers, become positively geotropic. They bend towards the soil and bury themselves in it to pass the winter. This has an important result, because, being lighter than water, these winter-buds would otherwise be liable to rise to the surface when set free by the decomposition of the parent plant[1]. Again, there are many cases of fruiting peduncles bending downwards and thus allowing the ovary to ripen under water; a similar curvature occurs not infrequently in terrestrial plants. Positive geotropism of the fruit stalk is characteristic of the Pontederiaceae[2] (Fig. 155, p. 240). *Limnobium Boscii* is a similar case; here it has been shown that the geotropic curvature is independent of fertilization[3].

Hochreutiner[4], who has paid special attention to the response of water plants to certain physical stimuli, has made some observations on 'rheotropism[5],' or reaction to current. He noticed that, in the case of *Zannichellia palustris*, where the water was still, the stem-branches rose erect, as would be expected of a negatively geotropic organ, but that, where there was a current, the axes adopted its direction. Hochreutiner observed this in the case of a current of such slight force that he was convinced that no mechanical compulsion was exerted, but that the stems responded to the stimulus by their own activity and might thus be called positively rheotropic. Roots, on the other hand, seem to show a tendency to grow against the current. It is suggested that this sensibility would be useful to the plant, since it would lead to the roots and stems taking up a position in which they would be unlikely to be damaged by the pulling force of the current. Further experimental work on rheotropism is obviously needed, however, before the subject lends itself to generalisation. The question is complicated by the fact that a rapid current alters the conditions of life of the plant very materially. Differences between the morphology of the same

[1] Hochreutiner, G. (1896).
[2] Müller, F. (1883).
[3] Montesantos, N. (1913).
[4] Hochreutiner, G. (1896).
[5] This term was suggested by Jönsson, B. (1883).

species, when grown in still or moving water, are possibly due, in some cases, to the better aeration of water which is in motion. Such differences are markedly exhibited by *Myriophyllum*[1], which in still, small pools may have leaves whose segments are very tender and almost hair-like, while in strongly flowing water they are shorter and firmer.

One of the most interesting problems connected with the tropisms of water plants, is the question of the influences which regulate the length of the petiole in the case of floating leaves. It is a matter of common observation that, in plants such as the Waterlilies, the length of the petiole varies with the depth of the water. The accommodation begins at the youngest stages, for, if the seeds of *Castalia alba*[2] are planted at different levels in the mud, the length of the first internode, the acicular first leaf, and the petiole of the second leaf, adapt themselves most remarkably to their circumstances, elongating until they are long enough to raise the leaves well into the water (Fig. 13, p. 28).

In free-floating plants, such as the Frogbit (*Hydrocharis Morsus-ranae*), this power of accommodation to depth is also in evidence, though it is naturally less conspicuous. The Frogbit has gained notoriety in the present connexion, since it was the subject of an oft-quoted series of experiments by Frank[3]. Its petioles are normally 6 to 8 cms. long, but when grown in shallow water they may not exceed 1 cm. If the plant is attached to the bottom of a deep glass vessel, on the other hand, very long petioles may be produced, a length of nearly 14 cms. being recorded in one case. Frank obtained a sensational contrast in petiole length, by growing a plant in a deep jar until its youngest leaf had succeeded in reaching the surface by elongating its petiole to 11 cms. It was then transferred to a shallow vessel in which the terminal bud was only just covered. The next leaf produced a petiole 1·5 cms. long, i.e. less than 14 per cent. of the length of the preceding leaf-stalk.

[1] Schenck, H. (1885).

[2] Massart, J. (1910).

[3] Frank, A. B. (1872).

Both common observation, and critical experiments such as these, leave no room for doubt about the fact that accommodation of petiole-length to water-depth does actually occur; but when we pass on to the question of the factors which bring about this accommodation, by causing cessation of growth at the appropriate moment, we find ourselves on controversial ground. One point seems to be uncontested—namely that, in the case of *Hydrocharis*, the regulation is not due to the change in light intensity, for even in darkness the petioles grow only to exactly the right length to bring the blade to the surface. Frank's experiments led him to the conclusion that, when the lamina reached the water-surface, the lowering of pressure, due to the absence of a superincumbent layer of water, was the physical factor which gave the signal to the petiole to cease growth. However, the repetition and critical analysis of Frank's experiments seem to have shown clearly that his deductions cannot be accepted. Karsten[1], using *Ranunculus sceleratus*, *Marsilea* and *Hydrocharis*, showed that if tubes of oxygen-free air were inverted over individual leaves, the growth of the petiole continued after the lamina had come in contact with the gas, instead of ceasing, as it did under normal conditions, as soon as the lamina reached the surface. His experiments seem to justify the conclusion that it is contact with the oxygen of the atmosphere which checks the further growth of the petiole, but we have no conception of the exact nature of the process by which this inhibition is brought about.

[1] Karsten, G. (1888); see also Vries, H. de (1873).

CHAPTER XXIII

THE ECOLOGY OF WATER PLANTS

THE study of the relation of plants to their habitats, of their different forms of association with one another, and of their applied physiology in general, is at the present day commonly included under the name of 'Ecology,' around which a complicated system of other technical terms has grown up. But, though the ecological language is new, the ecological standpoint and even the special ecology of water plants, are as old as the science itself. Theophrastus (370 B.C.–285 B.C.), whose writings form our earliest botanical classic, distinguishes water and marsh plants as a biological group and classifies them according to their varieties of habitat[1].

In a country such as Great Britain, where cultivation of the land, grazing of flocks and herds, and the numberless activities of man, have reduced the terrestrial plant population to a mere disheartening semblance of its former self, the vegetation of the waters has preserved, in many cases, a closer approximation to its original condition. Despite periodical disastrous clearances, ponds and streams, even in highly cultivated regions, sometimes show a fairly natural grouping of their inhabitants, while, on dry land, such a grouping can often only be discovered in remote districts, such as our few remaining areas of virgin fen and forest.

At the present day a voluminous literature has come into existence dealing with ecological topics, but it must be confessed that, as regards water plants, the results attained are, on the whole, scarcely of first-rate importance. On analysing the work in question, one is led to the conclusion that the chief service, which Ecology has rendered to the study of water plants, has probably been in emphasizing the influence of the substratum

[1] Greene, E. L. (1909).

and of the degree of aeration of the water in determining the distribution of aquatics[1]. It might have been supposed that the nature of the soil, underlying the water in which hydrophytes grow, would be relatively unimportant, but, on investigation, it proves to be a factor of almost as much significance as in the case of terrestrial plants. It is true that there are certain exceptions, such as the Podostemaceae, which seem indifferent to the chemical composition of the naked rocks on which they live[2], but this case may perhaps be explained by the fact that the rapidly flowing waters, to which they are confined, probably owe little of their dissolved constituents to the particular rocks over which they are passing at any given moment. The majority of hydrophytes, however, show definite preferences and aversions in the matter of the soil underlying the water in which they grow, and of the resulting differences in the nature of the solution in which they are immersed.

A case has been described in America, in which the dependence of water plants upon the substratum is shown with diagrammatic lucidity[3]. Lake Ellis in North Carolina is an area of shallow water, 2¼ by 3 miles across, and seldom more than two feet in depth; the entire floor is clothed with plants. Three distinct assemblages of vegetation occur in the Lake, the differentiation apparently depending wholly on the nature of the soil. The central region, where the soil is sandiest, is characterised by *Eriocaulon*, *Eleocharis* and *Myriophyllum*; a number of different plants, including one or two Waterlilies, frequent the intermediate muddy belt, while the marginal area of muddiest soil is chiefly clothed with Grasses and Sedges. The observation, made long ago by a German writer[4], that the variety of *Hydrilla verticillata* found in Pomerania is intolerant of sandy soil and is confined to muddy clay, is comparable with the facts just cited concerning Lake Ellis.

The two classes of substratum which offer the most marked contrast, as regards the flora which they support, are the cal-

[1] Tansley, A. G. (1911). [2] Willis, J. C. (1914[1]).
[3] Brown, W. H. (1911). [4] Seehaus, C. (1860).

careous and the peaty. Certain water plants are decidedly calcophil; *Stratiotes aloides*[1] is one of these cases, while another is *Scirpus lacustris*[2], which has been recorded as absent or rare in the Vosges, while it becomes common when the streams from this mountain region reach the Loess alluvium. When the substratum is peaty, on the other hand, the humous acids break up the calcium carbonate, thus rendering the water untenable for lime-loving plants but favourable for others, which are able to live in a solution poor in mineral salts, such as *Lobelia*, *Littorella* and *Isoetes*[3]. Those plants which can tolerate peaty water, enjoy the great advantage of freedom from the ravages of Water-snails[4].

Lists have been drawn up of the hydrophytes frequenting stagnant and slowly flowing waters in this country, showing that a different assemblage of plants is characteristic of each of these habitats[5]. This difference is probably due primarily to variations in the aeration. In extremely stagnant waters, which contain much decaying organic matter and are poorly aerated, the higher plants rarely appear. The Lemnaceae, however, form an exception to this rule, since they not only tolerate, but actually require, certain soluble products of organic decomposition. It has been shown that normal growth and multiplication cannot be sustained in *Lemna minor* for any length of time in the absence of certain organic, growth-promoting substances, or auximones[6].

A subject on which great stress is laid in descriptive ecological studies, is the "zonation" of the hydrophytes which characterises very many water areas. As a typical example we may refer to Magnin's[7] description of the Jura Lakes, where the plants are distributed with great regularity. Passing inwards from the shore, the following order is generally observed. There is, firstly, a littoral zone of plants standing out of the water—

[1] Davie, R. C. (1913). [2] Kirschleger, F. (1857).
[3] West, G. (1905), (1908), and (1910). [4] West, G. (1908).
[5] Tansley, A. G. (1911).
[6] Bottomley, W. B. (1917); see also p. 81. [7] Magnin, A. (1893).

Phragmites followed by *Scirpus lacustris*; next, a belt of plants with floating leaves, among which *Nymphaea lutea* is the dominant species, and, still farther from the shore, a zone of plants with leafy shoots reaching to the water surface, or nearly, consisting mainly of Potamogetons. To this succeeds a region in which the upper layers of the water are free from vegetation, while the grappling iron brings to light various plants which grow on the bottom, such as *Ceratophyllum*, *Naias*, *Chara* and *Nitella*. Fig. 165 shows, in the form of a section, the essentially similar zones of vegetation in the White Moss Loch in Perthshire[1].

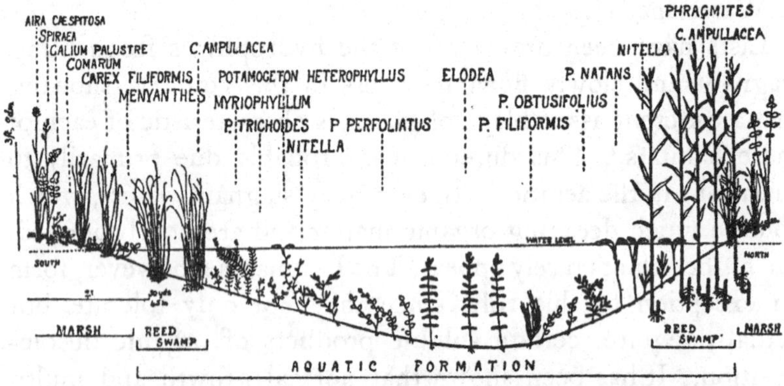

FIG. 165. Section nearly N. and S. across White Moss Loch, Perthshire, showing relations of plants to water environment. [Matthews, J. R. (1914).]

One of the chief reasons determining this zonation seems to be that plants with floating leaves can only flourish if guarded from the wind. For this reason they generally do not occur at a great distance from the shore, except in very sheltered basins, and often obtain the necessary protection by growing among reeds. It has been pointed out that in the larger English Broads, the "floating-leaf association" is almost coterminous with the "open reed-swamp[2]," while in Lake St Clair (Michigan) precisely the same thing occurs, the plants with large floating leaves all belonging to the "Phragmitetum[3]." In the case of the White

[1] For recent views on ecological classification of aquatics, see Pearsall, W. H. (1917–1918) and (1918).

[2] Pallis, M. in Tansley, A. G. (1911).

[3] Pieters, A. J. (1894).

Moss Loch, it has been recorded that the floating leaves of *Potamogeton natans* cover the surface in the parts of the loch which are protected from the prevailing winds; where the water is much exposed, however, such broad-leaved plants are absent, their place being taken by *Myriophyllum*, whose highly divided foliage is uninjured by wave-motion[1]. Submerged plants, as a rule, form a special zone farther from the shore than the floating-leaf association, because the latter shades the lower layers of the water so much that the subdued sunlight, that penetrates it, is insufficient to supply a deeper flora. An exception to this rule is afforded by *Aldrovandia vesiculosa*, a typical shade plant, which grows among reeds, or protected by the leaves of Waterlilies, in order to secure the dim light which suits its requirements[2].

In addition to the examination of well-established aquatic floras, another branch of the ecology of aquatics consists in the study of the process of colonisation of newly formed waters. We shall return in the next chapter to the methods by which this colonisation is achieved, but we may mention here an account, recently published by a Cambridge botanist[3], of an ecological experiment on a large scale which was carried out in the fen country, by Nature herself, not long ago. In January, 1915 an area of about 24 square miles became inundated, and remained under water for nine months, until re-drainage was accomplished; it was thus temporarily restored to something like its original aquatic conditions. Even in the brief period in question, water plants invaded the area, but, somewhat unexpectedly, the new flora was confined mainly, as far as flowering plants were concerned, to two species, *Alisma Plantago* and *Polygonum amphibium*. Those were present in abundance and tended locally to form "closed associations."

The effect of altitude above sea level upon the water vegetation, may be considered as coming within the purview of

[1] Matthews, J. R (1914). [2] Hausleutner, (1850[1]).
[3] Compton, R. H. (1916).

ecology[1]. It is a matter of common knowledge that the land flora suffers great changes in the passage from the lowlands to the mountains, until an Alpine flora is reached, whose facies is totally different from that of the plains below. The hydrophytes, on the other hand, show singularly little change, though the number of species diminishes rapidly as high altitudes are approached. In Scotland, West[2] has pointed out that, if a highland loch is well sheltered and possesses a good shore and water not too poor in mineral salts, its flora may scarcely be distinguishable from that of a lowland basin. In the Jura, to take a Continental example, sixty lakes were investigated by Magnin[3], who showed that out of thirty species of hydrophytes, twenty-four were common to all these basins, whose heights ranged from 200 to 1000 metres above sea level. Tansley[4], again, has drawn attention to the fact that the plants recorded by Graebner[5] from sandy pools in the barren heaths of North Germany—*Isoetes*, *Littorella*, *Lobelia*, etc.—are the same as those occurring in Britain in mountain lochs, and suggests that this indicates that the poverty in mineral salts, common to both types of locality, has more influence than the actual altitude in determining the flora.

In the Alps many aquatics reach considerable heights. In the Upper Engadine[6], *Ranunculus trichophyllus* has been found at above 2500 metres, and a *Potamogeton*, a *Callitriche* and *Hippuris vulgaris* at above 2000 metres. These plants have thus an astonishing range of altitude, since they abound, on the other hand, almost at sea level in the English fens. Outside Europe, the same great range is also observed. In South America near Chimborazo[7], *Myriophyllum*, *Lemna* and *Callitriche* have been recorded at a height of above 2400 metres. The

[1] Overton, E. (1899) has shown that the data on this point given by Schenck, H. (1885) have little value, since the altitudes which he names are, in reality, much exceeded.

[2] West, G. (1908).

[3] Magnin, A. (1893).

[4] Tansley, A. G. (1911).

[5] Graebner, P. (1901).

[6] Overton, E. (1899).

[7] Spruce, R. (1908).

genus *Isoetes*, like the flowering plants just mentioned, shows great indifference to altitude. One species, *I. amazonica*, Mgg., was found on the river margin at Santarem in the lowlands, while another occurred at about the same latitude on the cold Paramos of the Andes at nearly 3700 metres[1]. In India, *Lemna minor* has been recorded at Laboul at a height of above 2900 metres[2]. In Venezuela and Tibet, *Potamogeton pectinatus*, which flourishes at sea level in England, has been found at heights of above 5000 metres[3].

The term Ecology is used by some botanists in a sense so wide that it becomes almost co-extensive with out-of-door Botany in general. But, if we limit our consideration to that branch of plant study which strictly deserves the name, it does not appear, as far as the present writer is able to judge, that any general ideas of the first importance, bearing upon the study of water plants, have emerged from it, beyond those to which allusion has been made in this chapter. At present Ecology has scarcely passed the stage of a merely descriptive branch of the science; indeed one of its chief promoters[4] described it, a decade ago, as "still in its infancy." When it has become more closely linked up with Physiology, we may look to it for further help in solving the complex problems presented by the life of hydrophytes[5].

In conclusion, it may be suggested that there is room, in the case of aquatic plants, for ecological work of a rather different character from that usually attempted—namely, a study of the changes occurring from year to year in the Angiospermic flora

[1] Spruce, R. (1908). [2] Kurz, S. (1867).

[3] Ascherson, P. and Graebner, P. (1907).

[4] Warming, E. (1909).

[5] In addition to the references given in the course of this chapter, see Bruyant, C. (1914), Massart, J. (1910), Moss, C. E. (1913), Nakano, H. (1911), Pieters, A. J. (1902), Preston, T. A. (1895), Roux, M. le (1907), Schorler, B., Thallwitz, J. and Schiller, K. (1906), Schröter, C. and Kirchner, O. (1902), Thiébaud, M. (1908). On the cultivation of water plants see Mönkemeyer, W. (1897).

of the same waters. These changes seem to be much more notable and rapid than those occurring among terrestrial plants in corresponding periods. In certain dykes and ditches, which the present writer has had under more or less continuous observation for some years, various species appear, disappear, and reappear, in a fashion which seems at first glance wholly erratic, but which might, on thorough study, yield results which would throw some light upon the problems of dispersal and distribution.

PART IV

THE STUDY OF WATER PLANTS FROM THE PHYLOGENETIC AND EVOLUTIONARY STANDPOINTS

"The theorem of Organic Evolution is one thing; the problem of deciphering the lines of evolution, the order of phylogeny, the degrees of relationship and consanguinity, is quite another. Among the higher organisms we arrive at conclusions regarding these things by weighing much circumstantial evidence, by dealing with the resultant of many variations, and by considering the probability or improbability of many coincidences of cause and effect; but even then our conclusions are at best uncertain, our judgments are continually open to revision and subject to appeal, . . ."

D'Arcy Wentworth Thompson, *Growth and Form*, 1917.

CHAPTER XXIV

THE DISPERSAL AND GEOGRAPHICAL DISTRIBUTION OF WATER PLANTS

THE most striking character of the geographical distribution of water plants is, in general, their remarkably wide range[1]. Countless instances might be cited, but it may perhaps suffice to refer, as examples, to *Potamogeton crispus*, which occurs in Europe, Asia, Africa, America and Australia, and to *Ceratophyllum demersum* and *Lemna minor*, which are also found almost all over the world. Of the twenty-two genera of Lythraceae, again, only five are common to both hemispheres —*Rotala*, *Ammania*, *Peplis*, *Lythrum* and *Nesaea*—and these five all characteristically frequent water or marshy ground[2]. The wide distributions of aquatics often include occurrences on islands which are some distance from other land surfaces; *Lemna trisulca*, for example, which is found in Europe, Asia, North and South America, Australia and Africa, penetrates to Mauritius, Madeira, the Azores and the Canary Islands. The most marked exception to the rule of the wide distribution of hydrophytes is furnished by the Podostemads[3], many of which inhabit extremely restricted areas. The Brazilian river Araguay, for instance, has three sets of cataracts, each of which is populated by an almost entirely different group of species belonging to this family. Seven species of *Castelnavia* occur in this river, although the genus is almost unknown elsewhere.

If we except the Podostemads, the generalisation certainly holds good that aquatic Angiosperms have, as a rule, a wider distribution than the terrestrial members of the group. This

[1] Schenck, H. (1885), gives the ranges of a long series of aquatic plants, as far as they were known at that date.

[2] Gin, A. (1909). [3] Weddell, H. A. (1872).

is by no means what one would expect at first glance, since it might reasonably be supposed that salt-water areas, mountain ranges, and wide tracts of arid country would prove insuperable barriers to the migration of plants of fresh water[1]. This difficulty was so keenly felt by Alphonse de Candolle[2] that he was forced to the conclusion that the facts of the distribution of aquatic species were scarcely explicable except on the theory that there had been multiple centres of creation.

For the sake of simplicity we may first consider the distribution of hydrophytes within a single country such as our own, which, on a small scale, presents the same difficulties. A partial solution of the problem might be reached, if former connexions between the existing river basins could be postulated, in order to account for the uniformity of their floras. But the history of the land surfaces at once disposes of this possibility. In the words of Clement Reid[3], whose labours disinterred so much of the geological history of our present flora, "Each year's work at the subject makes it more clear, that ever since our climate became sufficiently mild to allow of the existence of our present fauna and flora, many of the river-basins of Britain have formed isolated areas." It is no doubt possible that floods may, in some cases, give a species the opportunity of introducing itself into fresh situations; an extension on a small scale, of the area of distribution of certain aquatic plants was induced by the great floods in East Anglia in 1912. Furthermore, floods may even, as Guppy[4] has suggested, occasionally bring about an exchange between plants belonging to different rivers traversing extensive level regions. But such effects can never be more than partial and they will not explain the passage of any species over a well-defined watershed[5].

[1] This paradox was noted by Darwin, C. (1859).

[2] Candolle, A. P. de (1855).

[3] Reid, C. (1892).

[4] Guppy, H. B. (1906).

[5] Dr Guppy has suggested to the writer "that the permanent head-springs of rivers in elevated regions where the sources of rivers may lie in proximity would serve as centres of dispersion for the same plants in

Even within a single river basin, the question of the seed-dispersal of aquatic plants is by no means a simple one. The expectation might perhaps be formed that aquatics would be characterised by floating seeds or fruits, capable of being water-borne for considerable distances. But, as is often the case, Nature fails to conform to the preconceived notions of the teleologist, and we find, as a matter of actual fact, that although many plants with *water-side* stations possess buoyant seeds, such seeds are relatively rare among *true aquatics*. Guppy[1], who gives the results of experiments on the floating powers of the seeds of more than 300 British plants, records that sinking occurred within a week in the case of 26 aquatics, e.g. *Ranunculus aquatilis*, *Hottonia palustris*, *Lobelia Dortmanna*, *Lemna gibba*, *Callitriche*, and others. He found that the seeds of *Limnanthemum nymphoides* would float for 1 to 4 weeks, while *Lemna minor*, *Sagittaria sagittifolia*, *Alisma Plantago* and certain species of *Potamogeton* were the only hydrophytes whose seeds and fruits were capable of floating for months at a time, and, of these, the Alismaceae should perhaps be reckoned, in this connexion, as water-side rather than as aquatic plants.

It is true that the seeds of those aquatics that sink very rapidly may yet sometimes be carried a short distance by the wind. For instance, the slender infructescences of *Hippuris vulgaris* are swayed to and fro by the breeze, and the fruits may be jerked a little way[2], but the migrations thus achieved can never be extensive.

If the dispersal of hydrophytes within a single river basin can only be explained with difficulty, this is still more the case when we come to consider migration from one country to another. As Guppy[3] has pointed out, *Ceratophyllum demersum*

different river-basins, and if that is right then the species held in common ought to include all those growing in the head-springs, e.g. in England, *Callitriche aquatica*, *Nasturtium officinale*, *Ranunculus aquatilis*, etc., etc." (By letter, February 3rd, 1918.)

[1] Guppy, H. B. (1906). See also Praeger, R. L. (1913).

[2] Fauth, A. (1903).

[3] Guppy, H. B. (1893).

possesses a fruit which sinks like a stone, and the plant is soon killed by sea water—yet it has established itself nearly all over the globe, reaching such islands as the Bermudas and Fijis. The Potamogetons, again, present little or no obvious capacity for dispersal by sea—yet such a species as *Potamogeton densus*, whose fruits sink at once in fresh or salt water, flourishes in Europe, Asia, Africa and America.

Water plants, as we have already pointed out, are particularly prone to reproduction by vegetative means, and any theory attempting to account for their dispersal must take into consideration the conveyance of detached fragments, and of various types of winter-buds or turions, which are probably more effective than fruits and seeds in the process of dissemination.

The hypothesis has been proposed that water-fowl are the chief agents in the dispersal of hydrophytes. This theory certainly explains a large proportion of the observed facts, and a considerable amount of indirect and circumstantial evidence has accumulated in its favour. Darwin[1] pointed out how readily wading birds, which are great wanderers, might convey seeds from one water basin to another, in the mud adhering to their feet. Clement Reid[2] came to conclusions bearing on this question in the course of his study of the colonisation of isolated ponds—such as pools which collect in old brick yards, quarries, etc., and the dew ponds dug on dry chalk downs to provide water for cattle and sheep. He found, in general, that the water plants which colonise isolated ponds are essentially the floating species with finely divided leaves. Their seeds and fruits are commonly such as would be digested and destroyed if eaten by birds, but their stems are brittle, and their leaves, on removal from the water, collapse and cling closely to any object they may touch. He therefore concluded that it was probable that these plants are transported in fragments that adhere to the feet of wading birds. This would also account for the constant presence of Limnaeids in these ponds, since their eggs might easily be carried, clinging to pieces of leaves or stems.

[1] Darwin, C. (1859). [2] Reid, C. (1892).

An interesting experiment in the colonisation of a pond was made at Garstang in Lancashire some years ago[1]. The pond in question was dug in a grass field and carefully railed off to prevent access of cattle. After about eighteen months certain aquatic Angiosperms had appeared in the pond—*Alisma Plantago*, *Callitriche* and *Glyceria fluitans*, as well as species of *Juncus*. In the course of the next five years no new hydrophytes appeared, but *Alisma*, *Glyceria* and *Juncus conglomeratus* developed so freely as practically to exclude any intruders. In connexion with the Garstang experiment, it is significant that fragments of *Alisma Plantago*, *Glyceria fluitans* and *Juncus* sp. were observed by a French botanist[2] many years ago attached to the feet and feathers of migrating birds. The only water birds, actually seen to visit the Garstang pond, were Moorhens, but other aquatic species were numerous in the district.

That water birds convey hydrophytes from place to place, is so far an accepted fact that it has been stated that it is "vain to make a shallow reservoir in the line of the constant migration of water fowl (i.e. between their resorts), and expect it to maintain a freedom from water plants[3]." On the other hand we occasionally meet with an apparent exception. A case was recorded in Germany[4] in which *Utricularia Bremii* grew in one locality, while in another, less than a mile away, *U. minor* was found. These marshes had been under observation for a century, and, during that time, no exchange of species had taken place, though, throughout the summer, numbers of Ducks and other water-fowl flew daily between the localities in question.

There is obviously no doubt that hydrophytes and water-fowl are constantly brought into intimate relations. One has only to watch Moorhens in summer, running for long distances over Waterlily leaves without wetting their feet, to realise that plant

[1] Wheldon, J. A. and Wilson, A. (1907). Information relating to this experiment has been most kindly supplied to me by letter by the authors, to supplement that recorded in their *Flora of West Lancashire*.

[2] Duval-Jouve, J. (1864). [3] West, G. (1910).

[4] Meister, F. (1900).

and bird play some part in one another's life. In British Guiana, Im Thurn[1] noticed Spurwings (*Parra jacana*) running about over the leaves of *Victoria regia*; one of them had even nested on a leaf. In the case of *Lawia zeylanica*, a Podostemad belonging to Ceylon, Willis[2] records that wading birds are often seen walking over the thalli. Very numerous fruits are produced, each containing a large number of seeds, whose epidermis swells up and becomes mucilaginous on wetting. This mucilage, when it dries, serves to fix the small seeds firmly to any object with which they come in contact, and Willis points out that in this way they may easily adhere to the feet of wading birds.

When we come, however, to the question of the *first-hand* evidence as to the part played by water-fowl in the dispersal of aquatic plants, we find that the facts actually recorded are relatively few. Our ignorance on this point was emphasized by Caspary[3] in 1870, and though almost half a century has elapsed since he propounded the question—"Welche Vögel verbreiten die Samen von Wasserpflanzen?"—very few observers have stepped into the breach. It is a question which might well engage the attention of local natural history societies, since it requires the co-operation of botanists and zoologists: an investigation conducted over a number of seasons could scarcely fail to produce interesting results.

Such direct evidence as we at present possess, relates partly to the unintentional conveyance of water plants attached to a bird's feet or feathers, and partly to the presence of undigested seeds and fruits in the alimentary canal. With regard to the Lemnaceae, Weddell[4] records that, when shooting in Brazil, he killed a water bird called "Camichi"; its feathers were soiled with greenish matter, and closer examination revealed the presence of a minute Duckweed, *Wolffia brasiliensis*, in full flower! At a later date Darwin[5] stated that, in this country, he had twice observed Duckweed adhering to the backs of Ducks

[1] Im Thurn, E. F. (1883).
[2] Willis, J. C. (1902).
[3] Caspary, R. (1870[2]).
[4] Weddell, H. A. (1849).
[5] Darwin, C. (1859).

on their suddenly emerging from the water. Guppy[1] found that a week in sea water killed the seeds of *Lemna minor*, while a day generally killed the fronds, but he considers that in damp weather the plants might, for a day or two, withstand exposure to the atmosphere and thus might be carried a few hundred miles entangled in a bird's plumage—a supposition to which the observations of Weddell and Darwin lend colour.

That seeds and fruits may be conveyed in mud, adhering to the beaks, feet, or feathers of birds, has long been known. Most of the records on this point relate to plants which are not strictly aquatic, but Kerner[2] mentions *Elatine hydropiper*, *Glyceria fluitans* and *Limosella aquatica* among the species which he has himself found in this situation. Duval-Jouve[3], who also paid attention to this subject, observed at different times the *débris* of twelve plant species adhering to the feet and breasts of the migrating web-footed birds exposed for sale in a market.

Our knowledge of the internal conveyance by birds of the seeds of aquatics, rests almost entirely on the work of Guppy[4], whose remarkable observations on the life-histories of water plants have been frequently cited in the foregoing chapters. Guppy dissected and examined thirteen wild Ducks purchased in the London markets, and found altogether 828 seeds and fruits, including those of *Sparganium* and *Potamogeton*. Seeds obtained from this source sprouted with such greatly increased rapidity that Guppy describes the wild Ducks as "flying germinators." He adds the observation that, of a large number of nutlets of *Potamogeton natans* which were eaten and passed by a domestic Duck in December, 60 per cent. germinated in the following spring, whereas, at the same date, sprouting had only occurred in the case of 1 per cent. of the nutlets left over in the vessel from which the Duck had been fed.

It is possible that, in certain cases, the seeds of water plants

[1] Guppy, H. B. (1893).
[2] Kerner, A. and Oliver, F. W. (1894–1895).
[3] Duval-Jouve, J. (1864).
[4] Guppy, H. B. (1894[1]), (1897) and (1906).

may—accidentally as it were—offer some lure to birds. When a fruit of *Castalia alba* bursts, some 1600 to 1700 seeds rise to the surface, where they float for a day or two in a mass, looking like a patch of fish spawn[1] and perhaps on this account attracting the attention of birds.

Ascherson[2], who has given much study to the distribution of marine Angiosperms, argues, from the occurrence of *Zostera nana* in the Caspian Sea, that this water area must have been in comparatively recent times connected with the Black Sea, where this species is also found. However, in the light of the part played by birds in the distribution of water plants, it is probable that little stress can be laid upon such evidence. It has been observed[3] in Britain that Brent Geese feed on *Zostera*, and that these birds are almost confined to the parts of the coast where the Grass-wrack is to be found. It is quite conceivable that they may occasionally carry seeds or fragments of the plant which would be able to take root on reaching salt water again: by analogy we may suppose that birds might also be competent to convey *Zostera nana* over the three hundred miles or so which separate the Black Sea from the Caspian.

Problems of plant distribution are often a good deal complicated by the interference of man. This is less the case with aquatic than with terrestrial vegetation, because, on the whole, water plants are of no great utility to the human race, and are seldom introduced intentionally. But the present distribution of certain aquatics cannot be understood unless allowance be made for the influence of mankind in their dispersal. *Trapa natans*, the Bull Nut or Water Chestnut, is an instance. This plant now occurs over a considerable part of Europe, the Caucasus and Siberia[4]. It has been used from early times for food, medicine and magic, and is supposed to have been introduced into Switzerland as long ago as the period of the lake dwellings[5]. It is now nearly exterminated in that country, and

[1] Guppy, H. B. (1893). [2] Ascherson, P. (1875).
[3] Walsingham, Lord and Payne-Gallwey, R. (1886).
[4] Areschoug, F. W. C. (1873[2]). [5] Jäggi, J. (1883).

has vanished from various localities in Belgium, Holland and Sweden, where there are records of its occurrence in comparatively recent times. It certainly seems to be a plant which is in process of extinction in various parts of its range, since it occurs in peat in a semi-fossil condition in places where it has never been known alive within the memory of man[1]. The exact reason for its disappearance is hard to find. Probably the lowering of the mean temperature has some bearing on the question, but it evidently does not provide a complete explanation, since the Bull Nut can live in the north of Scania, although that region is colder than Belgium and the Swiss lowlands, where the plant is now almost, if not entirely, extinct[2].

Just as certain terrestrial plants penetrate as weeds with the seeds of cereals into alien localities, so aquatics find a congenial home in swampy rice fields, and are disseminated to other countries in company with the rice. Thus the Lythraceous *Rotala indica* and several species of *Ammania*, belonging to the same family, have penetrated into Kurdestan, Transcaucasia and Astrakhan[3], while *Naias graminea* has reached Upper Italy[4] in the same fashion. The latter species has even been introduced into England, probably with Egyptian cotton, and grew at one time in a canal near Manchester, where the temperature happened to be artificially raised by the discharge of hot water from various mills[5]. Cotton is probably also responsible for the introduction into Yorkshire of *Potamogeton pennsylvanicus*, which is the only non-native Pondweed recorded from Britain[6]. In the Tropics, e.g. Fiji, a number of edible tubers, such as *Colocasia* and *Alocasia*, are cultivated at the borders of ponds and ditches. It has been suggested[7] that aboriginal man, in taking such moisture-loving food plants with him on his

[1] Reid, C. (1899).
[2] Areschoug, F. W. C. (1873[2]).
[3] Gin, A. (1909).
[4] Ascherson, P. (1874).
[5] Bailey, C. (1884) and Weiss, F. E. and Murray, H. (1909).
[6] Fryer, A., Bennett, A., and Evans, A. H. (1898–1915).
[7] Guppy, H. B. (1906) and (1917).

migrations, may often have assisted unintentionally in the dispersal of associated aquatics.

Turning from the detailed question of the modes of dispersal of hydrophytes, to the more general problem of their geographical distribution, we find that these plants furnish certain data bearing on the theories put forward in recent years by Guppy and Willis. The views of these two authors, though wholly independent, and in many ways quite distinct, seem in some respects to supplement one another.

The nature of Guppy's hypothesis—which he names the Differentiation Theory[1]—may be briefly indicated as follows. He supposes that the history of our present flora is "essentially the history of the differentiation of primitive world-ranging generalised types in response to the differentiation of their conditions." He expressly points out that his view does not attempt to explain the *origin* of these primitive generalised families, and he is careful to note that the present distribution is also "an expression of the influence of the arrangement of the continents during secular fluctuations of climate." For lack of space it is impossible here to do justice to Guppy's theory, but we may consider two cases among water plants, to each of which he draws attention as illustrating differentiation and distribution within a single genus. One of these is the genus *Naias*[2], which he treats in the light of Rendle's monograph[3]. Guppy considers that the polymorphic *Naias marina*, which occurs almost all over the whole area of the genus, is the primitive type, representing the stock from which the other species are derived. None of the remaining species are so widely distributed, and though some of them have a considerable range, others are extremely localised. In *Limnanthemum*[2], again, Guppy regards nearly all the tropical species as reducible to varieties of *L. indicum*, which he takes to be another typical polymorphic species of wide range; it has played a *rôle*, in the warm fresh waters of the globe, comparable with that of

[1] Guppy, H. B. (1917), etc. [2] Guppy, H. B. (1906).
[3] Rendle, A. B. (1901).

Naias marina, giving birth to new species in various parts of its range.

There is another case among water plants which, though Guppy does not allude to it, seems to the present writer to be readily interpreted on the differentiation theory. The case in question is that of the family Aponogetonaceae, with its one genus *Aponogeton*, the Arrowgrass, often cultivated in England[1]. Africa and Madagascar appear to be the headquarters of the genus; the species in this region consist almost entirely of plants with forked inflorescences, while the Indo-Australian species have simple inflorescences. The species can be placed, according to their geographical position, in a series extending from west to east which also represents their affinities. The African species lead on to the Madagascan; these show affinity with the Indian, while the North Australian are the most remote. It seems that we must interpret the genus *Aponogeton* as having reached a more advanced stage of differentiation than such genera as *Naias* and *Limnanthemum*. *Aponogeton* no longer contains any species whose range is approximately coterminous with that of the genus, but the original area has become "divided up into a number of smaller areas each with its own group of species[2]." However it must not be overlooked that this case might be interpreted in other ways by those who hold different views on plant evolution.

Willis[3] has in recent years put forward a remarkable hypothesis which is in many ways easily related to Guppy's theory—namely, the "Law of Age and Area," according to which the relative size of the geographical territory occupied by each species within a genus (or genus within a family) is, in general, proportional to the age of that species. According to this hypothesis, the most widely distributed genera and species—instead of being the best adapted, as is maintained by orthodox Darwinians—are in reality the most primitive, while those occupying limited areas are relatively modern. It is impossible here

[1] Krause, K. and Engler, A. (1906). [2] Guppy, H. B. (1906).
[3] Willis, J. C. (1914[2]), and a number of earlier and later papers.

to enter upon any discussion of the grounds upon which Willis bases his view, or of the criticisms to which it has been subjected. It must suffice to see whether it can be applied to any aquatic plants, and, if so, with what result. The only hydrophytes with which Willis himself deals are the Tristichaceae and Podostemaceae. He points out that, owing to the peculiar morphology of these plants, it seems possible to say with some degree of certainty which are the older forms. *Tristicha* and *Podostemon* are almost radially symmetrical, and do not diverge greatly from the ordinary type of submerged plant. *Lawia* and *Castelnavia*, on the other hand, show the most extreme dorsiventrality of structure and have highly modified flowers; most botanists would probably agree that *Tristicha* and *Podostemon* are the older types, while *Lawia* and *Castelnavia* represent a more recent evolutionary development. If this view be accepted, the families in question form a striking illustration of the principle of Age and Area, for *Tristicha* and *Podostemon* cover the whole range of distribution of the families, while *Lawia* and *Castelnavia* are both limited to comparatively small regions[1].

The difficulty of applying a morphological test to Willis's or to Guppy's theory lies in the fact that botanists seldom agree as to which members of any given family or genus are to be considered primitive and which are more specialised. The Water Starworts (*Callitriche*), however, seem to the present writer to present a case which is, in this regard, less problematical than most. Within *Callitriche* we have two sub-genera, one of which, *Eu-callitriche*, has the upper leaves floating and is characterised by aerial pollination, while the other, *Pseudo-callitriche*, is completely submersed throughout life. Most botanists would probably admit that the genus is descended from terrestrial ancestors, and that the submerged *Pseudo-callitriche* is hence a more highly specialised and recent type. In distribution, *Eu-callitriche* is almost cosmopolitan, while *Pseudo-callitriche* is confined to the North Temperate regions. The distinction holds even within our own country, where *C. verna* and its sub-species

[1] Willis, J. C. (1917).

representing *Eu-callitriche*, are abundant, whereas *C. autumnalis* (*Pseudo-callitriche*) is rare and local. The two sub-genera of Starworts are thus related to one another, as regards their distribution, in exactly the way that would be predicted, from their degree of specialisation, either on the Differentiation Theory, or on the Law of Age and Area. Further, it may be suggested that the Duckweeds afford another case in point. *Lemna minor*, which is the most widespread member of the family, also shows indications of being the least specialised. These instances are obviously too few for generalisation, but, as far as they go, they show that the evidence from hydrophytes is decidedly favourable to the views of Willis and Guppy. It is greatly to be wished that more test cases may come to light, in which certain species within a genus, or genera within a family, can be accepted with some degree of confidence as relatively primitive.

CHAPTER XXV

THE AFFINITIES OF WATER PLANTS AND THEIR SYSTEMATIC DISTRIBUTION AMONG THE ANGIOSPERMS

(I) THE AFFINITIES OF CERTAIN AQUATIC ANGIOSPERMS

IT is generally recognised that the primaeval forms of vegetable life were probably aquatic, and that it is only in the highly evolved group of Seed Plants that a terrestrial habit has become firmly established. It follows that any aquatics met with among the higher plants must be regarded as descendants of terrestrial ancestors, which have reverted in some degree to the hydrophytic habits of their remote forbears. That this view is tenable, and that the Aquatic Angiosperms cannot trace their ancestry in an unbroken aquatic line from some far-away algal progenitor, is demonstrated by the fact that their floral organs, in the vast majority of cases, belong to a decidedly terrestrial type[1].

Before discussing any significance which may be attributed to the systematic distribution of aquatics among the families and genera of terrestrial Angiosperms, it will be necessary briefly to review the natural affinities of various members of this biological group—affinities which are still in some cases decidedly problematical. The present writer accepts the theory that the Ranalean plexus includes the most primitive forms among the living Angiosperms[2], and also the view that from this plexus the Monocotyledons have been derived[3]; these theories provide the basis for the general order in which the plants are dealt with in this chapter, and they also form the bed-rock for the discussion arising out of the facts enumerated.

[1] See Chapter XVIII. [2] Arber, E. A. N. and Parkin, J. (1907).
[3] Sargant, E. (1908) and earlier papers.

Dealing first with those more primitive members of the Archichlamydeae, which are known as the Polypetalae, we find that, in the Ranalean plexus, the Nymphaeaceae offer a striking example of a family rich in genera and species, and consisting entirely of water and marsh plants. There is great variation in the structure of the flower, and the carpels range from superior to inferior. The variety of form occurring in the family suggests that it is an old one which has had a long time to evolve, since it adopted aquatic life. It should be noted that various observers[1] have regarded the Nymphaeaceae as Monocotyledons, but it seems more reasonable to suppose that they are truly Dicotyledonous, though descended from a stock closely related to that which gave rise to the Monocotyledons.

The curious genus *Ceratophyllum*, on whose affinities the most divergent claims have been made, seems best regarded as a reduced form, closely related to the Nymphaeaceae[2] and especially the Cabomboideae[3]. Thus this genus, which on account of its extreme specialisation is reasonably relegated to a distinct family, may be regarded as the ultimate term in the Nymphaeaceous series; its rootlessness, reduced anatomy and submerged pollination indicate how completely it has identified itself with aquatic life.

The Ranunculaceae are typically terrestrial, but the genus *Ranunculus* contains, besides purely terrestrial species, the subgenus *Batrachium* which is definitely aquatic, and also a number of species such as *R. Flammula*, which are amphibious.

The Cruciferae include certain types, e.g. *Nasturtium amphibium* and *Cardamine pratensis*, which are capable of living either in damp places or actually submerged. These form a link between the terrestrial Crucifers and such definitely aquatic forms as *Subularia aquatica*. This plant, which superficially resembles a tiny *Juncus*, lives entirely submerged and has been described as cleistogamic.

[1] Trécul, A. (1845) and (1854), Henfrey, A. (1852), Seidel, C. F. (1869), Schaffner, J. H. (1904), Cook, M. T. (1906).

[2] Brongniart, A. (1827), Strasburger, E. (1902). [3] Gray, A. (1848).

The Droseraceae possess one curious little floating water plant, *Aldrovandia vesiculosa*. Its flowers are aerial and of the type characteristic of the family, but it is rootless, and its anatomy is much simplified.

The Podostemaceae have been placed in the most various systematic positions, but botanists seem now to regard them as showing some affinity with such forms as *Nepenthes*[1] and the Saxifragaceae. The carpels present numerous points of similarity with those of the latter family, e.g. the gynaeceum is hypogynous, with a bicarpellary ovary, two free styles and a number of ovules on a thick placenta connected with the outer wall by a thin septum, while the ovule is anatropous, with a straight embryo and no endosperm[2]. The most modern view is to regard the Podostemads as an old phylum lying near the Rosales and Sarraceniales[3].

The Crassulaceae, which presumably belong to the same plexus as the Podostemaceae, though typically xerophytic, include certain aquatic forms belonging to the genus *Tillaea* (*Bulliarda*).

Several families containing a few aquatic plants are to be found in the same cycle of affinity as the Caryophyllaceae; the plants in question are characterised by their inconspicuous flowers, which suggest reduction from a more highly developed type. *Montia fontana* (Portulacaceae), which occurs in Britain, generally lives submerged. In the heat of summer, however, the shoots often become exposed, but the thickish stem and leaves do not collapse in drought in the manner characteristic of submerged plants. The Portulacaceae include many succulent xerophytes, and it has been suggested that *Montia* is descended from ancestors of this type, and that, in spite of adopting the water life, it has retained—to its own advantage—certain xerophytic characters[4]. As a water plant descended from a xerophilous stock, it may perhaps be compared with *Tillaea aquatica*. The Elatinaceae, which show affinities

[1] Gardner, G. (1847). [2] Warming, E. (1888).
[3] Willis, J. C. (1902). [4] Focke, W. O. (1893[1]).

both with the Caryophyllaceae and Hypericineae[1], contain the British species *Elatine hexandra* and *E. hydropiper*—small submerged herbs with minute flowers. *Illecebrum* (Illecebraceae), again, is so near to the Caryophyllaceae, that it is perhaps best included in this family.

Polygonum amphibium is an example of an aquatic species belonging to a terrestrial genus and family (Polygonaceae). It is amphibious, but only reaches its optimum growth in water.

The affinities of the little family Callitrichaceae have been much disputed. Robert Brown[2], followed by Hooker[3] and Hegelmaier[4], included it in the Haloragaceae. But it is better related to the Euphorbiaceae; in this family itself, aquatics are not unknown, e.g. the *Salvinia*-like *Phyllanthus fluitans*[5]. Richard[6] was the first to compare *Callitriche* with *Mercurialis*, and more recent work on the relation of its reduced flowers to those of various Euphorbiaceae has rendered it highly probable that he was right[7].

The Lythraceae contain a number of marsh plants, such as *Lythrum Salicaria*, the Water Loosestrife, and also a certain proportion of true aquatics, such as *Peplis Portula*, with its inconspicuous flowers.

The Onagraceae include genera occupying very varying habitats; some, such as the Willow Herbs, contain typically terrestrial species, while *Ludwigia* and *Jussiaea* are aquatic. A closely related group, generally separated under the name of Haloragaceae, includes *Myriophyllum*, the Water Milfoil and *Trapa*, the Bull Nut; *Trapa* is however sometimes placed in a distinct family, the Hydrocaryaceae[8]. The most problematic genus associated with the Onagraceae is *Hippuris*. By some

[1] Cambessedes, J. (1829) and Müller, F. (1877).
[2] Brown, R. (1814) [3] Hooker, J. D. (1847).
[4] Hegelmaier, F. (1864). [5] Spruce, R. (1908).
[6] Richard, L. C. (1808). [7] Baillon, H. (1858) and Lebel, E. (1863).
[8] The distinctness of *Trapa* from the Onagraceae has recently been emphasized by Täckholm, G. (1914) and (1915) on the ground of its embryo-sac characters.

authors it has been placed in the Haloragaceae[1], but it is excluded by others, and a remote position near the Santalaceae has even been assigned to it[2]. The most reasonable view seems to be the non-committal one of Juel[3], whose investigations led him to believe that the position of the genus must still be treated as uncertain, since it is by no means even proved that it belongs to the Archichlamydeae. So it is best, provisionally, to relegate it to a separate family, the Hippuridaceae, possibly allied to the Haloragaceae. The geographical distribution of the two families, as Schindler[2] has pointed out, lends colour to the idea of their distinctness. He shows that the Haloragaceae (including the two tribes Haloragideae and Gunnereae) form an "antarctic" group of plants, a few of which by virtue of their special dispersal-capacity as aquatics, extend into the north temperate zone; while the Hippuridaceae, on the contrary, are an "arctic" family, confined to the Northern Hemisphere.

At different times in the last century, botanical writers have grouped the following genera in pairs as members of the same family—*Ceratophyllum* with *Callitriche*, *Callitriche* with *Myriophyllum*, and *Myriophyllum* with *Hippuris*—but more recent research has led to the belief that these four genera may even belong to four different Cohorts; this example indicates the degree to which homoplastic convergence may prevail among aquatics, and the confusion which it is apt to introduce into systematic botany.

The Umbelliferae are primarily terrestrial, but certain genera and species have, to a greater or less degree, taken to aquatic life. In some, e.g. *Oenanthe Phellandrium*, var. *fluviatilis*, the vegetative organs are completely submerged.

Among the Sympetalae, water plants are more scattered, and there is a notable absence of wholly hydrophytic families.

In the Primulaceae there is the single aquatic genus, *Hottonia*, with one European and one American representative.

The Gentianaceae are mainly terrestrial, but such marsh

[1] Parmentier, P. (1897). [2] Schindler, A. K. (1904).
[3] Juel, O. (1910) and (1911).

plants as *Menyanthes*, the Bog Bean, form a transition to the typically aquatic genus *Limnanthemum*.

The Scrophulariaceae include several hydrophytic genera; heterophyllous species are found in *Ambulia* (*Limnophila*) and *Hydrotriche*. In Britain the water Scrophulariaceae are represented by *Limosella aquatica*, a small plant whose corolla scarcely exceeds the calyx in length, while its capsule sometimes fails to dehisce; these features are no doubt symptoms of the reduction so often associated with aquatic life.

The Bladderworts (*Utricularia*), belonging to the Lentibulariaceae, contain a number of species which are aquatic, besides others which live on dry land. The terrestrial Plantagos, whose anemophilous flowers are generally regarded as reduced from the Scrophulariaceous type, form a transition to the aquatic genus *Littorella* in which floral reduction has reached a still higher pitch; the flowers are unisexual with a one-seeded indehiscent fruit. *Limnosipanea* is an example of a Rubiaceous hydrophyte.

Among the otherwise terrestrial Campanulaceae, we find the submerged *Lobelia Dortmanna*, while the Compositae include a few hydrophytes[1] such as *Bidens Beckii*—a heterophyllous water plant from North America—and *Cotula myriophylloides*.

Passing to the Monocotyledons—which the present writer regards as a phylum comparable with the Dicotyledons in being ultimately derived from ancestral forms of the dicotylar Ranalean plexus—we are at once struck with the relatively high number of aquatic families. The Helobieae (or Fluviales) consist of a series of families which are generally grouped together, chiefly on account of the enlarged hypocotyl of their embryo, which forms a remarkable common character. The aquatic and marsh families generally included in the Cohort are the Alismaceae, Butomaceae, Hydrocharitaceae, Aponogetonaceae, Juncaginaceae, Potamogetonaceae and Naiadaceae. The Alismaceae, which appear to be the most primitive of the group, show striking similarities to the Ranunculaceae, which they re-

[1] Hutchinson, J. (1916).

semble in polycarpy, polyandry and the insertion of the ovules[1]. The scattered arrangement of the ovules on the carpellary wall of *Butomus* and *Vallisneria* is similar to that observed in certain Nymphaeaceae, while coalescence and epigyny occur in both Hydrocharitaceae and Nymphaeaceae[2]. The view that these resemblances are not indicative of affinity, and that the development of a similar type of flower in the two families is mere coincidence[3], seems to the present writer to have little to support it, except the fact that the flowers of the Ranunculaceae are generally more or less acyclic, while those of the Alismaceae have the parts whorled. This argument scarcely seems to carry much weight, when it is recalled that certain genera which are undoubtedly members of the Ranunculaceae, e.g. *Aquilegia*, have flowers which are verticillate throughout.

The Helobieae as a whole appear to be more nearly related to the Spathiflorae (Araceae and Lemnaceae) than to any other Cohort of Monocotyledons[4]—the Aponogetonaceae forming, in some respects, a link between the two Cohorts. This family recalls the Araceae in its sympodial growth and tuberous stem, its laticiferous tissue and its flower spike with a fleshy axis. In the perforation of the leaves, *Aponogeton fenestralis* may be compared with *Monstera*[5]. But the Aponogetonaceae show, in addition, certain distinctively Helobian characters, which have led to their association with the Alismaceae, Juncaginaceae and Potamogetonaceae[6].

The Potamogetonaceae and Naiadaceae seem to form a coherent group, while their affinity with the other members of the Cohort is by no means a close one. The Potamogetonaceae share with the Hydrocharitaceae one curious little character, which may be of systematic importance, the occurrence, namely, of peculiar teeth at the edge of the leaf, formed from elongated cells

[1] Buchenau, F. (1903[1]). [2] Schaffner, J. H. (1904).
[3] Rendle, A. B. (1904).
[4] Hegelmaier, F. (1868) and Engler, A. (1892).
[5] Serguéeff, M. (1907).
[6] Planchon, J. E. (1844) and Krause, K. and Engler, A. (1906).

with thickened walls[1]. The Potamogetonaceae, like the Aponogetonaceae, show certain features which suggest the Araceae. *Zostera*, in particular, was actually included among the Aroids by de Jussieu[2], while, nearly a century later, Engler[3] suggested that the carpels and anthers of this genus might possibly each represent a male or female flower, the arrangement thus being comparable with that prevailing in the Aroid *Spathicarpa*.

The position of the Naiadaceae is obscure, owing to the difficulty of interpreting the extremely simple flower. Rendle, in his authoritative work on *Naias*, regards it as an "apparently primitive type of Monocotyledon[4]." Such a view is of course entirely irreconcilable with the belief that the Monocotyledons are derived from some early member of the Ranalean plexus, and that the primitive Angiospermous flower was of the 'Eu-anthostrobilus' type[5], with a petaloid perianth of numerous members, and numerous free stamens and carpels. On this view *Naias* must be interpreted as a highly reduced form, representing perhaps the ultimate term of reduction in the Potamogetonaceae series. The female flower consists of a single ovule, around which a carpellary wall and integuments grow up in a rather belated fashion. The flower is sometimes naked, but sometimes surrounded by a membranous bottle-shaped envelope. The male flower consists of a single stamen, enclosed in most cases in two such envelopes, but sometimes in one only. According to Rendle, the outer envelope of the male flower, and the corresponding envelope which occasionally invests the female flower, are of the nature of spathes, comparable with the spathes occurring in other submerged water plants, e.g. *Hydrilla*, and with the membranous cup enclosing the female flowers of *Zannichellia*. The probabilities are perhaps in favour of this interpretation, but it is more difficult to agree with Rendle's explanation of the *inner* envelope of the male flower, which he regards as a perianth. The present writer would like to

[1] Ascherson, P. and Graebner, P. (1907); see also p. 133.
[2] Jussieu, A. L. de (1789). [3] Engler, A. (1879).
[4] Rendle, A. B. (1899). [5] Arber, E. A. N. and Parkin, J. (1907).

suggest that this envelope—and possibly the 'spathes' also—may be, not foliar organs at all, but structures more closely comparable with such outgrowths from the floral axis as the membranous cup which surrounds the essential organs in the male and female flower of the Poplar, and, more remotely, with the arillus of the seed of the Yew-tree. The Potamogetonaceae are characterised throughout by the absence of a perianth; if *Naias* be descended from the Pondweed stock, any 'perianth' which it possesses must have been acquired *de novo* and hence it is highly improbable that any such organ which it might form would be morphologically a normal perianth[1]. In *Althenia*, the 'perigonium' of the male flowers and the scarious 'bracts' associated with the female flowers, and, in *Zannichellia*, the membranous cup surrounding the female flowers (*m.c.* in Fig. 45, p. 70) may also be mere cupules of no phylogenetic importance, but in the case of the female flowers of these genera, the possibility that we are dealing with spathes is not excluded. The variable occurrence of the floral envelopes in different sections of the genus *Naias*, harmonises well with the theory that they are recently acquired organs of no historical significance. On this view we are absolved from making the forlorn attempt to recognise in this genus the counterparts of all the organs which characterise the typical Angiospermic flower.

The Lemnaceae have long been regarded as connected with the Arum family. More than eighty years ago Schleiden[2] propounded the view that *Pistia* and *Lemna* both belong to the Aroideae and are related to one another. He showed that in *Pistia* the axis is abbreviated instead of being elongated as in most Aroids, and he regarded the River Lettuce as forming, in this respect, a transition to the Duckweeds[3]. Certain dissimilarities between *Lemna* and *Pistia* have, however, been emphasised by Koch[4].

The aquatic family Pontederiaceae (Farinosae) is somewhat

[1] This follows from the 'Law of Loss' which will be discussed in Chapter XXVIII.

[2] Schleiden, M. J. (1838[1]).

[3] See also Arber, A. (1919[4]); and p. 74.

[4] Koch, K. (1852).

remote in affinity from those hitherto considered, and is probably best interpreted as ultimately descended from the stock from which the Liliiflorae were also derived. Solms-Laubach[1] regarded the genus *Eichhornia* as of older origin than *Pontederia*, an opinion which accords well with the fact that *Eichhornia* has a trilocular ovary with numerous ovules, while in *Pontederia* the ovary is reduced to a single loculus containing one ovule. Among the Farinosae we also find another entirely aquatic family in the small group of the Mayacaceae.

No other families among the Monocotyledons are exclusively aquatic, but there remain certain cases of hydrophytic genera and species, occurring among families which consist otherwise of terrestrial or marsh plants. Examples from the British flora are *Scirpus lacustris* and *S. fluitans* (Cyperaceae), *Glyceria aquatica* and *G. fluitans* (Gramineae) and *Sparganium natans* (Sparganiaceae). The resemblance of *Sparganium* to the Pandanaceae is so great that we may perhaps regard *S. natans* as representing an aquatic off-shoot from the stock which also gave rise to the Screw Pines.

(2) THEORETICAL CONSIDERATIONS[2]

From the foregoing section of this chapter certain general conclusions may be deduced. The most obvious and striking feature is the relative paucity of hydrophytes in comparison with terrestrial plants. Contrasted with those that live on land, the number of aquatic families is so small as to be almost negligible, and even when all the individual hydrophytic genera and species are added, the sum total is relatively insignificant. This result is however hardly surprising when we consider that the Phanerogams are essentially a terrestrial stock and are distinguished from the Cryptogams by their aerial mode of pollination, which has won for them the freedom of the land. Under these circumstances, the reversion to aquatic life could hardly be expected to occur on any great scale. It must also be remem-

[1] Solms-Laubach, H. Graf zu (1883).

[2] This section of the present chapter is based on a recent paper by the writer in the *Journal of Botany*. See Arber, A. (1919[1]).

bered that the entire area of the fresh waters of the globe is very small as compared with the land surfaces, and that thus the aquatic Angiosperms occupy a much more restricted field than their terrestrial compeers.

The mode of systematic distribution of aquatics among the Angiosperms shows every possible variety. In the earlier part of this chapter we have pointed out that among the Dicotyledons there are cases in which one species of a terrestrial genus is aquatic (e.g. *Polygonum amphibium*), and others in which a number of species in a genus are hydrophytic while some are terrestrial (e.g. *Ranunculus* with its aquatic sub-genus *Batrachium*). Again, an entire genus of an otherwise terrestrial family may be aquatic (e.g. *Hottonia* among the Primulaceae) or several genera of a land family may be aquatic (e.g. *Jussiaea*, *Ludwigia*, etc. among the Onagraceae, and *Limosella*, *Hydrotriche*, etc. among the Scrophulariaceae). Finally, an entire family may be aquatic and contain no terrestrial forms (e.g. Podostemaceae). A family given over wholly to aquatic life may include a number of genera (e.g. Nymphaeaceae and Podostemaceae) or a single genus (e.g. Ceratophyllaceae and Callitrichaceae). Among the Monocotyledons, on the other hand, we meet with more cases of entire families leading a water life (e.g. Lemnaceae, Pontederiaceae, and various families belonging to the Helobieae), but there are fewer instances of individual aquatic genera and species belonging to families which are mainly terrestrial, though these occasionally occur (e.g. *Glyceria aquatica* of the Gramineae).

When one genus or species in an otherwise terrestrial family has taken to aquatic life, this may well be held to indicate that the habit is a recent one; but when a whole family, containing a number of genera, is found to be hydrophytic, it is hardly possible to avoid the conclusion that the differentiation of the genera has occurred since the adoption of the aquatic habit, which, on this view, must be very ancient. The only other alternative, namely that all the genera have been evolved in the course of terrestrial life, and that they have all subsequently and independently taken to the water, seems too far-fetched to be considered seriously. A scrutiny of the characters of those aquatic families which contain a number of highly individualised genera, confirms the notion that such families adopted aquatic life at a relatively early stage in the course of evolution of the Angiosperms. The Nymphaeaceae show characters that are markedly primitive among the Dicotyledons, and the Podostemaceae, though not standing so low in the scale of floral evolution, yet appear to be a very old phylum related to the Rosales and Sarraceniales. That is to say, the only Dicotyledonous families which are both exclusively aquatic and also contain a number of distinct genera, belong to the more primitive groups among the Polypetalae, and hence may be regarded as ancient lines which took to the water before they had diverged widely from the ancestral type.

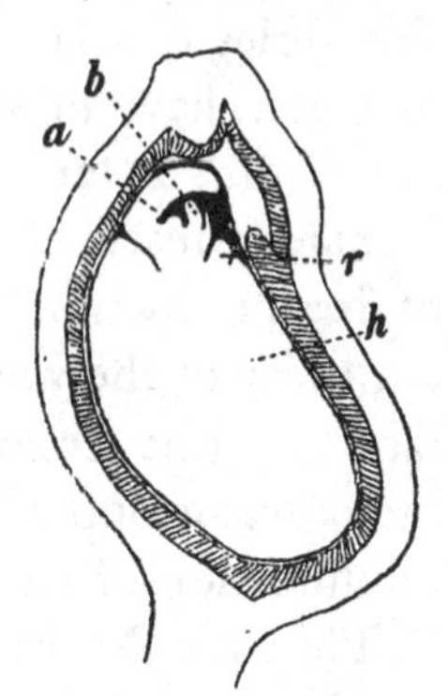

FIG. 166. *Ruppia brachypus*, J. Gay. Longitudinal section of fruit. (×15.) *a*, cotyledon; *b*, first leaf following cotyledon; *h*, hypocotyl; *r*, primary root. [Raunkiaer, C. (1896).]

Among the Helobieae, the Alismaceae are probably nearest to the ancestral stock. This family shows characters which are in many ways decidedly Ranalean, and which suggest that the Helobieae represent a branch that took to the water at a very early stage in the evolution of the Monocotyledons, while they still retained features recalling the Ranalean plexus from which they sprang. That they are descended from a geophytic ancestor is suggested by the characteristically abbreviated main axis, which in many cases does not elongate except to form the stalk of the inflorescence. It is also perhaps conceivable that the enlarged hypocotyl of the embryo (Fig. 166) recalls an ancestor which possessed a hypocotyledonary tuber, resembling that of *Eranthis hiemalis*, the chief difference being that in the Helobieae the storage of food in the hypocotyl has been shifted back to a pre-germination stage, owing perhaps to the exigencies of aquatic life[1]. It may be recalled, in this connexion, that tuberous

[1] See pp. 248, 249.

hypocotyls are common among Ranunculaceae with concrescent cotyledons, that is to say, among forms which supply indications of the characters of the original Monocotyledonous stock[1].

The idea that the Helobieae are descended from a very ancient group of Angiosperms, and have inhabited the water for a correspondingly long period, is ratified by the fact that this series consists of a whole plexus of related families, some of which have departed widely from the original type; it contains forms as far asunder, for instance, as *Alisma* with its many Ranalean features and *Naias* which represents the very acme of floral reduction. One minor piece of evidence favouring the antiquity of the water habit in the case of the Helobieae, is the fact that this Cohort includes all the marine Angiosperms—a biological group which probably originated through the further modification of fresh-water forms.

That the Nymphaeaceae and the related Ceratophyllaceae on the one hand, and the Helobieae on the other, have taken to aquatic life with such conspicuous success, suggests that the original Ranalean stock, from which they both sprang, may have been particularly well adapted to water life. In the Ranunculaceae the tendency to aquatic habits in the case of the genus *Ranunculus* is obvious; besides the definitely aquatic sub-genus *Batrachium*, the Buttercups include a number of forms, such as *R. sceleratus* and *R. Flammula*, which are capable both of land and water life. The singularly slight difference in general anatomy, between the terrestrial and aquatic species of *Ranunculus*, suggests that the land forms are of a type which does not require great changes of structure in order to succeed in water life.

It is a remarkable fact that the Sympetalae—the most highly evolved group of Angiosperms—has produced no entirely aquatic family, nor any single aquatic species which has become so far adapted to water life as to have acquired submerged hydrophilous pollination. The very large family of the Compositae, which may perhaps be classed as the ultimate term of the

[1] Sargant, E. (1903) and (1908).

Sympetalous series, contains apparently only four aquatic members[1]. Exactly the same is true of all the earlier Cohorts of Engler's Archichlamydeae, which, on the present writer's view, represent the more advanced and reduced forms of the Series. The families which are generally known as Polypetalae (the later Cohorts of Engler's Archichlamydeae), and which, on the view here adopted, include all the more primitive Dicotyledons, are markedly richer in aquatic types. It would hardly be going too far to say that independent aquatic families are chiefly characteristic of the Ranalean plexus, and of its derivatives—both Dicotyledonous and Monocotyledonous—while among the more advanced Polypetalae, and the Sympetalae, the sporadic occurrence of aquatic types and their close relation to terrestrial forms, indicate that the water-habit has been acquired comparatively recently. It is always possible that those individual genera and species among the Sympetalae which are hydrophytic at the present day, may each, in some future age, be represented by an entire aquatic family; for such groups as the Helobieae, Nymphaeaceae and Podostemaceae may owe their richness in genera and species partly to their ancient birth and to the length of time that has elapsed since they took to the water. But, on the other hand, a member of the Sympetalae embarking at the present day upon an aquatic career, may possibly be handicapped, as a potential ancestor, by the high degree of specialisation it has attained in its previous terrestrial life. The members of the primaeval Ranalean plexus may have possessed a greater plasticity in correlation with their lower degree of specialisation. It must also be remembered that the more primitive Angiosperms, which entered the water at an early period, had merely to take possession of a field undisputed by other seed plants, whereas species embarking on an aquatic life at the present day are exposed to acute competition from the numerous well-established hydrophytes with which the fresh waters of the world are already so fully stocked[2].

[1] Hutchinson, J. (1916).

[2] Since this chapter was written, I have learned that some of my conclusions were anticipated by Boulger, G. S. (1900).

CHAPTER XXVI

THE THEORY OF THE AQUATIC ORIGIN OF MONOCOTYLEDONS

THE high proportion of aquatic species among Monocotyledons, as compared with Dicotyledons, has been noticed in the preceding chapter. This, and other considerations, suggested to Professor Henslow his interesting theory of the aquatic origin of Monocotyledons[1], the broader aspects of which we may now briefly consider. He discusses the number of aquatic families to be found in each of the great groups, and concludes that only 4 per cent. of the Dicotyledonous families are aquatic, as compared with 33 per cent. of the Monocotyledonous. These figures probably have little *absolute* value—since it is difficult to decide, to begin with, exactly what we are to understand by the expression 'aquatic family'—but they serve a useful purpose in showing how much more numerous aquatics are among Monocotyledons than among Dicotyledons. This is indeed a matter of common observation. It is recorded[2] for instance, that in the case of the Bodensee, the plants living in the water or on the margin include forty Monocotyledons and thirty-eight Dicotyledons; this proportion is remarkable when we realise that the total number of species of Monocotyledons now existing on the face of the earth, bears to the total of Dicotyledons the ratio[3], very roughly, of 1 : 4·5. Henslow's general conclusion, with which most botanists will probably agree, is that marked numerical contrasts of this type "show that there is some decidedly important connexion between an

[1] Henslow, G. (1893) and (1911). It should be recalled that Gardiner, W. (1883) also regards Monocotyledons as essentially aquatic.

[2] Schröter, C. and Kirchner, O. (1902).

[3] The figures from which this ratio is deduced are taken from Coulter, J. M. and Chamberlain, C. J. (1904).

aquatic habit and endogenous structures." Further, Henslow points out that Monocotyledons and aquatic Dicotyledons have many characters in common, and he explains these resemblances, and the numerical preponderance of aquatic Monocotyledons, on the theory that Monocotyledons have arisen from a Dicotyledonous stock through "self-adaptation to an aquatic habit."

Henslow's theory has been criticised in some detail by Miss Sargant[1], who has shown that a large proportion of the characteristic features of Monocotyledons can be more readily interpreted on the supposition that the group was evolved through adaptation to a geophilous habit, than on the view that it was originally aquatic. The resemblances between aquatic plants and Monocotyledons are, on her view, largely due to the fact that both have suffered some reduction and degradation of structure, not necessarily arising from the same cause. It is true that the type of plant reconstructed by Miss Sargant, as representing the ancestral Monocotyledonous stock, would be, as she has pointed out, well adapted for subsequent aquatic life. Many aquatic Dicotyledons are formed more or less upon the geophilous plan, e.g. *Nymphaea* (Fig. 10, p. 25), *Castalia* (Fig. 11, p. 26), *Limnanthemum* (Fig. 22, p. 41), *Littorella* (Fig. 142, p. 218). It is worthy of note that, in an instance in which an aquatic flora—that of the Jura lakes—was analysed from this point of view[2], of the forty aquatic Phanerogams and Vascular Cryptogams recorded, thirty-one proved to have rhizomes.

The main lacuna in Henslow's theory appears to be that it treats the reduction of the cotyledons, from two to one, merely as a symptom of the general degeneracy of Monocotyledons, whereas Miss Sargant's theory of the geophilous origin of Monocotyledons offers a specific and convincing explanation of this peculiarity.

If Henslow's theory be not accepted, the onus rests upon his opponents of explaining the existence of so large a proportion of aquatic families within the Monocotyledons. Miss Sargant

[1] Sargant, E. (1908). [2] Magnin, A. (1893).

suggests as an explanation that Monocotyledons are on the whole a decadent race, of which some branches may have been driven to an aquatic habitat to escape the severer competition on land. She regards the existence of a large proportion of *small* families among the Monocotyledons as suggesting that the modern members of the group are survivals from a period when they were more numerous and widely spread, and she supposes that they have chiefly maintained themselves in such situations as fresh-water areas, in which competition is less keen than under more genial conditions. This view is obviously bound up with the assumption that the adoption of an aquatic life is a device by which a poorly equipped species may escape from the competition of its more favoured compeers[1], saving itself from extinction by retirement into a quiet back-water of existence. In other words, water life is regarded as a refuge for the destitute among plants. The present writer, having begun the study of aquatics ten years ago with a full conviction of the truth of this picturesque theory, has gradually and reluctantly been forced to the conclusion that there is no sound evidence in its favour. On the hypothesis in question, water plants are more or less comparable with the remnant of a defeated race among mankind, which preserves its existence by retreating into some forbidding and inaccessible region, into which its conquerors have little temptation to pursue it. But this analogy is probably quite misleading; it would perhaps be more illuminating to compare water plants with the pioneers who are to be found leading hard and difficult lives in barbaric regions on the frontiers of civilisation—not forced thereto by failure to 'make good' in the excessive competition prevailing in regions more anciently inhabited, but impelled to the frontiersman's life by a natural, inborn affinity for the adventurous conditions which it offers. In the same way, water plants appear to the present writer to have adopted this mode of life, not as a last resource, but *because it happened to suit their particular constitution and*

[1] Darwin, C. (1859), Goebel, K. (1891–1893), Hutchinson, J. (1916), etc.

character. There is little doubt that, after they had once entered upon an aquatic career, they must have evolved along lines which gradually harmonised them more and more completely with their surroundings, but the initial step or steps, which led to the adoption of the water-habit, must have been due to an innate affinity for the environment, rather than to the negative quality of incapacity for success in terrestrial life; to pursue our metaphor—the man, who fails in the struggle for existence at home, is not of the type that makes the successful colonist.

West's[1] critical study of the vegetation of certain Scottish lakes, led him to a similar conclusion, which is best expressed in his own words. "It seems to me," he writes, "that aquatic plants have not always had their origin from terrestrial forms that had been forced into the water by more robust competitors on the land, as is sometimes stated, but, more probably, because certain suitable forms have exhibited a tendency, as some do even now, to take on the aquatic habit, that mode of living being more agreeable to their requirements,...never have I observed the case of a plant being forced into the water by a stronger competitor."

If the preponderance of aquatic families among Monocotyledons is neither to be explained as due to the aquatic origin of the Class, nor to the part played by the waters in offering a harbour of refuge to a decadent and unsuccessful race, it remains to be seen whether any other interpretation can be offered. In scrutinising more closely the numerical preponderance of aquatic Monocotyledonous families, it becomes obvious that this does not depend so much upon the constitution of the group in general, as upon the existence of the very large and highly differentiated Cohort of the Helobieae. Apart from the Helobieae, there is no particular disparity between the proportion of aquatic families in the two Classes, and, if the number of species is to be taken into consideration, the theory that there is a decided aquatic tendency among the Monocotyledons becomes hard to maintain. It has been pointed out[2], for instance,

[1] West, G. (1910). [2] Coulter, J. M. and Chamberlain, C. J. (1904).

that the half-dozen purely hydrophytic families of Monocotyledons, though they have a world-wide distribution, contain altogether less than two hundred species, whereas the four great world-wide terrestrial families—Gramineae, Cyperaceae, Liliaceae and Iridaceae—contain ten thousand species.

As we attempted to show in Chapter xxv, the Helobieae carry every indication of being an ancient group which took to the water very early in the history of the Monocotyledons, and in which the existence of the macropodous embryo has possibly played a considerable part in favouring aquatic life. The Cohort seems in the main monophyletic, though it is conceivable that certain families, therein included, are really offshoots from other Cohorts, which have come by secondary modification to resemble the true Helobieae.

The two factors that have led to the great development of the Helobieae, and hence to the prevailing impression that there is a strong aquatic tendency among Monocotyledons in general, may be held to be—firstly, the long period which has elapsed since the ancestral stock of the Cohort became aquatic[1], thus allowing time for its differentiation into a wide variety of forms—and secondly, the fortunate provision of an embryo with its food stored in the swollen hypocotyl, which has possibly been one of the chief instruments in determining the remarkable success of the group in aquatic life[2].

[1] See pp. 319, 320.

[2] See pp. 248, 249.

CHAPTER XXVII

WATER PLANTS AND THE THEORY OF NATURAL SELECTION, WITH SPECIAL REFERENCE TO THE PODOSTEMACEAE[1]

FROM a study of the Podostemaceae, Dr Willis[2] has arrived at certain views as to their evolution which, if accepted, have a peculiarly wide bearing. The great variety and anomalous character of the features exhibited by this family have been touched upon in Chapter IX. There is little doubt that these plants have been derived from some terrestrial group, since the structure of the flower and fruit is typically adapted to land life. Willis suggests that a possible origin for the family is from plants already growing on the banks of mountain streams, with creeping adventitious roots, upon which secondary shoots were regularly developed; these secondary shoots might provide the opportunity for an entrance into aquatic life. Most of the peculiarities of the group, as Willis points out, can be traced to the remarkable plasticity of the skeletonless root, and to the parallel dorsiventrality of the vegetative organs and flowers. This dorsiventrality is associated with "their plagiotropic method of growth, forced upon them by the fact that they live only upon an unyielding substratum; they have not, and can never have had, primary roots going downwards into the rock, and are thus, one might almost say, cut in half, or deprived of one-half of their polarity[3]." "No other family above the liverworts shows so marked and far-reaching a dorsiventrality in organisation[4]." The dorsiventrality of the flowers, Willis

[1] For the sake of brevity the term Podostemaceae will be used in this chapter in the old sense, to include both the Podostemaceae proper, and the closely related Tristichaceae.

[2] Willis, J. C. (1902), (1914[1]), and (1915[2]).

[3] Willis, J. C. (1914[1]).

[4] Willis, J. C. (1902).

regards as forced upon them, so to speak, by that of the vegetative organs, "without any reference to advantages or disadvantages to be derived from it in the performance of the functions of the floral organs themselves[1]." He believes that the dorsiventrality was first impressed upon the vegetative organs, whence it spread, as it were, to the reproductive regions, affecting the bracts, spathe and flower. The stamens most commonly exhibit it, but, in the cases in which it is carried furthest, the gynaeceum also conforms to it. The zygomorphy of the flower develops concurrently with a tendency towards anemophily and autogamy, whereas in most families it is associated with adaptation to entomophily. Willis looks upon the zygomorphy of the more specialised Podostemaceae as a character without survival-value, which thus cannot owe its presence to Natural Selection, but which originates as an inevitable corollary to the dorsiventrality of the vegetative organs. In fact, he even goes so far as to regard the zygomorphy of the flower as a positive disadvantage, whose influence the plant seems to attempt to neutralise. "However dorsiventral the flower becomes it still stands erect as long as it possesses a stalk, and when at last we come to the forms without the stalk we find the flower curving its ovary and stamens so as to get them as erect as possible. It seems as if the flower were, so to speak, struggling against the dorsiventrality to the last[1]."

The aspect of the zygomorphy of the Podostemad flower upon which Willis dwells with the greatest emphasis, is its apparent uselessness. This is one of the points which he brings forward to show that, though the family as a whole is probably more completely transformed than almost any other from the average mesophytic type, the great variety in morphological structure presented by the individual members cannot be explained as due to adaptation to their individual surroundings. For, though the family has become differentiated into at least thirty genera and one hundred species of the most varied morphological structure, the conditions under which they live

[1] Willis, J. C. (1902).

are uniform in the extreme. "By no stretch of imagination can the variety in the conditions of life be made to fit one quarter of the variety of structure[1]." Even the dorsiventrality, which is obviously associated with the mode of growth, must not, according to Willis, be interpreted as an advantageous adaptation, for he points out that the least modified species, in which dorsiventrality hardly occurs at all, can and do live in nearly all the places occupied by the family. As a conspicuous example of the lack of adaptation among these plants, Willis[1] instances the fact that, in the great majority of species, there is no device to enable the seeds to cling to the rocks upon which they find themselves shed; he thinks it probable that it often takes from five hundred to one thousand seeds to produce three or four seedlings.

The Podostemaceae thus exhibit great variety and marked specific differentiation, but the features in which the genera and the species differ from one another cannot, according to Willis, be explained as adaptational. Further, the particular situations in which they thrive are such as almost to preclude competition with other plant forms, and there is also relatively little struggle for existence even between members of the same species. On these grounds Willis concludes that the evolution of the group cannot be explained as due to the natural selection of infinitesimal variations.

In scrutinising Willis's criticism of selectionist views, no progress can be made unless Natural Selection be analysed in accordance with the two distinct claims which have been made on its behalf—firstly, that it is the cause of the origin of species, and secondly, that it is one of the factors conditioning adaptation. Unfortunately the distinctness of these two functions is not clearly recognised in Darwin's own work, and the confusion thus initiated has given rise to much obscurity in later writings. Willis's observations certainly strike a severe blow at Natural Selection considered from the first point of view, i.e. as *the originator of specific types*. In the Podostemaceae we

[1] Willis, J. C. (1914[1]).

undoubtedly have a case in which Natural Selection can scarcely be a factor of any great importance, and yet there is a quite extraordinary variety of specific forms, many of which are confined to extremely limited areas.

That specific forms may be markedly definite and distinct, and that yet the differences between them may be such that it is scarcely possible to imagine that they have any special survival value, is also indicated in the case of a number of aquatics outside the Podostemaceae. Water plants in general have the character of being Protean, and there is undoubtedly great individual variability associated with varying conditions of life, but, at the same time, the opinions of those best qualified to judge, tend to the conviction that there is great fixity rather than plasticity of specific characters. It is probable that the general impression as to the specific variability of aquatics is partly attributable to the fact that, owing to the prevalence of vegetative reproduction, local races readily come into being, since any variation may be perpetuated by this means for a considerable time. But there is no reason to suppose that such local races would come true from seed. In the case of the Eu-callitriches, great variation may often be observed in the form of the leaves and the size of the floating rosette. Little groups of plants growing together often conform to one type in these respects; but it is probable that such homogeneous groups are merely the vegetative progeny of one individual. The Potamogetons are proverbially variable, and their specific identification presents almost insuperable difficulties to the tyro, yet a great authority on this group was led, by a critical study of some of these puzzling forms, to write: "All I have observed during the past summer induces me to believe that, at the present time, each form of the *lucens* group is so far constant that seed of each form produces its like. Their imitation of one another under variation, induced by abnormal circumstances, may betray a comparatively recent common origin, but at the present day our fenland pondweeds certainly seem to be 'fixed quantities[1].'"

[1] Fryer, A. (1887).

The idea that the specific distinctions among the Potamogetons are somewhat fluid, may be partly attributed to a too exclusive use of external features in systematic work; there is no logical reason for the exclusion of anatomical characters from taxonomic study and their importance is fortunately now becoming recognised[1]. It has been demonstrated, for instance, that, though the flower and fruit characters of the Potamogetons show very small differences in the different species, and the external characters of the vegetative parts which can be used in diagnosis are few and variable, the anatomical characters of the vegetative organs prove to be much more constant[2]. That the majority of specific differences observed among the Pondweeds could be of any survival-value, seems almost incomprehensible, and the lack of any apparent utility in certain specific characters is seen almost more clearly when we turn to the marine Potamogetonaceae. In the case of these plants, the anatomy of the leaves, taken by itself, furnishes data for exact specific determination[3]. Dealing with *Cymodocea* and *Halodule*, Sauvageau[4] remarks, "It is an interesting fact that plants which in general are of relatively simple structure, present such a variation from one species to another, and, at the same time, such constancy in specific anatomical characters." It can scarcely be imagined that the majority of the specific differences, observed in the anatomy of the vegetative organs of the marine Potamogetonaceae, can serve any purpose in connexion with the relatively uniform conditions of their submerged life, and, unless these differences are advantageous, it is impossible to suppose that they are due to Natural Selection. It is most remarkable that in so simple a genus as *Naias*, in which some, at least, of the external specific differences can hardly, by any stretch of imagination, be supposed to fit their possessors in

[1] The excellent method advocated by R. C. McLean (*New Phyt.* Vol. XV, 1916, p. 103) for rendering herbarium material available for anatomical work, makes the use of internal characters in systematic study more practicable than hitherto.

[2] Raunkiaer, C. (1903).

[3] Sauvageau, C. (1891[1]); see also p. 131.

[4] Sauvageau, C. (1891[3]).

any special way for their environment, these specific or varietal characters are exceedingly constant. *Naias graminea*, var. *Delilei*, for example, has been known in Egypt to have the same characters for about a century, and when introduced into England these characters remain wholly unchanged[1]. These considerations seem to the present writer to confirm the conclusion drawn by Willis from his study of the Podostemaceae—a conclusion which has also been arrived at by various workers in other fields—that Natural Selection is incompetent to explain the origin of the sharply defined entities which we call species.

But when we turn to Natural Selection in its second aspect—as one of the various factors to which adaptation may be due—Willis's conclusions seem to need some revision. Accepting the view that we have, among the Podostemaceae, a case of evolution untrammelled by the limiting influence of Natural Selection, we find associated with this freedom, the development of a large number of well-defined species, remarkable for their lack of definite adaptation to the conditions of their life. The view may well be taken that the lack of adaptation which Willis finds so striking, is actually in part *attributable* to the absence of competition and hence to the elimination of Natural Selection. From this point of view, the Podostemaceae furnish evidence—negative but forcible—for the importance of Natural Selection in the development of adaptation, since here we have a case of the *absence* of Natural Selection correlated with the *absence* of special adaptations. Among the Podostemads, presumably, all variations—good, bad, or indifferent—have had an almost equal chance of perpetuation, provided they did not interfere with those general features which gave the group its special capacity for growth in the rapidly running water, which is so inimical to most forms of plant life. Perhaps the present condition of the Podostemaceae may be broadly compared with that of certain of our domestic animals, consisting at the present day of many sharply defined breeds, which could not have survived the

[1] Magnus, P. (1883).

stringent ordeal of Natural Selection, to which they would have been subjected in the feral state.

But Natural Selection is, after all, merely a negative force. That in the struggle for existence the less fit go to the wall, is a truism which all must admit; but, curiously enough, we do not seem to possess many records among aquatics of this process having been observed in actual operation. It has been noticed, however, that, in the lake district of Pico in the Azores, *Potamogeton polygonifolius* is playing the part of an aggressive species and is ousting such plants as *Littorella* and *Isoetes* from the ponds[1]. Possibly the chief work of Natural Selection consists in sorting out species into the environments most suited for them; it has, for instance, suffered plants which can tolerate aquatic conditions to embrace that mode of life, while annihilating any others, with a constitution unfavourable for the purpose, which may also have attempted it. In the same way a Labour Exchange may distribute men into appropriate situations, and may also be responsible for the elimination of the unfit, by setting some of them to tasks not within their capacity, but yet it has no claim to be the *originator* of any skill which they display in their respective crafts.

If we can no longer whole-heartedly accept the facile Darwinian explanation, we must be content to confess that adaptation remains one of the outstanding mysteries of biology. It seems impossible to arrive at any glimmer of a comprehension of its nature, without accepting, in some form, the notion of the inheritance of acquired characters, with which the inheritance of unconscious memory is probably bound up. Many biologists to-day seem disposed, at the best, to regard the inheritance of acquired characters as both unproven and improbable, but it seems to the present writer to be an almost inevitable article of belief, if it is understood *in a broad and general sense*. Whether the offspring of a mutilated Guinea-pig derives abnormal characters from its injured parent, is quite beside the point. If we suppose that the whole organic world

[1] Guppy, H. B. (1917).

has arisen from a single primaeval form of life, those complex powers of reaction to the environment, and the structures subserving them, which distinguish man from the primordial speck of protoplasm, must, in a broad sense, be regarded as 'acquired characters,' and, unless such characters were heritable, we should not have advanced to-day beyond the unicellular stage. This contention remains true even if we accept the suggestion, which Bateson has recently made, that the course of evolution may conceivably be represented by "an unpacking of an original complex which contained within itself the whole range of diversity which living things present." This striking, and at first sight paradoxical, notion contains fundamentally nothing new. Erasmus Darwin, for instance, who believed in the origin of the whole animal and vegetable world from "one living filament, which the great First Cause endued with animality," must have had in mind, as the essential attribute of this primordial living stuff, its inherent potentiality of development. Every evolutionist must suppose that, as the descendants of the primaeval speck of protoplasm multiplied and advanced along diverse lines of development, what they gained in specialisation they lost in plasticity. In other words, while the original living matter contained within itself the power of development in the direction of any and every class of the organic world as we now know it, one of its descendants which has gone far along the path towards becoming, let us say, an Angiosperm or a Rodent, has only done so by closing the gates upon itself in countless other directions: it no longer retains the power of developing, for instance, into a Bryophyte or a Bird. In this sense all evolution is accompanied by a succession of losses, and the highly evolved descendant of the original "living filament" pays the price of its specialisation in losing the power to develop in countless other directions. A human analogy, though obviously imperfect, may perhaps make this point clearer. The future of a new-born infant presents a wide variety of possibilities. In one case, for instance, he may contain within himself the powers—at this stage, necessarily, in a latent form—for ultimately becoming, let us say, an artist, a bishop, or a stock-broker. But we know that if he achieves any one of these aims in later life, it will almost inevitably be at the expense of the power to arrive at the other two. If, in the course of his ontogeny, the stock-broker triumphs, we may regard him as built up upon the ashes of the potential bishop and artist. The man, though superior to the baby in actual achievement, is inferior to it in the qualities which may be summed up in the word "promise," just as the Angiosperm, though its degree of differentiation so greatly exceeds that of the primordial protoplasmic speck, is inferior to it when judged by its power to produce descendants of widely varying types.

CHAPTER XXVIII

WATER PLANTS AND THE 'LAW OF LOSS' IN EVOLUTION[1]

IT is a well-known fact—indeed almost a truism—that structural reduction is one of the most marked characteristics of water plants. In the preceding chapters of this book we have alluded to numerous cases in which aquatics are reduced, both in their vegetative and reproductive organs, as compared with their terrestrial relations. The consideration of this reduction, and of some of its sequels, led the present writer to formulate, under the name of the 'Law of Loss,' a certain minor principle which seems to be operative in various phases of plant evolution. The expression 'Law of Loss' is meant to indicate the general rule that a structure or organ once lost in the course of phylogeny can never be regained; if the organism subsequently has occasion to replace it, it cannot be reproduced, but must be constructed afresh in some different mode.

In the very nature of the case, such a law is not susceptible of formal proof. We can only here consider whether, if accepted as a working hypothesis, it throws any light on structural features observed among water plants, which would otherwise be obscure. We may begin with a case which happened to be the first to arrest the present writer's attention in connexion with the 'Law of Loss,' and to which allusion has already been made[2]. *Ceratophyllum demersum* and certain species of *Utricularia* are entirely rootless at all stages of their life-history—even the primary root of the seedling being either altogether absent

[1] The greater part of this chapter has already appeared in two papers by the present writer, Arber, A. (1918) and (1919[2]), to which reference can be made for a fuller treatment of the subject. The 'Law of Loss' is closely related to Dollo's 'Law of Irreversibility.'

[2] See pp. 88, 89, 96, 97.

or remaining quite rudimentary. These plants are undoubtedly descended from ancestors of the normal Angiospermic type, characterised by possessing a root system; but they have themselves entirely lost the ancestral capacity for producing roots. Nevertheless, in both these unrelated genera, the need for an absorbing organ seems to have re-asserted itself, and to have been met, not by the re-establishment of root-formation, but by the development of special subterranean shoots which—though not of the morphological nature of roots—perform root-like functions. This behaviour, in the case of *Ceratophyllum* and *Utricularia*, may be interpreted to mean that a plant which has entirely given up root-formation and afterwards again experiences the need of roots, cannot re-acquire them, but can only press some existing organ into the service, modifying it as best it may. It is possible that the root-like water leaves of *Salvinia* indicate a similar history.

As another instance of the working of the 'Law of Loss,' we may take the phylogenetic history of the leaves of the Alismaceae, Pontederiaceae, or Potamogetonaceae—or indeed of any other Monocotyledons which possess 'laminae.' But whether or no this illustration be accepted, depends upon the standpoint adopted regarding the general morphology of the leaves of Monocotyledons. The typical Monocotyledonous leaf is of simple, linear to ovate form, with a sheathing base and parallel veins; how is such a leaf to be compared with that of a Dicotyledon, consisting, in its fullest expression, of leaf-base and stipules, petiole, and net-veined lamina? This question has naturally attracted the attention of morphologists, and an interpretation, which has become known as the 'phyllode theory' was first put forward with some reservations by de Candolle[1], not much less than a century ago. According to this view, the typical Monocotyledonous leaf does not correspond to the complete Dicotyledonous leaf, with its leaf-base and stipules, petiole and lamina, but is merely the equivalent of a petiole with a sheathing base. It seems to the present writer probable

[1] Candolle, A. P. de (1827).

that in many cases reduction may have gone yet further, so that the leaf-base alone is represented.

The phyllode theory is supported by the existence of a number of examples among Dicotyledons, in which organs not dissimilar to typical Monocotyledonous leaves can be shown to be equivalent either to leaf-bases, or to leaf-bases and petioles. Such cases are numerous and familiar—those in which the reduced leaves correspond to leaf-bases alone, being decidedly the commoner. In *Cabomba caroliniana*[1], to take an instance from among aquatics, two or three pairs of *lanceolate simplified leaves with no laminae* are followed by transitional forms in which a lamina occurs but is much reduced. These are succeeded by the normal submerged leaves with finely divided laminae.

It is a commonplace of every text-book that one of the most distinctive features of Monocotyledons is the parallel venation of the leaves. But no theory hitherto propounded regarding the origin of Monocotyledons has offered any satisfactory explanation of this well-marked character of the Class. To the present writer it appears that one of the chief merits of de Candolle's theory is that it explains the parallel venation of Monocotyledonous leaves in a perfectly unstrained way. For some form of parallel veining is one of the most obvious characters of Dicotyledonous leaf-bases, petioles and phyllodes. Hence, on de Candolle's theory, the venation of the Monocotyledonous leaf ceases to present any problem; it shows precisely those characters which might have been anticipated from the morphological nature of the organ.

So far we have only considered those Monocotyledonous leaves in which no lamina is differentiated, but we must now return to the question with which we started—what are the homologies of the lamina in the Alismaceae and other families with a corresponding foliar morphology? If the Monocotyledons are monophyletic, two explanations are open to us; it is either a revival of the lamina as it occurs among the Dicotyledons, or an organ which has arisen *de novo* as a modification

[1] Raciborski, M. (1894[2]).

of the distal region of a pre-existing phyllode. In deciding between these two alternatives, the Law of Loss comes to our assistance. On this law, the blade *once lost cannot be regained*, and it is therefore clear that the 'lamina' of the Monocotyledon is, as Henslow[1] has suggested, an expansion of the petiole, imitative of, but not identical with, the blade of a Dicotyledon: the present writer proposes to distinguish such a blade as a 'pseudo-lamina.' This interpretation certainly accords well

FIG. 167. *Potamogeton lucens*, L. Apical part of a shoot showing range of leaf form. (Reduced.) [Raunkiaer, C. (1896).]

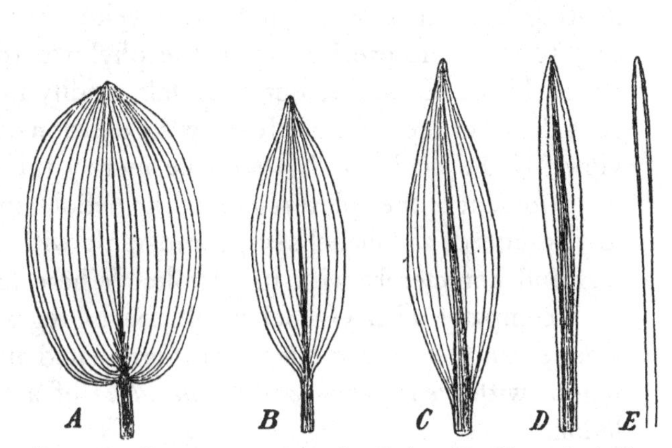

FIG. 168. *Potamogeton natans*, L. Series of leaf forms including *A*, the normal floating 'lamina.' (*A* and *B*, reduced; *C–E*, nat. size.) [Raunkiaer, C. (1896).]

with the venation of many Monocotyledonous leaves. The transitional leaf forms produced in *Sagittaria* between the band and arrow-shaped types (Fig. 5, p. 14) have all the appearance of merely representing different degrees of expansion of the upper region of the petiole, with correspondingly varying degrees of outward curvature and apical detachment of the veins. A somewhat similar series can be traced in certain Potamogetons (Figs. 167 and 168). These series afford an

[1] Henslow, G. (1911).

illustration of the way in which the development of the 'pseudo-lamina' may have occurred in the course of phyletic history.

The phyllode theory has met with lively opposition at the hands of Goebel[1]. He discusses the question chiefly in connexion with *Sagittaria*, and takes the view that the band-like submerged leaves of this plant are not *reduced* leaves in which the lamina has disappeared, but *rudimentary* leaves in which no differentiation of blade from petiole has occurred. He supports this view by recalling that, in the ontogeny of the individual arrow-head leaf, stages are passed through corresponding, firstly, to the band-shaped submerged leaf, and secondly to the oval floating leaf. It is true that these developmental facts are not easy to reconcile precisely with the phyllode theory as enunciated by de Candolle, but they fall readily into place when considered in the light of Henslow's extension of de Candolle's view. If the blade of *Sagittaria* be merely the expansion and development of the apical region of the petiole, the band-shaped leaf is indeed, as Goebel says, comparable with a complete air-leaf and not merely with its petiole. Where Henslow would part company with Goebel would be in regarding both the simple band-leaf and the highly differentiated air-leaf as homologous with *the leaf-base and petiole alone* of a typical Dicotyledon.

The present writer had felt for many years that it ought to be possible to apply anatomical evidence to the phyllode theory, and at length a path leading in this direction was disclosed. Solereder[2], in the course of a general anatomical study of the Hydrocharitaceae, reported the discovery of vascular bundles of inverted orientation in the leaves of various members of the family (Fig. 28, p. 46). He compared the structure thus revealed to that of petioles, *Acacia* phyllodes and *Iris* leaves, but he did not, apparently, attach any theoretical importance to it. It seemed, however, to the present writer that these inverted bundles were an indication of the phyllodic nature of the leaves in question. In the light of this idea, a general examination of

[1] Goebel, K. (1891–1893). [2] Solereder, H. (1913).

the leaves of Monocotyledons was undertaken, with the result that 'phyllodic' anatomy was found to occur frequently in this Class. In many cases the existence of this type of structure had already been recognised, but it had not been interpreted as 'phyllodic.' In other instances the existence of inverted bundles had apparently been overlooked. This was the case in the Pontederiaceae—an aquatic family belonging to the Farinosae; it may therefore be worth while to describe the leaf structure of this group in some detail. The leaves, as a rule, have a sheathing leaf-base, a petiole, which is sometimes much

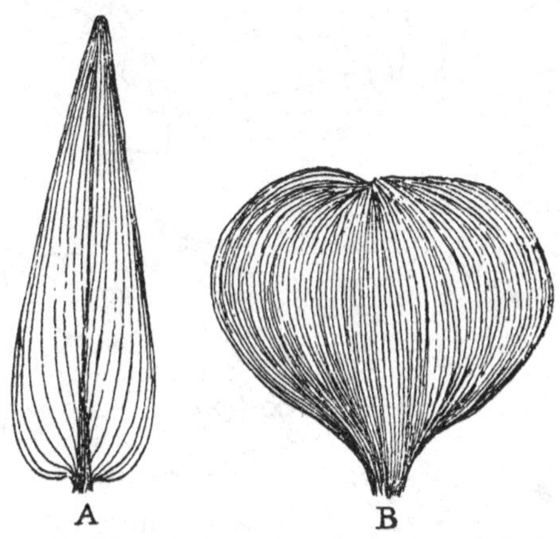

FIG. 169. *A*, 'lamina' of *Pontederia cordata*, L.; *B*, small 'lamina' of *Eichhornia speciosa*, Kunth. (Reduced.) [Arber, A. (1918).]

swollen, and a 'lamina.' In external appearance and venation the leaves of *Pontederia* (Fig. 169 *A*) and *Eichhornia* (Fig. 169 *B*) distinctly suggest that the 'laminae' are produced by expansion of the apical region of the petiole, and that they are thus 'pseudo-laminae' and not equivalent to the blades of Dicotyledonous leaves. The anatomy confirms this idea in a striking fashion. Fig. 170 *D*, p. 342, shows the transverse section of a petiole of *Pontederia cordata*, L. with inverted bundles towards the upper side. When the 'lamina' is cut transversely, its structure is found to be exactly such as might have been anticipated on the theory that it is produced by extreme

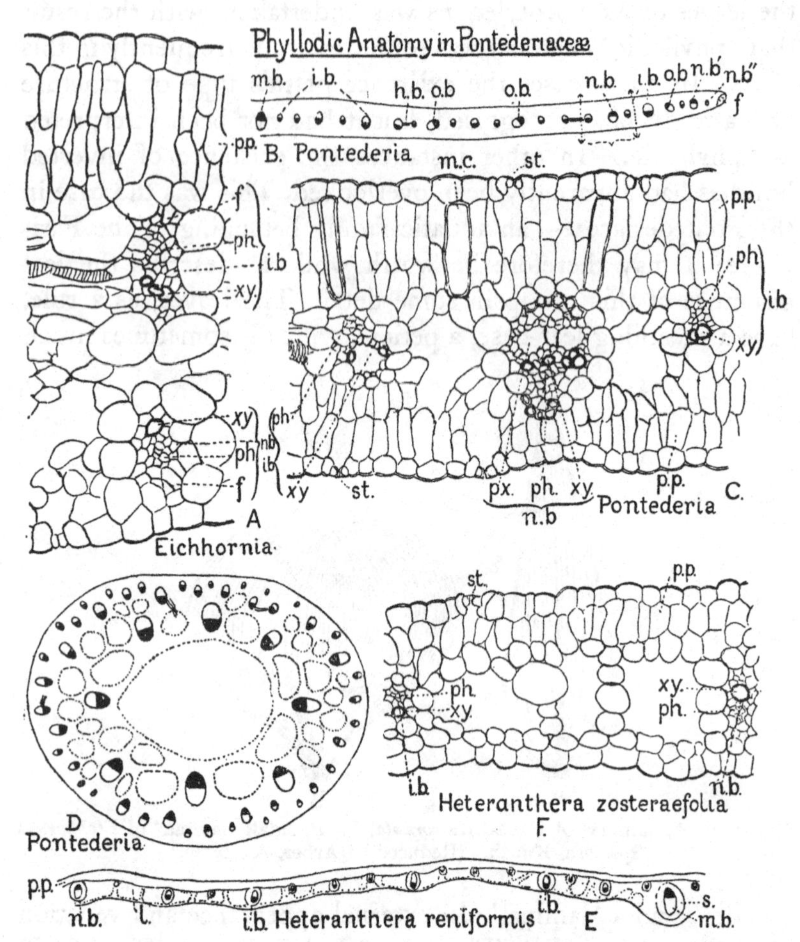

FIG. 170. Leaf anatomy of Pontederiaceae. *A*, *Eichhornia speciosa*, Kunth, T.S. lateral vein of 'lamina.' One small normal bundle (*n.b.*). One larger inverted bundle (*i.b.*) higher in leaf is giving off a branch, also inverted. *B*, *Pontederia cordata*, L., half T.S. lamina near apex. All bundles inverted (*i.b.*) or oblique (*o.b.*) except the median bundle (*m.b.*) and the three bundles *n.b.*, *n.b.'*, and *n.b.''*. Fibres (*f*) at margin; *h.b.* = horizontal branch. *C*, *Pontederia cordata*, L., the part of the T.S. shown in *B* which is included between the dotted arrows. One normal bundle (*n.b.*) and two inverted bundles (*i.b.*), one with an inverted branch. *m.c.* = cells containing a secretion, probably myriophyllin. *D*, *Pontederia cordata*, L., T.S. petiole near its upper end, outlines of lacunae dotted. *E*, *Heteranthera reniformis*, Ruiz. and Pav., part of T.S. of lamina, including midrib (*m.b.*). All the bundles shown are inverted, except the midrib and main lateral. *F*, *Heteranthera zosteraefolia*, Mart., T.S. part of ribbon-leaf to show one normal and one inverted bundle [Arber, A. (1918).]

flattening and expansion of the petiole in the horizontal plane (Fig. 170 *B* and *C*). For, instead of the normal arrangement of bundles, all orientated with the xylem upwards, which we are accustomed to find in laminae, the vascular strands in this case, though in a single series, are orientated, some normally (*n.b.*), including the median bundle (*m.b.*), the majority inversely (*i.b.*), and a few obliquely placed (*o.b.*). A small part of the transverse section is shown in greater detail in Fig. 170 *C*. In this drawing, the central and largest bundle is seen to be normally orientated, but the bundles on either side of it have the xylem below and phloem above.

In the heart-shaped 'lamina' of *Heteranthera reniformis*, Ruiz. and Pav., a very similar bundle arrangement is found (Fig. 170 *E*). Here, only the midrib and main laterals are normally placed, the remaining bundles being inverted.

The 'lamina' of *Eichhornia speciosa*, Kunth (Fig. 170 *A*) differs from that of the other members of the family here considered, in its much greater thickness. Inverted bundles occur, not only in the thick basal region in which the transition from petiole to 'lamina' takes place quite gradually, but also near the margin. Here, there is only a single series of vascular strands, among which inversely orientated bundles are very numerous. Some of the lateral veins in the 'lamina' consist of a single, normally orientated bundle, while others include a pair of bundles, one normal and one inverted.

Among the Pontederiaceae, we not only find leaves, such as those just described, in which there is a differentiation between petiole and 'lamina,' but others, which are ribbon-like, with no distinction of blade and stalk. For comparison with the more highly differentiated leaves, sections were cut of the ribbon-leaf of *Heteranthera zosteraefolia*, Mart. Here the midrib and main laterals proved to be normal, but the others—i.e. the majority of the laterals—were inverted. Fig. 170 *F* shows two adjacent bundles orientated in opposite ways. The structure of this ribbon-leaf is closely similar to that of the 'lamina' in *H. reniformis*.

It may be worth noting that a peculiar submerged member of this family, *Hydrothrix Gardneri*, Hook. f., described by Goebel[1], has leaves with a sheathing base and hair-like upper region, whose external morphology distinctly suggests a phyllodic origin. In this case anatomical evidence cannot be sought, since the extremely slender leaves are said to be traversed by a single bundle only.

The presence of inverted bundles in all species of Pontederiaceae of which material has been available to the present writer, is a remarkable anomaly which calls for some explanation. It is difficult to see how such a structural peculiarity can be explained as an adaptation, since it is common to leaves otherwise differing notably in type and mode of life. It is equally conspicuous in the very delicate ribbon-leaf of *Heteranthera zosteraefolia* and in the well-defined, thick 'lamina' of *Eichhornia speciosa*; it occurs both in *Heteranthera reniformis*, in which palisade parenchyma is confined to the upper side and in *Pontederia cordata*, in which this tissue occurs towards both surfaces. In the present writer's opinion, this anatomical anomaly is best interpreted on the view that the 'laminae' of the Pontederiaceae, instead of being homologous with the blades of Dicotyledons, are merely the expanded apices of pre-existing phyllodes: the inverted bundles are thus an indication of the petiolar nature of the organ, and are regarded as an ancestral feature rather than as an adaptation.

The Pontederiaceae are not the only family in which we meet with phyllodic anatomy of the 'lamina.' The present writer has found, in the arrow-head blade of *Sagittaria montevidensis*, Cham. and Schlecht. (Fig. 171 *B*), that, besides the normal main bundles ($n.b_1$) and a series of smaller bundles running near the lower surface ($n.b_2$), there is a third series of small *inverted* bundles near the upper surface (*i.b.*). In *Sagittaria sagittifolia*, L., inverted bundles are a less striking feature, but the lateral ribs, one of which is represented in Fig. 171 *A*, show both normal and inverted bundles.

[1] Goebel, K. (1913).

If the view here advocated regarding the nature of the blades of Monocotyledonous leaves be accepted, it forms a particularly salient instance of the working of the 'Law of Loss,' since we have here an instance of a discarded organ (the lamina) being replaced by a modification of another (the petiole) in lieu of being re-acquired.

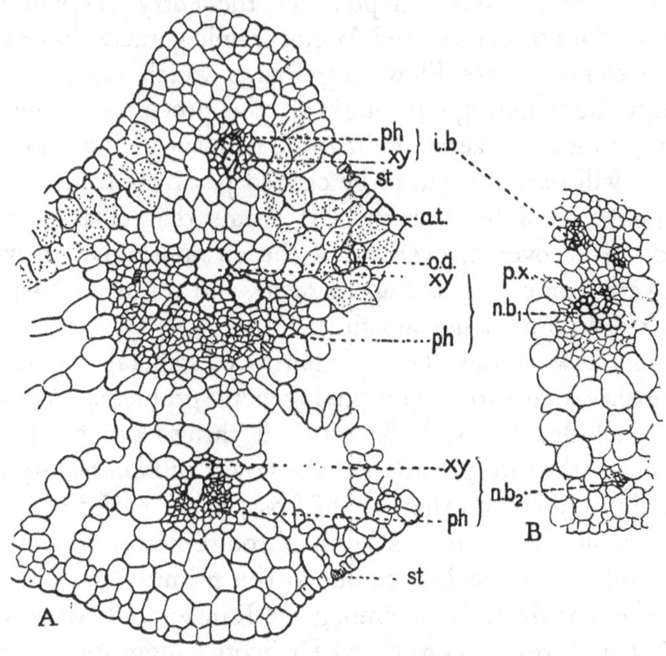

FIG. 171. *A*, *Sagittaria sagittifolia*, L., T.S. lateral vein of lamina, next but one to midrib. *B*, *Sagittaria montevidensis*, Cham. and Schlecht., small part of T.S. of leaf near margin. The lower of the two bundles belonging to the normal series ($n.b_2$) is irregularly placed. ($n.b_1$ = bundle of main normal series; *i.b.* = inverted bundle; *xy* = xylem; *ph* = phloem; *a.t.* = assimilating tissue; *st* = stomate; *o.d.* = oil duct.) [Arber, A. (1918).]

The pollination methods of submerged Angiosperms may also possibly be regarded as illustrating the Law of Loss. The ciliation of the male gamete in the great group of the Pteridophyta—from which it is supposed that Flowering Plants are ultimately derived—is associated essentially with aquatic fertilisation; with the adoption of terrestrial life this feature was lost, and is now unknown either in the higher Gymnosperms or the

Angiosperms. It might well have been expected that when certain Angiosperms adopted water-life so completely as even to revert to the remotely ancestral habit of submerged fertilisation, they would also simultaneously revert to ciliated sperms, associated with a broad stylar canal and open micropyle. Such a trumpet-shaped stigma as that possessed by *Zannichellia* seems, indeed, exactly adapted for the entry of swimming sperms. But no such ciliated Angiospermic gametes have come into existence; those Flowering Plants which are pollinated beneath the water, go through all the processes of making pollen-grains as for aerial pollination, with such slight modifications as will permit them to be carried passively to the stigma by gravity or water currents. It seems that cilia once lost cannot be recovered, even when the circumstances in which they were formerly of use again recur, and the plant has, as it were, to patch up some substitute.

If the Law of Loss be accepted as of general application, it furnishes a clue to certain phylogenetic problems. We have already alluded to the light which it throws on the difficult question of the interpretation of the flower of *Naias*[1]. Again it is highly unlikely, on the Law of Loss, that a naked unisexual flower could evolve into a hermaphrodite flower with a perianth, and hence the law points to the primitiveness of such floral types as those found among the Ranales and Alismaceae.

We have already[2] considered Dr Scott's suggestion that the anatomical peculiarities of the polystelic genus *Gunnera* might lie in an ancestral history in which an original terrestrial period, followed by an aquatic phase, has been succeeded by a second terrestrial period. Expressing this example in terms of the Law of Loss, we may say that the cambial system, once discarded under the influence of water-life, could not be regained even when the plant reverted to terrestrial conditions; the expedient of adding to the number of the existing reduced steles represents a device for repairing this irrevocable loss of means by such substitutes as are to hand.

[1] See p. 315. [2] See p. 180.

Some time after the present writer had deduced the Law of Loss from a consideration of the structure of the water plants living to-day, she learned that zoologists had already arrived, on fossil evidence, at very similar conclusions regarding animals. The Law of Loss covers part of the same ground as Dollo's 'Law of Irreversibility.' That this law should have been arrived at independently for plants and for animals is perhaps an indication of its probable validity.

With current Mendelian conceptions, the 'Law of Loss' harmonises without apparent difficulty. If evolution has proceeded by variations due to successive losses of factors, we should certainly expect that the complete loss of an organ might be associated with inability to recall it, even when circumstances seem to put a premium upon its reappearance.

If we accept the views of Samuel Butler so far as to admit that there is at least an analogy of a highly intimate nature between heredity and unconscious memory, each example of the 'Law of Loss' may perhaps be visualised as representing a lapse or failure of memory. If an organ be lost, the remembrance of it presumably in course of time becomes more and more remote, until finally, even if circumstances renew the need for it, the memory has so entirely faded that the plant cannot, as it were, recall how to reconstruct it. It is thrown, so to speak, on its own resources, and is thus compelled to discover for itself some method of responding upon new lines to the ancient need.

ALPHABETICAL LIST OF BOOKS AND MEMOIRS BEARING ON THE STUDY OF AQUATIC ANGIOSPERMS

[This list is far from exhaustive, being merely intended to indicate the principal sources. Each title is followed by a brief note on the contents and scope of the memoir. In the case of works cited in the body of the text, or from which figures have been reproduced, references to the pages in question will be found beneath the authors' names.]

Agardh, C. A. (1821) [p 123]
Species Algarum, Vol. I. 1821, 531 pp. Gryphiswaldiae.
(*Amphibolis zosteraefolia* [= *Cymodocea antarctica*] included under the Algae.)

Anon., (1828) [p. 17]
Honzo Zufu (*Phonzo Zoufou*). Yedo, 1828.
(A large series of volumes with fine illustrations of Japanese plants. Vols. 69–76 contain coloured figures of *Nymphaeaceae*, *Trapa*, *Trapella* and other water plants. There is a copy in the Library of the Kew Herbarium.)

Anon., (1895) [p. 17]
Useful Plants of Japan, described and illustrated. Agricultural Society of Japan, Tokyo, 1895.
(*Trapa*, *Nelumbo*, *Euryale*, *Sagittaria* and *Scirpus tuberosus* are figured and their uses described.)

Arber, A. (1914) [pp. 50, 186 and Figs. 31, p. 49 and 121, p. 186]
On Root Development in *Stratiotes aloides* L. Proc. Camb. Phil. Soc. Vol. XVII. 1914, pp. 369–379, 2 pls.
(The development of the adventitious roots is discussed in this paper, and attention is called to the frequently bi-lobed character of the nuclei in their stelar tissues.)

Arber, A. (1918) [pp. 52, 336 and Figs. 169, p. 341, 170, p. 342 and 171, p. 345]
The Phyllode Theory of the Monocotyledonous Leaf, with Special Reference to Anatomical Evidence. Ann. Bot. Vol. XXXII. 1918, pp. 465–501, 32 text-figs.
(In this paper the nature of the leaves in the Pontederiaceae, *Sagittaria* and other aquatic Monocotyledons is discussed.)

Arber, A. (1919[1]) [p. 317]
Aquatic Angiosperms and their Systematic Distribution. Journ. Bot. Vol. 57, 1919, pp. 83–86.
(See Chapter 25 of the present book.)

Arber, A. (1919[2]) [pp. 182, 336]
The 'Law of Loss' in Evolution. Proc. Linn. Soc. Session 131, 1918–1919, pp. 70–78.
(See the last chapter of the present book.)

Arber, A. (1919[3]) [p. 143]
Heterophylly in Water Plants. Amer. Nat. Vol. 53, 1919, pp. 272–278.
(A general discussion of this question.)

Arber, A. (1919[4]) [pp. 74, 82, 316]
On the Vegetative Morphology of *Pistia* and the Lemnaceae. Proc. Roy. Soc. B, Vol. 91, 1919, pp. 96–103, 8 text-figs.
(It is here shown that the leaf of *Pistia* is phyllodic in anatomy, and that its sheath forms a lateral pocket in which a bud is produced, in a position comparable with that of a young frond of *Lemna*.)

Arber, E. A. N. and Parkin, J. (1907) [pp. 308, 315] On the Origin of Angiosperms. Linn. Soc. Journ. Bot. Vol. 38, 1907, pp. 29–80, 4 text-figs.

(This paper is partly devoted to a reconstruction of the primitive type of Angiospermic flower. Among aquatics, the Nymphaeaceae, Alismaceae and Butomaceae are regarded as showing certain primitive features of flower structure.)

Arcangeli, G. (1890) [pp. 27, 159] Sulle foglie delle piante acquatiche e specialmente sopra quelle della Nymphaea e del Nuphar. Nuovo Giornale Botanico Italiano, Vol. XXII. 1890, pp. 441–446.

(A study of heterophylly in these genera.)

Areschoug, F. W. C. (1873[1]) Om *Trapa natans* L. och dess i Skåne ännu lefvande form. Öfversigt af k. vet. akad. Förhandl. XXX. 1874 (for 1873), No. 1, pp. 65–80, 1 pl.

[An account of this Swedish paper was given in the same year in the Journ. of Bot. See **Areschoug, F. W. C. (1873[2])**.]

Areschoug, F. W. C. (1873[2]) [pp. 302, 303] On *Trapa natans* L., especially the form now living in the southernmost part of Sweden. Journ. Bot. Vol. XI. N.S. Vol. II. 1873, pp. 239–246, 1 pl.

[This paper is a translation, revised by the author, of **Areschoug, F. W. C. (1873[1])**.]

Armand, L. (1912) [p. 166] Recherches morphologiques sur le *Lobelia Dortmanna* L. Revue gén. de Bot. T. XXIV. 1912, pp. 465–478, 18 text-figs.

(A description of the anatomy of this species, and, for comparison, of the terrestrial species, *L. urens* and *L. erinus*.)

Ascherson, P. (1867) [pp. 123, 124] Vorarbeiten zu einer Uebersicht der phanerogamen Meergewächse. Linnaea, Bd. 35, N.F. Bd. I. 1867–1868, pp. 152–208.

(A systematic account of the marine Hydrocharitaceae and Potamogetonaceae, the synonymy and distribution being dealt with in detail.)

Ascherson, P. (1870) [p. 135] Über die Phanerogamen des rothen Meeres, besonders *Schizotheca Hemprichii* Ehrb., *Phucagrostis rotundata* Ehrb. und *Phucagrostis ciliata*. Sitzungs-Berichte d. Gesellsch. naturforsch. Freunde zu Berlin, Dec. 20, 1870, pp. 83–85.

[This brief descriptive account of the marine Phanerogams of the Red Sea should be read in conjunction with **Magnus, P. (1870[2])**.]

Ascherson, P. (1873) [p. 146] Ueber Schwimmblätter bei *Ranunculus sceleratus*. Sitzungs-Ber. d. Gesellsch. naturforsch. Freunde zu Berlin, May 20, 1873, pp. 53–55.

(The first record of the occurrence of floating leaves in this species.)

Ascherson, P. (1874) [p. 303] Vorläufiger Bericht über die botanischen Ergebnisse der Rohlfs'schen Expedition zur Erforschung der libyschen Wüste. (Schluss.) Bot. Zeit. Jahrg. 32, 1874, pp. 641–647.

(In this paper mention is made of the occurrence of *Naias graminea*, Del. in the rice fields both of Egypt and Upper Italy.)

Ascherson, P. (1875) [pp. 135, 302] Die geographische Verbreitung der Seegräser, in Dr G. von Neumayer's Anleitung zu wissenschaftlichen Beobachtungen auf Reisen, 1875, pp. 358–373 (also later editions).

(A detailed and suggestive account of the distribution of the marine members of the Potamogetonaceae and Hydrocharitaceae.)

Ascherson, P. (1883) Bemerkungen über das Vorkommen gefärbter Wurzeln bei den Pontederiaceen, Haemodoraceen und einigen Cyperaceen. Ber. d. deutsch. Bot. Gesellsch. Bd. 1. 1883, pp. 498–502.

(The author describes the blue or pale lilac colouring of the roots of several genera of Pontederiaceae.)

Ascherson, P. and Graebner, P. (1907) [pp. 133, 291, 315] Potamogetonaceae, in Das Pflanzenreich, IV. 11 (herausgegeben von A. Engler), 184 pp., 221 text-figs. Leipzig, 1907.

(An authoritative account of all the species, Ascherson being responsible for the marine forms.)

Ascherson, P. and Gürke, M. (1889) Hydrocharitaceae, in Die Natürlichen Pflanzenfamilien, II. 1 (Engler, A. and Prantl, K.). Leipzig, 1889, pp. 238–258, 11 text-figs.

(A systematic treatment of the family.)

Ascherson, P. See **Delpino, F. and Ascherson, P.** (1871).

Askenasy, E. (1870) [pp. 144, 228 and Fig. 126, p. 196] Ueber den Einfluss des Wachsthumsmediums auf die Gestalt der Pflanzen. Bot. Zeit. Jahrg. 28, 1870, pp. 193–201, 209–219, 225–231, 2 pls.

(An account of the structure and development of *Ranunculus aquatilis*, L. and *R. divaricatus*, Schr. The chief feature of the work is the experimental investigation into the effect of land or water conditions on these two species.)

Aublet, F. (1775) [p. 113] Histoire des plantes de la Guiane Françoise, T. I. London and Paris, 1775.

(On pp. 582–584 there is the first account of the Podostemaceous genus *Mourera*. The author notes that the plant grows on rocks in rapidly running water and is entirely submerged with the exception of the flowers.)

Augé de Lassu (1861) [p. 109] Analyse du mémoire de Gaetan Monti sur l'*Aldrovandia*, suivie de quelques observations sur l'irritabilité des *follicules* de cette plante. Bull. de la Soc. bot. de France, T. VIII. 1861, pp. 519–523.

(An analysis of Monti's original memoir on this plant, published between 1737 and 1747, followed by the first record of the closure of the leaves when irritated.)

Bachmann, H. (1896) [pp. 32, 195] Submerse Blätter von *Nymphaea alba*. Landformen von *Nymphaea alba*. Ber. d. Schweiz. bot. Gesellsch. Heft VI. 1896 (Jahresber. d. zürcher. bot. Gesellsch.), pp. [11] and [12].

[The author describes certain cases of the occurrence of the submerged leaves of *Castalia* (*Nymphaea*) *alba*, and also of a land form which he found in three localities in the dry summer of 1895.]

Bailey, C. (1884) [pp. 237, 275, 303] Notes on the Structure, the Occurrence in Lancashire, and the Source of Origin, of *Naias graminea* Delile, var. *Delilei* Magnus. Journ. Bot. Vol. XXII. 1884, pp. 305–333, 47 text-figs., 4 pls.

[This account of an Egyptian species, which has been introduced into Lancashire, in some points supplements **Magnus, P.** (1870[1]). **Magnus, P.** (1883), **Ascherson, P.** (1874) and **Weiss, F. E. and Murray, H.** (1909) deal with the same plant.]

Bailey, C. (1887) [p. 145] Forms and Allies of *Ranunculus Flammula* L. Journ. of Bot. XXV. 1887, pp. 135–138.

(In this paper the existence of a form of *Ranunculus Flammula* with floating leaves is recorded.)

Baillon, H. (1858) [p. 311] Recherches sur l'organogénie du *Callitriche* et sur ses rapports naturels. Bull. de la Soc. bot. de France, T. V. 1858, pp. 337–341.

(A defence of the Euphorbiaceous affinity of *Callitriche*, based upon the structure and development of the gynaeceum.)

Balfour, I. B. (1879) [p. 129 and Fig. 87, p. 130] On the Genus *Halophila*. Trans. and Proc. Bot. Soc. Edinburgh, Vol. XIII. 1879, pp. 290–343, 5 pls.

[A full account of two species of this genus collected by the author on the reefs surrounding the island of Rodriguez; **Solereder, H.** (1913), pp. 46, 47, discusses Balfour's material from the systematic standpoint.]

Barbé, C. (1887) See **Dangeard, P. A. and Barbé, C.** (1887).

Barber, C. A. (1889) [pp. 36, 225 and Fig. 19, p. 37] On a change of Flowers to Tubers in *Nymphaea Lotus*, var. *monstrosa*. Ann. Bot. Vol. IV. 1889–1891, pp. 105–116, 1 pl.

(An account of a case of the replacement—under cultivation—of flowers by tubers, which, when detached were capable of reproducing the plant.)

Barnéoud, F. M. (1848) [p. 207] Mémoire sur l'anatomie et l'organogénie du *Trapa natans* (Linn.). Ann. d. sci. nat. Sér. III. Bot. T. IX. 1848, pp. 222–244, 4 pls.

(This early description of *Trapa natans* includes a study of the germination, anatomy and floral development.)

Barratt, K. (1916) [p. 185 and Fig. 120, p. 185] The Origin of the Endodermis in the Stem of Hippuris. Ann. Bot. Vol. XXX. 1916, pp. 91–99, 6 text-figs.

(The author's results regarding the apical anatomy of the stem of *Hippuris* are in general agreement with those of Schoute.)

Barthélemy, A. (1883) Sur la respiration des plantes aquatiques ou des plantes aquatico-aériennes submergées. Comptes rendus de l'acad. des sciences, Paris, T. 96, 1883, pp. 388–390.

(An account of experiments on the assimilation and respiration of aquatic plants, from which the author concludes that "la respiration spéciale des organes verts ne peut avoir l'importance cosmique qu'on lui attribue.")

Batten, L. (1918) [p. 188]
Observations on the Ecology of *Epilobium hirsutum*. Journ. Ecology, Vol. 6, 1918, pp. 161–177, 15 text-figs.
[A fully illustrated account of the "aerenchyma" of this species—a tissue whose existence had previously been recorded by **Lewakoffski, N. (1873[1])** and **Schenck, H. (1889)**.]

Bauhin, G. (1596) [p. 9]
Phytopinax seu Enumeratio Plantarum...Basileae per Sebastianum Henricpetri 1596.
(Bauhin describes the germinating tuber of *Sagittaria* as "Gramen bulbosum," p. 21.)

Bauhin, G. (1620) [p. 9 and Fig. 3, p. 11]
Prodromos Theatri Botanici...Francofurti ad Moenum, Typis Pauli Jacobi, impensis Joannis Treudelii, 1620.
[Bauhin gives a figure (p. 4) of "Gramen bulbosum aquaticum" to which he has already referred in **Bauhin, G. (1596)**.]

Bauhin, G. (1623) [p. 27]
Pinax Theatri Botanici...Basileae Helvet. Sumptibus et typis Ludovici Regis, 1623.
[The submerged leaves of *Nymphaea* (*Castalia*) *alba* are described on p. 193.]

Belhomme, (1862) [p. 219]
Note sur les bourgeons reproducteurs du *Ranunculus Lingua*. Bull. de la Soc. bot. de France, T. IX. 1862, p. 241.
(A note on the wintering of this species.)

Benjamin, L. (1848) [pp. 97, 99, 101]
Ueber den Bau und die Physiologie der Utricularien. Bot. Zeit. Jahrg. 6, 1848, pp. 1–5, 17–23, 45–50, 57–61, 81–86.
(This paper, which contains some interesting observations, was written before the insectivorous nature of the bladders was recognised.)

Bennett, A. (1896)
Fortschritte der schweizerischen Floristik. *Potamogeton*. Ber. d. Schweiz. bot. Gesellsch. Heft VI. 1896, pp. 94–99.
(A systematic enumeration of the results obtained by the author in the course of a revision of the principal Swiss herbaria.)

Bennett, A. (1913)
Remarks on Some Aquatic Forms and Aquatic Species of the British Flora. Trans. Bot. Soc. Edinb. Vol. XXVI. 1917 (for 1911–1915), Part II. 1913, pp. 21–27.
(Notes relating to the occurrence and nomenclature of some of the aquatic forms and species described by West, Glück, etc.)

Bennett, A. (1914) [p. 55]
Hydrilla verticillata Casp. in England. Journ. Bot. Vol. LII. 1914, pp. 257–258, 1 pl.
(This plant, which is new to the British flora, has been found growing at Estwaite Water associated with *Naias flexilis*, etc.)

Bennett, A.
See **Fryer, A., Bennett, A. and Evans, A. H. (1898–1915)**.

Berry, E. W. (1917) [p. 38 and Fig. 21, p. 39] Geologic History indicated by the Fossiliferous Deposits of the Wilcox Group (Eocene) at Meridian, Mississippi. U.S. Geol. Survey. Professional Paper 108 E. Shorter contributions to general geology, 1917, Washington, pp. 61–72, 3 pls., 1 text-fig., 1 map.
(This memoir contains an account with map of the past and present distribution of the genus *Nelumbo*.)

Blake, J. H. (1887) The Prickle-pores of *Victoria regia*. Ann. Bot. Vol. I. 1887–1888, pp. 74–75.
[The author criticises the account of these structures given by **Trécul, A.** (1854), and concludes that the function of the spines is probably merely protective.]

Blanc, M. le (1912) [p. 183 and Figs. 8, p. 19, and 118, p. 184] Sur les diaphragmes des canaux aérifères des plantes. Revue gén. de Bot. T. 24, 1912, pp. 233–243, 1 pl.
(In this paper the diaphragms crossing the intercellular spaces of the stems and leaves of certain aquatics are described, and they are figured in the cases of *Sagittaria sagittifolia, Pontederia cordata* and *Potamogeton natans*.)

Blenk, P. (1884) [p. 37] Ueber die durchsichtigen Punkte in den Blättern. Flora, N. R. Jahrg. XLII. (G. R. Jahrg. LXVII.) 1884, pp. 49–57, 97–112, 136–144, 204–210, 223–225, 275–283, 291–299, 339–349, 355–370, 371–386.
(The transparent dots on the leaves of Nymphaeaceae are referred to on pages 100–102.)

Bois, D. See **Paillieux, A. and Bois, D.** (1888).

Bokorny, T. (1890) [p. 261] Weitere Mittheilung über die wasserleitenden Gewebe. Pringsheim's Jahrb. f. wissen. Bot. Bd. XXI. 1890, pp. 505–519.
(An account of an experimental investigation of the transpiration stream in *Myriophyllum proserpinacoides*, when the plant is growing with its leafy shoots above water.)

Bolle, C. (1861–1862) Notiz über die Alismaceenformen der Mark. Verhandl. d. bot. Vereins Provinz Brandenburg, Heft. III. and IV. 1861–1862, pp. 159–167.
[An account of certain forms of *Sagittaria* and *Alisma* found by the author. A more modern discussion of the subject will be found in **Glück, H.** (1905).]

Bolle, C. (1865) [p. 210] Eine Wasserpflanze mehr in der Mark. Verhandl. d. bot. Vereins Provinz Brandenburg, Jahrg. 7, 1865, pp. 1–15.
[See note on **Bolle, C.** (1867).]

Bolle, C. (1867) [p. 210] Weiteres über die fortschreitende Verbreitung der *Elodea canadensis*. Verhandl. d. bot. Vereins Provinz Brandenburg, Jahrg. 9, 1867, pp. 137–147.
[This paper and **Bolle, C.** (1865) record the way in which *Elodea*, at that date a comparative rarity, was spreading over Germany.]

Bonpland, A. See **Humboldt, A. de, and Bonpland, A.** (1808).

Boresch, K. (**1912**) [p. 154] Die Gestalt der Blattstiele der *Eichhornia crassipes* (Mart.) Solms in ihrer Abhängigkeit von verschiedenen Faktoren. Flora, N.R. Bd. 4 (Ganze Reihe, Bd. 104), 1912, pp. 296–308, 1 pl., 3 text-figs.

(This paper describes a series of experiments which show that the inflated form of petiole in *Eichhornia crassipes* can be induced by full light, low temperature and a free-swimming life, whereas the converse conditions tend to be associated with the elongated form of petiole.)

Bornet, E. (**1864**) [p. 125 and Fig. 83, p. 124] Recherches sur le *Phucagrostis major* Cavol. Ann. d. sci. nat. Sér. v. Bot. T. 1. 1864, pp. 5–51, 11 pls.

(This finely illustrated memoir gives a singularly complete account of the structure and life-history of the plant now called *Cymodocea aequorea*, Kon.)

Borodin, J. (**1870**) [pp. 86, 169 and Fig. 163, p. 268] Ueber den Bau der Blattspitze einiger Wasserpflanzen. Bot. Zeit. Jahrg. 28, 1870, pp. 841–851, 1 pl.

[A description of the stomates which occur in small numbers near the apices of the submerged leaves of *Callitriche* and *Hippuris*. Mention is also made of the peculiar oil-containing processes at the tips of the leaves of *Myriophyllum* and *Ceratophyllum*. For a criticism of this paper see **Magnus, P.** (**1871**).]

Bottomley, W. B. (**1917**) [p. 287] Some Effects of Organic Growth-Promoting Substances (Auximones) on the Growth of *Lemna minor* in Mineral Culture Solutions. Proc. Roy. Soc. B, Vol. 89, 1917, pp. 481–507, 2 pls.

(By means of comparative cultures it is shown that Duckweed cannot be kept healthy in solutions with only mineral salts—soluble organic matter is essential.)

Boulger, G. S. (**1900**) [p. 321] Aquatic Plants. Journ. Roy. Hort. Soc. Vol. 25, 1900, pp. 64–77.

(A suggestive general account of hydrophytes, with a systematic appendix showing the independent origin of the aquatic habit in a comparatively small number of Cohorts.)

Brand, F. (**1894**) [pp. 27, 159] Ueber die drei Blattarten unserer *Nymphaeaceen*. Bot. Centralbl. Bd. LVII. 1894, pp. 168–171.

(A brief account of the submerged, floating and air leaves of *Nymphaea lutea* and *Castalia alba*.)

Brongniart, A. (**1827**) [p. 309] Mémoire sur la Génération et le Développement de l'Embryon dans les végétaux phanérogames. Ann. des sci. nat. Vol. 12, 1827, pp. 14–53, 145–172, 225–296, 11 pls.

(On p. 253 et seq. the author compares the embryo of *Ceratophyllum* with that of *Nelumbo*.)

Brongniart, A. (**1833**) [p. 76] Note sur la structure du fruit des *Lemna*. Archives de Botanique, T. II. 1833, pp. 97–104.

(An account of the structure of the seed and fruit in *Lemna minor* and *L. gibba*.)

Brongniart, A. (**1834**) [p. 164] Nouvelles recherches sur la structure de l'Épiderme des Végétaux. Ann. d. sci. nat. Sér. II. T. I. Bot. 1834, pp. 65–71, 2 pls.

[On p. 68 the author records the discovery of chlorophyll in the epidermis of the leaves of *Potamogeton lucens* and the existence of "une pellicule tout-à-fait incolore" (=cuticle) on the surface of the epidermal layer. In Pl. III, Fig. 5, the characters of the epidermis are clearly demonstrated.]

Brown, C. Barrington (1876) [p. 119] Canoe and Camp Life in British Guiana. xi+400 pp., 10 pls. and map. London, 1876.

(On p. 11 some Podostemaceae occurring in the Cuyuni River are described under the name of *Lacis* spp.)

Brown, R. (1814) [p. 311] General remarks on the Botany of Terra Australis. 89 pp. Reprinted in the Miscellaneous Botanical Works of Robert Brown, Vol. I. 1866.

(The author includes *Callitriche* in the Halorageae; see p. 22.)

Brown, W. H. (1911) [p. 286] The Plant Life of Ellis, Great, Little, and Long Lakes in North Carolina. Contributions from the U.S. National Herbarium, Vol. 13, Part 10 (Misc. Papers), Washington, 1911, pp. 323–341, 1 text-fig.

(An account from the ecological standpoint of the plant life of these lakes, special attention being paid to the relation of soils to aquatic vegetation.)

Brown, W. H. (1913) [pp. 253, 264, 265] The Relation of the Substratum to the Growth of *Elodea*. The Philippine Journal of Science, C, Botany, Vol. VIII. 1913, pp. 1–20.

(An important experimental study on the factors affecting the growth of *Elodea*, especially the CO_2 supply.)

Bruyant, C. (1914) [p. 291] Les Tourbières du massif Mont-Dorien. Annales de Biologie Lacustre, T. VI. Fasc. 4, 1914, pp. 339–391, 1 map, 14 text-figs.

(This memoir contains an ecological study of the peat bogs of this region.)

Buchenau, F. (1857) Ueber die Blüthenentwickelung von *Alisma* und *Butomus*. Flora, N.R. Jahrg. XV. (G.R. Jahrg. XL.) 1857, pp. 241–254, 1 pl.

(A description of the development of the parts of the flower in *Alisma Plantago* and *Butomus umbellatus*, with a briefer mention of *Sagittaria sagittifolia*.)

Buchenau, F. (1859) [pp. 217, 232] Zur Naturgeschichte der *Littorella lacustris* L. Flora, N.R. Jahrg. XVII. (G.R. Jahrg. XLII.) 1859, pp. 81–87, 464, 705–706, 1 pl.

(A study of the external morphology of the flowering land form and the sterile water form of this species.)

Buchenau, F. (1865) Morphologische Studien an deutschen Lentibularieen. Bot. Zeit. Jahrg. 23, 1865, pp. 61–66, 69–71, 77–80, 85–91, 93–99, 2 pls.

(In the 3rd and later parts of this memoir the branching and flower development of *Utricularia* are dealt with.)

Buchenau, F. (1866) [p. 245] Morphologische Bemerkungen über *Lobelia Dortmanna* L. Flora, N.R. Jahrg. 24 (G.R. Jahrg. 49), 1866, pp. 33–38, 1 pl.

(An account of the germination and general morphology of this species.)

Buchenau, F. (1882) [p. 17] Beiträge zur Kenntniss der Butomaceen, Alismaceen und Juncaginaceen. Bot. Jahrbücher (Engler's), Bd. II. 1882, pp. 465–510.

[This paper is intended to supplement and correct Micheli's monograph of the same group; see **Micheli, M. (1881)**.]

Buchenau, F. (1903[1]) [pp. 9, 314] Alismataceae, in Das Pflanzenreich, IV. 15 (herausgegeben von A. Engler), Leipzig, 1903, 66 pp., 19 text-figs.

(The standard systematic account of this family.)

Buchenau, F. (1903[2]) Butomaceae, in Das Pflanzenreich, IV. 16 (herausgegeben von A. Engler), 12 pp., 5 text-figs. 1903.

(An authoritative account of the species of this family which includes water plants such as *Hydrocleis nymphoides*.)

Burgerstein, A. (1904) [pp. 266, 267] Die Transpiration der Pflanzen. x + 283 pp., 24 text-figs. Jena, 1904.

[This critical compilation contains a chapter (XXVI. "Guttation, Hydathoden") dealing with the elimination of liquid water from the leaves. The case of water plants is discussed on pp. 195–197.]

Burkill, I. H. See **Willis, J. C. and Burkill, I. H. (1895).**

Burns, G. P. (1904) [pp. 160, 161] Heterophylly in *Proserpinaca palustris*, L. Ann. Bot. Vol. XVIII. 1904, pp. 579–587, 1 pl.

[An account of experimental work on the conditions determining the formation of leaves of the "land-type" and "water-type." This paper should be read in conjunction with **McCallum, W. B. (1902)**, on which it is based.]

Burrell, W. H. and Clarke, W. G. (1911) [p. 215] Botanical Rambles in West Norfolk, with notes on the genus *Utricularia*. Trans. Norfolk and Norwich Naturalists' Society, Vol. IX. 1914 (Pt II. 1911), pp. 263–268.

(These notes contain a reference to remarkably luxuriant growth observed in *Utricularia*.)

Büsgen, M. (1888) [pp. 93, 94, 95] Ueber die Art und Bedeutung des Thierfangs bei *Utricularia vulgaris* L. Ber. d. deutsch. bot. Gesellsch. Bd. VI. 1888, pp. lv–lxiii.

(The author discusses the function of the bladders in this species and shows experimentally that the carnivorous habit is an advantage.)

Caldwell, O. W. (1899) [p. 76] On the Life-history of *Lemna minor*. Bot. Gaz. Vol. XXVII. 1899, pp. 37–66, 59 text-figs.

(In this memoir special attention is paid to the gametophytes and fertilisation.)

Cambessedes, J. (1829) [p. 311] Note sur les Élatinées, nouvelle famille de plantes. Mém. du muséum d'histoire nat. T. XVIII. 1829, pp. 225–231.

(The author proposes to remove *Elatine*, *Bergia* and *Merimea* from the Caryophyllaceae and to place them in a separate family. He remarks on certain resemblances which they show to the Hypericineae.)

Campbell, D. H. (1897) A Morphological Study of *Naias* and *Zannichellia*. Proc. Cal. Acad. Sci. Ser. III. Botany, Vol. I. 1897–1900, pp. 1–70, 5 pls.

(In this memoir special attention is paid to the anatomy and the gametophytes.)

Candolle, Alphonse P. de (1855) [p. 296]
Géographie Botanique. Paris, T. II. 1855.
(Pages 998–1006 deal with the distribution of aquatic species. After showing how widely these plants are distributed, the author concludes that the facts are scarcely explicable except on the ground that there have been multiple centres of creation.)

Candolle, Auguste P. de (1827) [pp. 12, 337]
Organographie végétale. Paris, 1827.
(Vol. I. Book 2, Chap. III. contains the first enunciation of the phyllode theory of the Monocotyledonous leaf.)

Cario, R. (1881)
Anatomische Untersuchung von *Tristicha hypnoides* Spreng. Bot. Zeit. Jahrg. 39, 1881, pp. 25–33, 41–48, 57–64, 73–82, 1 pl.
[The author obtained material of this plant in Guatemala. The present paper forms an anatomical monograph of the species which was incompletely treated in **Tulasne, L. R. (1852)**. The part of the plant which Cario describes as the "thallus" is now generally regarded as representing the root-system.]

Caspary, R. (1847)
Ueber *Elatine Alsinastrum* und *Trapa natans*. Verhandl. des naturhistorischen Vereines der preuss. Rheinlande, Jahrg. 4, 1847, pp. 111, 112.
(A brief note on a new locality for *Elatine*, and on the absence of *Trapa* in the neighbourhood of Bensberg.)

Caspary, R. (1856[1])
Les Nymphéacées fossiles. Ann. des sci. nat. Sér. IV. Bot. T. VI. 1856, pp. 199–222, 2 pls.
(An account of the remains of this family found in Tertiary beds.)

Caspary, R. (1856[2]) [p. 214]
Ueber die tägliche Periode des Wachsthums des Blattes der *Victoria regia* Lindl. und des Pflanzenwachsthums überhaupt. Flora, N.R. Jahrg. XIV. (G.R. Jahrg. XXXIX.) 1856, pp. 113–126, 129–143, 145–160, 161–171.
(A detailed study of the growth of the leaves of *Victoria regia* in a hot-house. The maximum growth in 24 hrs was 30·8 cms. in length, and 36·7 cms. in breadth.)

Caspary, R. (1857)
Note sur la division de la famille des Hydrocharidées, proposée par M. Chatin. Bull. de la Soc. bot. de France, T. IV. 1857, pp. 98–101.
[A criticism of views expressed in **Chatin, A. (1856)**.]

Caspary, R. (1858[1])
Eine systematische Übersicht der Hydrilleen. Monatsber. d. König. Preuss. Akad. d. Wiss. Berlin, 1858 (for 1857), pp. 39–51.
(A systematic account of the tribe of the Hydrocharitaceae which includes *Elodea*, etc.)

Caspary, R. (1858[2]) [pp. 55, 56, 173, 210, 211]
Die Hydrilleen (Anacharideen Endl.). Pringsheim's Jahrb. f. wiss. Bot. Bd. I. 1858, pp. 377–513, 5 pls.
(A very important monograph of that tribe of the Hydrocharitaceae which includes *Hydrilla*, *Elodea* and *Lagarosiphon*. The standpoint is systematic, but a good deal of anatomical work is included.)

Caspary, R. (1858[3])
Die Blüthe von *Elodea canadensis* Rich. Bot. Zeit. Jahrg. 16, 1858, pp. 313–317, 1 pl.
(A description of the female flower based on living material.)

Caspary, R. (1858[4]) [p. 111] Sur l'*Aldrovanda vesiculosa*. Bull. de la Soc. bot. de France, T. v. 1858, pp. 716–726.
[The observations in this paper are expanded and illustrated in **Caspary, R.** (1859 and 1862).]

Caspary, R. (1859 and 1862) [pp. 110, 239 and Fig. 75, p. 111] *Aldrovanda vesiculosa* Monti. Bot. Zeit. Jahrg. 17, 1859, pp. 117–123, 125–132, 133–139, 141–150, 2 pls. *Aldrovandia vesiculosa*. Bot. Zeit. Jahrg. 20, 1862, pp. 185–188, 193–197, 201–206, 1 pl.
[These papers form a monograph of this species. An abstract of part of Caspary's work on the subject is also to be found in Flora, N.R. Jahrg. XVII. (G.R. Jahrg. XLII.) 1859, pp. 140–143.]

Caspary, R. (1860) [p. 234] *Bulliarda aquatica* D.C. Schriften d. köniġ. phys.-ök. Gesellsch. zu Königsberg, Jahrg. I. 1861 (for 1860), pp. 66–91, 2 pls.
(A monograph of this aquatic member of the Crassulaceae, now known as *Tillaea aquatica* L.)

Caspary, R. (1861) [p. 276] *Nuphar luteum* L. var. *rubropetalum*. Schriften d. königl. phys.-ök. Gesellsch. zu Königsberg, Jahrg. II. 1862 (for 1861), pp. 49–50, 1 pl.
(A description, illustrated with a coloured plate, of a variety of *Nymphaea lutea* with red petals.)

Caspary, R. (1870[1]) Neue und seltene Pflanzen Preussens. Schriften d. königl. phys.-ök. Gesellsch. zu Königsberg, Jahrg. XI. 1871 (for 1870), pp. 61–64.
(These field notes include an account of certain varieties of *Castalia alba*.)

Caspary, R. (1870[2]) [p. 300] Welche Vögel verbreiten die Samen von Wasserpflanzen? Schriften d. königl. phys.-ök. Gesellsch. zu Königsberg, Jahrg. XI. 1871 (for 1870), Sitzungsber. p. 9.
(This note emphasizes our ignorance of the part played by water birds in the distribution of water plants.)

Caspary, R. (1875) [p. 54] Die geographische Verbreitung der Geschlechter von *Stratiotes aloides* L. Sitzungs-Ber. d. Gesellsch. Naturforsch. Freunde zu Berlin, 1875, pp. 101–106.
(An account of the distribution of this species, supplementing and criticising previous work, and showing that though in some regions female plants alone are present, no region is known in which male plants appear exclusively.)

Cavolini, F. (**Caulinus, P.**) (1792[1]) [p. 125] Zosterae Oceanicae Linnei ΑΝΘΗΣΙΣ. Contemplatus est Philippus Caulinus Neapolitanus. Annis 1787 et 1791, 20 pp., 1 pl. Neapoli, 1792.
[An account of the flowering and vegetative organs of *Posidonia Caulini* = "*Zostera oceanica*." This paper and **Cavolini, F.** (**1792**[2]) are analysed in **Delpino, F. and Ascherson, P.** (**1871**).]

Cavolini, F. (**Caulinus, P.**) (1792[2]) [p. 125] Phucagrostidum Theophrasti ΑΝΘΗΣΙΣ. Contemplatus est Philippus Caulinus Neapolitanus. Anno 1792, 35 pp., 3 pls. Neapoli, 1792.
(An account with good figures of the vegetative and flowering structure of *Cymodocea aequorea* = "*Phucagrostis major*" and *Zostera nana* = "*Phucagrostis minor*.")

Chamberlain, C. J. See **Coulter, J. M. and Chamberlain, C. J.** (1904).

Chatin, A. (1855[1]) [pp. 164, 166] Note sur la présence de matière verte dans l'épiderme des feuilles de l'*Hippuris vulgaris*, du *Peplis portula*, des *Jussiaea longifolia* et *J. lutea*, de l'*Isnardia palustris* et du *Trapa natans*. Bull. de la Soc. bot. de France, T. II. 1855, pp. 674–676.

(The object of this note is to draw attention to the existence in many water plants of an epidermis supplied with stomates and also containing chlorophyll. The author points out that this type of epidermis is well adapted to amphibious life.)

Chatin, A. (1855[2]) [pp. 134, 235] Mémoire sur le *Vallisneria spiralis*, L. 31 pp., 5 pls. Paris, 1855.

(The morphology, anatomy and floral structure are dealt with in detail, and there is a habit drawing showing male and female plants.)

Chatin, A. (1856) Anatomie comparée des végétaux, Livraison 1 and 2, pp. 1–96, 20 pls. Paris, 1856.

(The first part of this work deals with Monocotyledonous water plants. It is fully illustrated but singularly inaccurate.)

Chatin, A. (1858[1]) Note sur le cresson de fontaine (*Sisymbrium Nasturtium* L., *Nasturtium officinale* R. Br.) et sur sa culture. Bull. de la Soc. bot. de France, T. V. 1858, pp. 158–166.

(This economic paper deals with the cultivation of the Watercress.)

Chatin, A. (1858[2]) Faits d'anatomie et de physiologie pour servir à l'histoire de l'*Aldrovanda*. Bull. de la Soc. bot. de France, T. V. 1858, pp. 580–590.

(This paper is of less importance than those of Caspary dealing with the same subject. Chatin and Caspary obtained the main part of their material from the same source.)

Chrysler, M. A. (1907) [pp. 63, 65, 135 and Fig. 39, p. 62] The Structure and Relationships of the Potamogetonaceae and allied Families. Bot. Gaz. Vol. XLIV. 1907, pp. 161–188, 3 text-figs., 5 pls.

(A discussion of the affinities of these families is based upon a study of the anatomy of *Potamogeton*, *Ruppia*, *Zostera*, *Phyllospadix*, *Cymodocea* and *Zannichellia*, etc.)

Clarke, W. G. See **Burrell, W. H. and Clarke, W. G.** (1911).

Clavaud, A. (1876) [p. 78] Sur une particularité du *Lemna trisulca* L. Actes de la Soc. Linn. de Bordeaux, T. XXXI. (Sér. IV. T. I.) 1876, pp. 309–311.

(A note on the occurrence of raphides in this species and their possible biological significance.)

Clavaud, A. (1878) [p. 127] Sur le véritable mode de fécondation du *Zostera marina*. Actes de la Soc. Linn. de Bordeaux, T. XXXII. (Sér. IV. T. II.) 1878, pp. 109–115.

(An account of the pollination of *Zostera* growing *in situ*, from observations made from a boat.)

Cloëz, S. (1863) [p. 256] Observations sur la nature des gaz produits par les plantes submergées sous l'influence de la lumière. Comptes rendus de l'acad. des sciences, Paris, T. LVII. 1863, pp. 354–357.

(The author describes experiments showing that the gas given off by aquatic plants exposed to light is a mixture of oxygen and nitrogen: he holds that this nitrogen is produced by decomposition of the substance of the plant.)

Cloëz, S. and Gratiolet, P. (1850) [p. 256] Recherches sur la végétation. Comptes rendus de l'académie des sciences, Paris, T. XXXI. 1850, pp. 626–629.

(An early account of the gaseous exchange in submerged plants.)

Clos, D. (1856) [p. 67] Mode de propagation particulier au *Potamogeton crispus* L. Bull. de la Soc. bot. de France, T. III. 1856, pp. 350–352.

(The first account of the peculiar turions of this plant. According to the author, they are unique among organs of vegetative reproduction in their horny consistency, and also in the fact that the detached shoot grows no further, but its whole vitality is concentrated in its axillary buds.)

Cohn, F. (1850) [p. 110] Ueber *Aldrovanda vesiculosa* Monti. Flora, N.R. Jahrg. VIII. (G. R. Jahrg. XXXIII.) 1850, pp. 673-685, 1 pl.

[A description of the anatomy and morphology of this species, less detailed than that of **Caspary, R. (1859 and 1862)**. A brief account of early references to the plant is given in an appendix.]

Cohn, F. (1875) [pp. 93, 96, 110, 270] Ueber die Function der Blasen von *Aldrovanda* und *Utricularia*. Cohn's Beiträge zur Biologie der Pflanzen, Bd. I. Heft 3, 1875, pp. 71–92, 1 pl.

(The earliest memoir in which the existence of the carnivorous habit in these two genera is fully established.)

Coleman, W. H. (1844) [pp. 150, 204] Observations on a new species of *Œnanthe*. Annals and Mag. of Nat. Hist. Vol. XIII. 1844, pp. 188–191, 1 pl.

(The author makes out what appears to be a good case for regarding *Oenanthe fluviatilis* as a species distinct from *Oe. Phellandrium*, Lamk., instead of as a mere variety of it.)

Compton, R. H. (1916) [pp. 200, 289] The Botanical Results of a Fenland Flood. Journ. of Ecology, Vol. IV. 1916, pp. 15–17, 2 pls.

(This paper gives an account of the effect of a nine months' period of submergence upon the flora of an area of fenland in E. Anglia, 24 square miles in extent.)

Cook, M. T. (1906) [p. 309] The Embryology of some Cuban Nymphaeaceae. Bot. Gaz. Vol. 42, pp. 376–392, 3 pls.

(The author's study of several genera leads him to the conclusion that the Nymphaeaceae are anomalous Monocotyledons.)

Costantin, J. (1884) [pp. 192, 200, 201, 259] Recherches sur la structure de la tige des plantes aquatiques. Annales des sci. nat. VI. Sér. Bot. T. XIX. 1884, pp. 287–331, 4 pls.

(A comparison of the anatomy of stems of different individuals of the same species, or of different parts of the same stem, grown in water, in air, or embedded in soil beneath water A very important contribution to the experimental anatomy of water plants.)

Costantin, J. (1885[1]) [pp. 165, 166] Observations critiques sur l'épiderme des feuilles des végétaux aquatiques. Bull. de la Soc. bot. de France, T. XXXII. (Sér. II. T. VII.) 1885, pp. 83–88 (followed by an account of a discussion in which É. Mer and P. Duchartre took part, pp. 88–92).

(The author attempts to show that the influence of the aquatic medium is one of the causes of the loss of stomates in submerged leaves. He also maintains that submerged plants possess a true epidermis, even if stomates are absent and chlorophyll present in this layer.)

Costantin, J. (1885[2]) [p. 155] Recherches sur la Sagittaire. Bull. de la Soc. bot. de France, T. XXXII. (Sér. II. T. VII.) 1885, pp. 218–223.

(Observations on the heterophylly of *Sagittaria sagittifolia* and a comparison of the anatomy of the submerged and aerial leaves.)

Costantin, J. (1885[3]) [p. 51] Influence du milieu aquatique sur les stomates. Bull. de la Soc. bot. de France, T. XXXII. (Sér. II. T. VII.) 1885, pp. 259–264.

[This paper forms a continuation of **Costantin, J.** (1885[1]). The author criticises Mer's view that the presence or absence of stomates is partly an hereditary character and partly due to variations in illumination and nutrition, and brings forward further evidence to show that the milieu has a great influence on the distribution of stomates.]

Costantin, J. (1886) [pp. 12, 28, 30, 51, 145, 151, 155, 156] Études sur les feuilles des plantes aquatiques. Ann. d. sci. nat. Sér. VII. Bot. T. 3, 1886, pp. 94–162, 5 pls.

(A memoir on the morphology and anatomy of the leaves of water plants, with special reference to the effect of the environment upon their structure.)

Cöster, B. F. (1875) [p. 67] Om *Potamogeton crispus* L. och dess groddknoppar. Botaniska Notiser, Lund, 1875, pp. 97–102, 1 text-fig.

[This paper, which deals with the winter buds of *Potamogeton crispus*, is reviewed in Bot. Jahresber. (Just) Jahrg. III. 1877 (for 1875), p. 425.]

Coulter, J. M. and Chamberlain, C. J. (1904) [pp. 322, 325] Morphology of Angiosperms. x + 348 pp., 113 text-figs. London and New York, 1904.

(This general work contains a number of references to water plants.)

Coulter, J. M. and Land, W. J. G. (1914) [p. 15] The Origin of Monocotyledony. Bot. Gaz. Vol. 57, 1914, pp. 509–519, 2 pls., 2 text-figs.

(In this paper the seedling of *Sagittaria variabilis* is described.)

Crocker, W. (1907) [p. 243] Germination of Seeds of Water Plants. Bot. Gaz. Vol. 44, 1907, pp. 375–380.

[The author shows experimentally that the delay in germination of the seeds of water plants, which have not been subjected to a period of desiccation, is due to the impossibility of absorbing sufficient water through the intact seed coats. Drying followed by a soaking seems to induce rupture of the coats, and thus to allow growth to begin. The paper contains a criticism of **Fischer, A.** (1907).]

Crocker, W. and Davis, W. E. (1914) [pp. 242, 243] Delayed germination in seed of *Alisma Plantago*. Bot. Gaz. Vol. 58, 1914, pp. 285–321, 8 text-figs.

(A detailed study of one case, *Alisma Plantago*, illustrating the delay in germination so common among water plants; the dormancy of the achenes is here due to the mechanical restraint exercised by the seed coats.)

Crouan (Frères) (1858) [p. 93] Observations sur un mode particulier de propagation des *Utricularia*. Bull. de la Soc. bot. de France, T. v. 1858, pp. 27–29.

(These notes on *U. minor* are written without knowledge of the previous literature.)

Cunnington, H. M. (1912) [p. 135] Anatomy of *Enhalus acoroides* (Linn. f.), Zoll. Trans. Linn. Soc. Lond. Ser. II. Bot. Vol. VII. Pt 16, 1912 (1904–1913), pp. 355–371, 1 pl., 13 text-figs.

(A detailed account of the anatomy of this marine Angiosperm, in which special attention is paid to the development of the various tissues.)

Dangeard, P. A. and Barbé, C. (1887) [p. 181] La Polystélie dans le genre *Pinguicula*. Bull. de la Soc. bot. de France, T. 34, 1887, pp. 307–309.

(The authors show that the old axes of *Pinguicula vulgaris* may contain four or five steles, each surrounded by a well-marked endodermis.)

Darwin, C. (1859) [pp. 296, 298, 300, 324] On the Origin of Species. ix + 502 pp. London, 1859.

(Chapter XII. contains a section dealing with the distribution of fresh-water animals and plants, pp. 383–388.)

Darwin, C. (1875) [pp. 93, 95, 111] Insectivorous Plants. x + 462 pp. 30 text-figs. London, 1875.

(Chapter XIV. deals with *Aldrovandia* and Chapters XVII. and XVIII. with *Utricularia*.)

Darwin, C. (1888) [p. 95] Insectivorous Plants. Second Edition revised by Francis Darwin. xiv + 377 pp., 30 text-figs. London, 1888.

[This edition contains a certain number of additional facts and references not found in **Darwin, C.** (**1875**).]

Darwin, C. (1890) [p. 181] Journal of Researches into the Natural History and Geology of the...voyage of...H.M.S. 'Beagle.' London, 1890.

(See reference to *Gunnera* on p. 298.)

Darwin, C. (1891) [p. 206] The Movements and Habits of Climbing Plants. ix + 208 pp., 13 text-figs. London, 1891.

(Darwin's references to climbing roots are of interest in connection with the tendril roots of certain water plants.)

Darwin, C. and F. (1880) [pp. 90, 161, 206] The Power of Movement in Plants. x + 592 pp., 196 text-figs. London, 1880.

[On p. 211 the observations made by Rodier on the movements of *Ceratophyllum* are discussed. See **Rodier, É.** (**1877**[1]) and (**1877**[2]).]

Davie, R. C. (1913) [pp. 50, 287] *Stratiotes Aloides*, Linn., near Crieff. Trans. and Proc. Bot. Soc. Edinb. Vol. XXVI. 1913, pp. 180–183, 1 pl.

(The author regards this plant as introduced in all Scottish localities. Water more or less richly charged with lime seems to suit it best.)

Davis, W. E. See **Crocker, W. and Davis, W. E.** (1914).

Delpino, F. (1870) [p. 135] Ulteriori osservazioni et considerazioni sulla dicogamia nel regno vegetale II. Atti della Soc. Ital. di Scienze Naturali, Vol. XIII. 1870, pp. 167–205.

[Pp. 168–187 deal with hydrophilous plants, giving a résumé of the work on their pollination up to 1870. For a German version with some additions see **Delpino, F. and Ascherson, P.** (1871).]

Delpino, F. (1871) [p. 110] Sulle Piante a Bicchieri. Nuovo Giornale Botanico Italiano, Vol. III. 1871, pp. 174–176.

(A footnote on p. 175 deals with the carnivorous habits of *Aldrovandia*.)

Delpino, F. and Ascherson, P. (1871) [pp. 84, 135, 236] Federico Delpino's Eintheilung der Pflanzen nach dem Mechanismus der dichogamischen Befruchtung und Bemerkungen über die Befruchtungsvorgänge bei Wasserpflanzen. Mitgetheilt und mit einigen Zusätzen versehen von P. Ascherson. Bot. Zeit. Jahrg. 29, 1871, pp. 443–445, 447–459, 463–467.

(This paper is based on **Delpino, F.** (1870) with certain additions: it consists of a critical compilation from the literature dealing with the pollination of *Posidonia, Cymodocea, Halodule, Zostera, Halophila, Ruppia, Vallisneria, Ceratophyllum* and *Enhalus*.)

Desmoulins, C. (1849) [p. 27] Feuilles du *Nymphaea* et du *Scirpus lacustris*. Actes de la Soc. Linnéenne de Bordeaux, T. XVI. (Sér. II. T. VI.) 1849, pp. 63–64.

(A record of the fact that the submerged leaves of *Castalia* were known to Gaspard Bauhin, and that the floating leaves of *Scirpus lacustris* were described by Scheuchzer.)

Devaux, H. (1889) [pp. 253, 254, 256] Du mécanisme des échanges gazeux chez les plantes aquatiques submergées. Ann. d. sci. nat. Sér. VII. T. 9, 1889, pp. 35–179, 8 text-figs.

(This may be regarded as the classic memoir on the physics of the gaseous exchange in submerged plants. It includes a discussion of earlier works on the subject.)

Dodoens, R. (1578) [p. 144] A Nievve Herball, or Historie of Plantes:...nowe first translated out of French into English, by Henry Lyte Esquyer. At London by me Gerard Dewes... 1578.

(This herbal contains an account of the heterophylly of the Water Buttercup.)

Dollo, L. (1912) [p. 39] Les Céphalopodes adaptés à la Vie Nectique Secondaire et à la Vie Benthique Tertiaire. Zool. Jahrb. Suppl. 15, Bd. I. 1912, pp. 105–140.

[In this paper Dollo applies the Law of Irreversibility to certain aquatic plants; see also **Arber, A.** (1919[2]).]

Douglas, D. (1880) [p. 55] Notes on the Water Thyme (*Anacharis alsinastrum*, Bab.). Science Gossip (Hardwicke's), Vol. XVI. 1880, pp. 227–229, 4 text-figs.

(The male flowers of *Elodea canadensis*, hitherto unknown in Britain, are here recorded from Scotland and are described and figured.)

Duchartre, P. (1855) Quelques mots sur la fécondation chez la Vallisnérie. Bull. de la Soc. bot. de France, T. II. 1855, pp. 289–293.

(An historical account of the different views which have been held on the question whether the male flowers of *Vallisneria* do or do not become detached from their pedicels and float to the surface of the water.)

Duchartre, P. (1858) [p. 261] Recherches expérimentales sur la transpiration des plantes dans les milieux humides. Bull. de la Soc. bot. de France, T. V. 1858, pp. 105–111.

(The author concludes from his experiments that the transpiration of a terrestrial plant can continue when it is grown in a saturated atmosphere or even when it is completely immersed in water.)

Duchartre, P. (1872) [p. 131] Quelques observations sur les caractères anatomiques des *Zostera* et *Cymodocea*, à propos d'une plante trouvée près de Montpellier. Bull. de la Soc. bot. de France, T. XIX. 1872, pp. 289–302.

[The author shows that, in the absence of the organs of fructification, *Zostera* and *Cymodocea* can be distinguished by their anatomy. This analysis of the anatomical characters of marine Angiosperms was carried much further by another French observer about twenty years later; see **Sauvageau, C.** (**1890**[1]) and following titles.]

Dudley, W. R. (1894) [p. 123] Phyllospadix, its systematic characters and distribution. Zoe, San Francisco, Vol. IV. 1894, No. 4, pp. 381–385.

(A revised diagnosis of this genus, and of the two species, *P. Scouleri*, Hook. and *P. Torreyi*, Wats.)

Dutailly, G. (1878) [p. 73] Sur la nature réelle de la "fronde" et du "cotylédon" des *Lemna*. Bull. mens. de la Soc. Linnéenne de Paris, T. I. 1874–1889, No. 19, 1878, pp. 147–149.

(This author regards the thallus of *Lemna* as "un sympode d'embryons disposés à la suite les uns des autres.")

Dutailly, G. (1892) [p. 85] La fécondation chez les *Ceratophyllum*. Bull. mens. de la Soc. Linnéenne de Paris, No. 132, 1892, p. 1056.

(The author describes the rising to the surface of the detached anthers, and the descent of the pollen through the water.)

Duval-Jouve, J. (1864) [pp. 299, 301] Lettre sur la découverte du *Coleanthus subtilis* en Bretagne. Bull. de la Soc. bot. de France, T. XI. 1864, pp. 265, 266.

(Notes on the part played by birds in the dispersal of aquatic plants.)

Duval-Jouve, J. (1872) [pp. 167, 183] Diaphragmes vasculifères des monocotylédones aquatiques. Académie des Sciences et Lettres de Montpellier. Mém. de la section des sciences, T. VIII. 1872–1875, pp. 157–176, 1 pl.

(The author of this paper shows that the occurrence of diaphragms crossing the lacunae of the leaves of aquatic Angiosperms is more general than has hitherto been supposed, and that transverse vascular connexions between the longitudinal veins are commonly associated with such diaphragms.)

Ehrhart, F. (1787) Wiedergefundene Blüte der dicken Wasserlinse (Lemna gibba L.). Ehrhart's Beiträge zur Naturkunde, Bd. I. 1787, pp. 43–51.

[An account of the finding of the flowers of *Lemna gibba* which had not been seen since they were described in **Micheli, P. A.** (**1729**).]

Engler, A. (1877) [pp. 74, 82] Vergleichende Untersuchungen über die morphologischen Verhältnisse der Araceae. II. Theil. Ueber Blattstellung and Sprossverhältnisse der Araceae. Nova Acta der Ksl. Leop.-Carol. Deutschen Akad. der Naturforscher, Bd. 39, No. 4, 1877, pp. 159–232, 6 pls.

(The author explains the nature of the shoot of the Lemnaceae on the basis of a close comparison with *Pistia*, after an exhaustive discussion of the morphology of the Araceae in general.)

Engler, A. (1879) [pp. 135, 315] Notiz über die Befruchtung von *Zostera marina* und das Wachsthum derselben. Bot. Zeit. Jahrg. 37, 1879, pp. 654–655.

[A criticism of **Hofmeister, W.** (**1852**), with remarks on the method of pollination, the branching of the sterile and fertile shoots, etc.]

Engler, A. (1892) [p. 314] Die systematische Anordnung der monokotyledoneen Angiospermen. Abhandl. d. k. Akad. d. Wiss. Berlin, 1892, Abh. II. 1892, 55 pp.

(The systematic relationships of the Helobieae are dealt with on pp. 11–20.)

Engler, A. See **Krause, K. and Engler, A.** (**1906**).

Ernst, A. (1872[1]) Ueber Stufengang und Entwickelung der Blätter von *Hydrocleis nymphoides* Buchenau (*Limnocharis Humboldtii* C. L. Richard). Bot. Zeit. Jahrg. 30, 1872, pp. 518–520.

(A brief account of heterophylly in this species.)

Ernst, A. (1872[2]) [p. 191] Ueber die Anschwellung des unter Wasser befindlichen Stammtheiles von *Aeschynomene hispidula* H. B. K. Bot. Zeit. Jahrg. 30, 1872, pp. 586–587.

(A description of the aerenchyma found in this Leguminous shrub—a native of Venezuela.)

Esenbeck, E. (1914) [pp. 151, 157 and Figs. 104, p. 158, and 105, p. 159] Beiträge zur Biologie der Gattungen *Potamogeton* und *Scirpus*. Flora, N.F. Bd. 7 (G.R. Bd. 107), 1914, pp. 151–212, 59 text-figs.

(An account of experimental and anatomical work on the land forms of *Potamogeton* and on leaf development in *Scirpus lacuster* and other Cyperaceae which are normally leafless. The author follows Goebel in regarding the water leaves of all these plants as youth leaves, to which the plant reverts under conditions of poor nutrition, rather than as direct adaptations to the medium.)

Evans, A. H. See **Fryer, A., Bennett, A. and Evans, A. H.** (1898–1915).

Fauth, A. (1903) [pp. 15, 18, 241, 242, 246, 248, 271, 297] Beiträge zur Anatomie und Biologie der Früchte und Samen einiger einheimischer Wasser- und Sumpfpflanzen. Beihefte zum Bot. Centralblatt, Bd. XIV. 1903, pp. 327–373, 3 pls.

(The fruit and seeds of *Alisma, Elisma, Sagittaria, Butomus, Callitriche, Hippuris, Myriophyllum, Limnanthemum, Menyanthes* and *Littorella* are dealt with, and certain land plants are included for comparison.)

Fenner, C. A. (1904) [p. 111] Beiträge zur Kenntnis der Anatomie, Entwicklungsgeschichte und Biologie der Laubblätter und Drüsen einiger Insektivoren. Flora, Bd. 93, 1904, pp. 335–434, 16 pls.

(One section of this paper is devoted to *Aldrovandia*.)

Ferrero, F. See **Gibelli, G. and Ferrero, F.** (1891).

Fischer, A. (1907) [p. 243] Wasserstoff- und Hydroxylionen als Keimungsreize. Ber. d. deutsch. Bot. Gesellsch. Bd. XXV. 1907, pp. 108–122.

[A study of the delayed germination characteristic of many water plants, which the author attributes to the lack of certain chemical stimuli. For a criticism see **Crocker, W.** (1907).]

Fischer, G. (1907) Die bayerischen Potamogetonen und Zannichellien. Ber. d. Bayer. Bot. Gesellschaft, München, Bd. XI. 1907, pp. 20–162.

(A detailed systematic monograph of the Bavarian Potamogetonaceae, without illustrations.)

Focke, W. O. (1893[1]) [p. 310] Eine Fettpflanze des süssen Wassers. Abhandl. naturwiss. Vereine zu Bremen, Bd. XII. Heft III. 1893, p. 408.

(This paper deals with *Montia rivularis* Gm. and its possibly xerophytic ancestry.)

Focke, W. O. (1893[2]) Fehlen der Schläuche bei Utricularia. Abhandl. naturwiss. Vereine zu Bremen, Bd. XII. 1893, p. 563.

(In this brief note the author reports the discovery of a form of *Utricularia vulgaris* without bladders. He considers that it cannot be a hybrid between *U. vulgaris* and *U. intermedia* because it resembles *U. vulgaris* in all points except the absence of bladders.)

Foerste, A. F. (1889) [p. 216] Botanical Notes. Bull. Torr. Bot. Club, Vol. XVI. 1889, pp. 266–268, 1 pl.

(On p. 266 there is a note on the adventitious buds which arise from the base of the submerged leaves in *Nasturtium lacustre*. In this species marked heterophylly occurs, the submerged leaves being pinnately dissected and the air leaves simple.)

Forel, F. A. (1901) [p. 255] Handbuch der Seenkunde. Allgemeine Limnologie (Bibl. Geog. Handbücher herausgegeben von F. Ratzel). Stuttgart, 1901.

[This general treatise on Limnology contains a chapter (pp. 161–241) on the biology of lakes.]

Forel, F. A. (1892–1904) [pp. 253, 278] Le Léman. Monographie limnologique. 3 vols. Lausanne, 1904.

(This elaborate monograph of the Lake of Geneva throws much light on the physics and chemistry of fresh waters. The Biology of the Lake is dealt with in Vol. III. pp. 1–408.)

Frank, A. B. (1872) [pp. 281, 283] Ueber die Lage und die Richtung schwimmender und submerser Pflanzentheile. Cohn's Beiträge zur Biologie der Pflanzen, Bd. I. (1870–1875) Heft 2, 1872, pp. 31–86.

[This memoir is the record of a series of experiments which the author undertook in order to examine the influences which regulate the position and direction of floating and submerged leaves. He chiefly employed *Hydrocharis*, *Trapa* and *Callitriche*. For criticisms of the work see **Karsten, G. (1888)** and **Vries, H. de (1873)**.]

Freyn, J. (1890) [p. 228] Beiträge zur Kenntniss einiger Arten der Gattung *Ranunculus*. Bot. Centralbl. Bd. XLI. 1890, pp. 1–6.

(On p. 5 the author gives some observations on the pollination of the aquatic species of *Ranunculus*.)

Fries, E. (1858) [p. 276] Kürzere briefliche Mittheilungen. Ueber Avena, Datura und Nymphaea. Bot. Zeit. Jahrg. 16, 1858, p. 73.

[These notes contain the record of the occurrence in a lake in Sweden (Fagersjö in Nerike) of a (*Nymphaea*) *Castalia* with rose-purple flowers, which is regarded by the author as a variety of *C. alba*.]

Fryer, A. (1887) [pp. 195, 330] Notes on Pondweeds. 6. On Land-forms of *Potamogeton*. Journ. of Bot. Vol. XXV. 1887, pp. 306–310.

(This paper forms one of a series of contributions made by the author to the study of this group, the majority of which are not included in this bibliography, as their interest is almost exclusively systematic. In the present paper the land forms of *Potamogeton natans*, *P. fluitans*, *P. plantagineus*, *P. heterophyllus* and *P. Zizii* are described.)

Fryer, A., Bennett, A. and Evans, A. H. (1898–1915) [pp. 58, 195, 303] The Potamogetons (Pond Weeds) of the British Isles. x + 94 pages, 60 pls., 2 text-figs. London, 1898–1915.

(A systematic monograph of the genus, as far as it is represented in Britain, with fine coloured plates by R. Morgan.)

Gardiner, W. (1883) [pp. 267, 322] On the Physiological Significance of Water Glands and Nectaries. Proc. Camb. Phil. Soc. Vol. v. 1886 (for 1883–1886). Paper read, Nov. 12, 1883, pp. 35–50, 1 pl.

[In the course of this paper the author suggests (p. 43) that Dicotyledons are typically land plants while Monocotyledons are of an essentially aquatic nature.]

Gardner, G. (1846) [p. 108] Travels in the Interior of Brazil. xvi + 562 pp., 1 map, 1 pl. London, 1846.

(This volume of travels by the Superintendent of the Royal Botanic Gardens of Ceylon contains an account on pp. 527, 528 of the curious *Utricularia nelumbifolia*.)

Gardner, G. (1847) [pp. 112, 310] Observations on the Structure and Affinities of the Plants belonging to the natural order Podostemaceae, together with a Monograph of the Indian species. Calcutta Journ. of Nat. Hist. Vol. VII. 1847, pp. 165–189.

(This paper is chiefly systematic, but points connected with the life-history are also touched upon. The author suggests that there is an affinity between the Podostemaceae and *Nepenthes*.)

Gaudichaud, C. (1826) [p. 130] Voyage autour du monde, par Louis de Freycinet. Botanique. vii + 522 pp.

[On p. 430 the filamentous pollen of *Halophila ovata* and *Ruppia antarctica* (= *Cymodocea antarctica*) is mentioned.]

Geldart, A. M. (1906) [pp. 50, 54] *Stratiotes Aloides* L. Trans. Norfolk and Norwich Naturalists' Society, Vol. VIII. 1905, pp. 181–200, 1 pl.

[This paper forms a useful account of the Water Soldier, partly drawn from **Nolte, E. F.** (1825) and other sources, but also containing original observations on the life-history of the plant.]

Géneau de Lamarlière, L. (1906) [pp. 163, 260] Sur les membranes cutinisées des plantes aquatiques. Revue gén. de Bot. T. 18, 1906, pp. 289–295.

(A micro-chemical study of the epidermis and of the cells in contact with the internal lacunae in the cases of *Ranunculus fluitans, Caltha palustris, Castalia alba, Myriophyllum spicatum, Hottonia palustris, Elodea canadensis, Potamogeton densus Glyceria spectabilis* and *Equisetum limosum*.)

Gibelli, G. and Ferrero, F. (1891) Intorno allo sviluppo dell' ovolo e del seme della *Trapa natans* L. Ricerche di anatomia e di morfologia. Malpighia, v. 1891, pp. 156–218, 11 pls.

(An elaborate and fully illustrated monograph dealing with the ovary, ovule and seed of *Trapa natans*. The vascular anatomy of the ovary is fully described, and the development of the embryo. The authors regard the embryo as a degraded structure which cannot be homologised with normal embryos.)

Gin, A. (1909) [pp. 234, 295, 303] Recherches sur les Lythracées. 166 pages, 13 pls., 28 text-figs. Thèse Doct. Univ. Paris, 1909.

(This memoir contains information about the structure, distribution, etc. of the aquatic Lythraceae.)

Glück, H. (1901) [p. 44]
Die Stipulargebilde der Monokotyledonen. Verhandl. d. Naturhist.-Med. Vereins zu Heidelberg, N.F. Bd. 7, Heft 1, 1901, pp. 1–96, 5 pls., 1 text-fig.
(In this work the morphology and biology of the stipular structures of many Monocotyledons are described, including those found in a number of aquatic forms such as Potamogetonaceae, Hydrocharitaceae, etc.)

Glück, H. (1902)
Ueber die systematische Stellung und geographische Verbreitung der *Utricularia ochroleuca* R. Hartman. Ber. d. deutsch. bot. Gesellsch. Bd. XX. 1902, pp. 141–156, 1 pl.
(This paper contains a good deal of information about the submerged species of *Utricularia* in general.)

Glück, H. (1905) [pp. 9, 19, 195, 223, 280 and Figs. 147, p. 224, 148 and 149, p. 225]
Biologische und morphologische Untersuchungen über Wasser- und Sumpfgewächse. I. Die Lebensgeschichte der europäischen Alismaceen. xxiv + 312 pp., 7 pls., 25 text-figs. Jena, 1905.
[The species studied were *Alisma Plantago*, (L.) Michalet, *A. graminifolium*, Ehrh., *Elisma natans*, Buchenau, *Echinodorus ranunculoides*, (L.) Engelm., *E. ranunculoides* var. *repens*, (Lam.), *Caldesia parnassifolia*, (Bassi) Parl., *Damasonium stellatum*, (Rich.) Pers., and *Sagittaria sagittifolia*, L. An elaborate series of culture experiments was carried out, to determine the effect of external conditions upon these plants.]

Glück, H. (1906) [Passim and Figs. 44, p. 69, 57, p. 89, 58, p. 89, 59, p. 92, 63, p. 96, 64, p. 96, 66, p. 99, 69, p. 102, 146, p. 223]
Biologische und morphologische Untersuchungen über Wasser- und Sumpfgewächse. II. Untersuchungen über die mitteleuropäischen *Utricularia*-Arten, über die Turionenbildung bei Wasserpflanzen, sowie über *Ceratophyllum*. xvii + 256 pp., 28 text-figs., 6 pls. Jena, 1906.
(An admirable account of the genus *Utricularia*, of 'winter-bud' formation in general, and of the biology of the genus *Ceratophyllum*, with special reference to the formation of 'rhizoids.')

Glück, H. (1911) [pp. 145, 188, 198, 199, 200, and Figs. 95, p. 147, 128, p. 198, 129, 130 and 131, p. 199, 134 and 135, p. 203]
Biologische und morphologische Untersuchungen über Wasser- und Sumpfgewächse. III. Die Uferflora. xxxiv + 644 pp., 8 pls., 105 text-figs. Jena, 1911.
[A detailed study of the manner of life of those plants which grow on the margin of fresh waters and have adopted an amphibious habit. As in his previous work, the author combines cultural experiments with observations in the field. He shows that a large number of shore plants have aquatic forms which have remained hitherto undescribed. Like **Glück, H. (1905)** and **(1906)** the book is beautifully illustrated and provided with a useful index.]

Glück, H. (1913)
Contributions to our Knowledge of the Species of *Utricularia* of Great Britain with Special Regard to the Morphology and Geographical Distribution of *Utricularia ochroleuca*. Ann. Bot. Vol. XXVII. 1913, pp. 607–620, 2 pls., 7 text-figs.
(The author records *Utricularia ochroleuca* from a number of stations in Great Britain and discusses the morphology, biology and distribution of this species.)

Goebel, K. (1879) [p. 225]
Ueber Sprossbildung auf Isoëtesblättern. Bot. Zeit. Jahrg. 37, 1879, pp. 1–6, 4 text-figs.
(A record of the replacement of sporangia by young plants in the case of certain examples of *Isoëtes lacustris* and *I. echinospora* from the Vosges.)

Goebel, K. (1880) [p. 12]
Beiträge zur Morphologie und Physiologie des Blattes. (Schluss.) Bot. Zeit. Jahrg. 38, 1880, pp. 833–845, 1 pl.
(On pp. 833–836 the heterophylly of *Sagittaria sagittifolia* is described. In opposition to de Candolle, Goebel takes the view that the band-shaped leaf of *Sagittaria* represents the entire leaf, not merely a modified petiole.)

Goebel, K. (1889[1])
Ueber die Jugendzustände der Pflanzen. Flora, Neue Reihe, Jahrg. 47, 1889, pp. 1–45, 6 text-figs., 2 pls.
(Pp. 40–43 contain an account of the germination of *Utricularia montana*.)

Goebel, K. (1889[2]) [pp. 93, 99]
Der Aufbau von Utricularia. Flora, Neue Reihe, Jahrg. 47 (G. R. Jahrg. 72), 1889, pp. 291–297, 1 pl.
(This paper forms a continuation of the author's previous work on *Utricularia*; *U. affinis*, *U. longifolia*, and *U. bryophila* are figured.)

Goebel, K. (1889[3]) [p. 117]
Pflanzenbiologische Schilderungen. Teil I. 239 pp., 9 plates, 98 text-figs. Marburg, 1889.
(Pp. 166–169 deal with one of the Podostemaceae, a species of *Terniola*.)

Goebel, K. (1891) [pp. 40, 100, 103, 104, 106, and Fig. 68, p. 100]
Morphologische und Biologische Studien. V. *Utricularia*. VI. *Limnanthemum*. Ann. du Jardin Bot. de Buitenzorg, Vol. IX. 1891, pp. 41–126, 11 pls.
(In these papers certain extra-European species of *Utricularia* and *Limnanthemum* are dealt with; the vexed question of the morphology of the *Utricularia* shoot receives special consideration.)

Goebel, K. (1891–1893) [Passim and Figs. 14, p. 29, 20, p. 38, 60, p. 92, 65, p. 98, 92, p. 144, 103, p. 154, 143, p. 220, 150, p. 229, 160, p. 247]
Pflanzenbiologische Schilderungen. Teil II. iv + 386 pp., 31 pls., 121 text-figs. Marburg, Lief. 1, 1891, Lief. 2, 1893.
[This work contains sections dealing with *Utricularia* (pp. 127–160, 173–181, pls. XIV, XV) and the Podostemaceae (pp. 331–354, pls. XXVI–XXX). There is also a very important general discussion of water plants from the biological standpoint (pp. 217–373, pls. XXIV, XXV, etc.).]

Goebel, K. (1895)
Ueber die Einwirkung des Lichtes auf die Gestaltung der Kakteen und anderer Pflanzen. Flora, Bd. 80, 1895, pp. 96–116, 5 text-figs.
[This paper includes a short account (pp. 110, 111) of certain experiments upon *Sagittaria* which show that want of light induces this plant to return to the 'youth form' in which only band-shaped leaves are developed. Its behaviour is thus analogous to that of *Phyllocactus* which, under similar conditions, also reverts to the youth form.]

Goebel, K. (1896) [p. 156] Ueber Jugendformen von Pflanzen und deren künstliche Wiederhervorrufung. Sitzungsber. d. math.-phys. Classe d. k. b Akademie d. Wissensch. zu München, Bd. XXVI. 1897 (for 1896), pp. 447–497, 16 text-figs.

(Pp. 487–491 are devoted to heterophylly in water plants. The author regards the band-shaped submerged leaves of many Monocotyledons, not as representing a direct adaptation to the medium, but as a juvenile form of leaf which may also be produced at later stages in the life of the plant, if the external conditions are unfavourable.)

Goebel, K. (1904) [p. 104 and Fig. 70, p. 104] Morphologische und biologische Bemerkungen. 15. Regeneration bei *Utricularia*. Flora, Bd. 93, 1904, pp. 98–126, 17 text-figs.

(Includes an account of the formation of adventitious shoots from the leaves of the water Utricularias.)

Goebel, K. (1908) [pp. 161, 281] Einleitung in die experimentelle Morphologie der Pflanzen. viii + 260 pp., 135 text-figs. Leipzig and Berlin, 1908.

(In this book heterophylly in amphibious plants is dealt with at some length, with special reference to *Myriophyllum proserpinacoides* and *Limnophila heterophylla*.)

Goebel, K. (1913) [pp. 234, 344] Morphologische und biologische Bemerkungen. 22. *Hydrothrix Gardneri*. Flora, N.F. Bd. 5 (Ganze Reihe, Bd. 105), 1913, pp. 88–100, 9 text-figs.

(An investigation of a peculiar submerged member of the Pontederiaceae with 'long' and 'short' shoots and cleistogamic flowers.)

Göppert, H. R. (1847) Ueber die Schläuche von *Utricularia vulgaris* und einen Farbestoff in denselben. Bot. Zeit. Jahrg. 5, 1847, pp. 721–726.

(An account of the structure and development of the bladder, which the author regards as a metamorphosed "Fiederblättchen." He records the occurrence of blue pigment in the cells of the bladder.)

Göppert, H. R. (1848) [p. 86] Ueber den rothen Farbestoff in den Ceratophylleen. Bot. Zeit. Jahrg. VI. 1848, pp. 147, 148.

(A record of the occurrence of a violet colouring matter, turning brown with age, in the cellular processes at the tips of the leaf segments in *Ceratophyllum*.)

Graebner, P. (1901) [p. 290] Die Heide Norddeutschlands. (Engler, A. und Drude, O. Die Vegetation der Erde, V.) xii + 320 pages, 1 map. Leipzig, 1901.

(This book contains some information about the flora of lowland heath pools.)

Graebner, P. See **Ascherson, P. and Graebner, P.** (1907).

Gratiolet, P. See **Clöez, S. and Gratiolet, P.** (1850).

Gray, A. (1848) [p. 309] Remarks on the Structure and Affinities of the Order Ceratophyllaceae. Annals of the Lyceum of Nat. Hist., New York, Vol. IV. 1848, pp. 41–50 (read Feb. 20, 1837).

(The author regards *Ceratophyllum* as allied to the Cabombaceae and Nelumbiaceae and supports this conclusion by a comparison of the seed characters.)

Greene, E. L. (1909) [p. 285] Landmarks of Botanical History. Part I. Prior to 1562 A.D. Smithsonian Misc. Coll. Vol. 54, 1909, pp. 1–329.
(On pp. 126, 127, attention is drawn to the opinions of Theophrastus upon the ecology of water plants.)

Grew, Nehemiah (1682) [p. 154] The Anatomy of Plants. 1682. 304 pp., 83 pls.
(This classic account of structural botany contains occasional references to aquatics or to subjects bearing on their study.)

Griset, H. E. (1894) Circulatory Movements of Protoplasm. Science-Gossip, Vol. I. New Series, 1894, pp. 132–133, 2 text-figs.
(The author draws attention to the stipules of *Hydrocharis Morsus-ranae* and the diaphragms of the petiole and peduncle of *Alisma Plantago* as affording excellent material for the observation of intracellular protoplasmic movements.)

Grönland, J. (1851) [p. 127] Beitrag zur Kenntniss der *Zostera marina* L. Bot. Zeit. Jahrg. IX. 1851, pp. 185–192, 1 pl.
[This account of the ovules and anthers of *Zostera* is supplemented and corrected by **Hofmeister, W. (1852)**.]

Guppy, H. B. (1893) [pp. 35, 220, 243, 244, 297, 301, 302] The River Thames as an Agent in Plant Dispersal. Journ. Linn. Soc. Bot. Vol. XXIX. 1893, pp. 333–346.
(An account of observations upon river drift in the Thames, Lea and Roding, with a discussion of the part played by birds in the dispersal of aquatic plants.)

Guppy, H. B. (1894[1]) [pp. 85, 88, 273, 274, 275, 301 and Fig. 55, p. 86] Water-Plants and their Ways. Science-Gossip, Vol. I. New Series, 1894. Their Dispersal and its Observation, pp. 145–147. Their Thermal Conditions, pp. 178–180. *Ceratophyllum demersum*, pp. 195–199, 1 text-fig.
(These short papers, though published in a popular journal, contain original observations of great importance.)

Guppy, H. B. (1894[2]) [pp. 75, 77, 275] On the Habits of *Lemna minor*, *L. gibba*, and *L. polyrrhiza*. Journ. Linn. Soc. Lond. Bot. Vol. XXX. 1895 (for 1894), pp. 323–330.
[Observations on the life-history of these forms, including a detailed study of the temperature conditions necessary for germination, flowering, etc. The paper may be regarded as supplementary to **Hegelmaier, F. (1868)**.]

Guppy, H. B. (1894[3]) [p. 274] River Temperature. Part I. Its Daily Changes and Method of Observation. Proc. Roy. Phys. Soc. Edinburgh, Vol. XII. 1892–1894, pp. 286–312.
[A more detailed consideration of the subject than in **Guppy, H. B. (1894[1])**.]

Guppy, H. B. (1896) [p. 274] River Temperature. Part III. Comparison of the Thermal Conditions of Rivers and Ponds in the South of England. Proc. Roy. Phys. Soc. Edinb. Vol. XIII. 1894–1897, pp. 204–211.
[The comparison of the temperatures of ponds with that of the Thames is treated more fully in this paper than in **Guppy, H. B. (1894[1])**.]

Guppy, H. B. (1897) [pp. 243, 244, 280, 301] On the Postponement of the Germination of the Seeds of Aquatic Plants. Proc. Roy. Phys. Soc. Edinburgh, Vol. XIII. 1894–1897, pp. 344–359.

(An account of experimental work on delayed germination of the seeds of water plants kept in water, with notes on the effect of drying, freezing and exposure to light or darkness.)

Guppy, H. B. (1906) [pp. 88, 162, 241, 296, 297, 301, 303, 304, 305] Observations of a Naturalist in the Pacific between 1896 and 1899. Vol. II. Plant-dispersal. xxviii + 627 pp., 1 pl. London, 1906.

(*The water-side plants of the British flora are considered in* Chapters III. and IV. Note 10, pp. 535–538, records the degree of buoyancy of the seeds and seed vessels of more than 300 British plants, including a large number of aquatics. The book also contains numerous other notes on water plants, e.g. distribution of *Naias*, p. 367.)

Guppy, H. B. (1917) [pp. 303, 304, 333] Plants, Seeds, and Currents in the West Indies and Azores. x + 531 pages, 3 maps, 1 pl. London, 1917.

(This book contains further developments of the author's "differentiation" hypothesis. A number of references to water plants are included.)

Gürke, M. See **Ascherson, P. and Gürke, M.** (1889).

Gwynne-Vaughan, D. T. (1897) [pp. 33, 37, 38, 182] On some Points in the Morphology and Anatomy of the Nymphaeaceae. Trans. Linn. Soc. Lond. Ser. II. Vol. V. 1895–1901, Part 7, 1897, pp. 287–299, 2 pls.

(The most important discovery recorded in this paper is that of the occurrence of clear cases of polystely in certain stem structures of the Nymphaeaceae.)

Haberlandt, G. (1914) [pp. 45, 183] Physiological Plant Anatomy, translated from the fourth German edition by Montagu Drummond. xv + 777 pages, 291 text-figs., 1914.

(This standard work contains many references to the structure of water plants and its interpretation.)

Hall, J. G. (1902) An Embryological Study of *Limnocharis emarginata*. Bot. Gaz. Vol. XXXIII. 1902, pp. 214–219, 1 pl.

(An account of the embryo-sac and embryo in this species.)

Hallier, E. (1859) [p. 192] *Aedemone mirabilis* Kotschy. Ein neues Schwimmholz vom weissen Nil, anatomisch bearbeitet. Bot. Zeit. Jahrg. 17, 1859, pp. 153–156, 1 pl.

[The anatomy of *Aedemone mirabilis*, Kotschy (=*Herminiera Elaphroxylon*, Guill. et Perr.) is described and its close resemblance is pointed out to that of *Aeschynomene paludosa*, Roxb. (=*Sesbania aculeata*, Poir.).]

Hannig, E. (1912) [pp. 260, 266] Untersuchungen über die Verteilung des osmotischen Drucks in der Pflanze in Hinsicht auf der Wasserleitung. Ber. d. deutschen bot. Gesellsch. Jahrg. XXX. 1912, pp. 194–204.

[On p. 200 the author gives an account of the differences between the osmotic pressure in leaf and root in certain water plants. For a criticism of his interpretation of his results see **Snell, K.** (1912).]

Hansgirg, A. (1903) [pp. 143, 151, 154] Phyllobiologie. xiv + 486 pp., 40 text-figs. Leipzig, 1903.

[This book includes (pp. 52–84) a summarised account, chiefly based upon previous work, of the various types of leaf met with among aquatic plants.]

Hauman-Merck, L. (1913[1]) [p. 239 and Fig. 155, p. 240] Sur un cas de géotropisme hydrocarpique chez *Pontederia rotundifolia* L. Recueil de l'Institut Bot. Léo Errera, T. IX. 1913, pp. 28–32, 1 text-fig.

(The author shows that after fertilisation the inflorescences of this plant, which have been previously held erect above the water, bend down through an angle of 180° and dip into the water where the fruits ripen.)

Hauman-Merck, L. (1913[2]) [pp. 55, 57, 236] Observations éthologiques et systématiques sur deux espèces argentines du genre *Elodea*. Recueil de l'Instit. Bot. Léo Errera, T. IX. 1913, pp. 33–39.

(An account of the morphology and mode of pollination of *Elodea densa* and *E. callitrichoides*.)

Hauman, L. (1915) **(formerly Hauman-Merck)** [p. 57] Note sur *Hydromystria stolonifera* Mey. Anales del Museo Nac. de Hist. Nat. de Buenos Aires, T. 27, 1915, pp. 325–331.

(The author draws attention to root dimorphism and hydro-anemophily in this species.)

Hausleutner, (1850[1]) [pp. 111, 289] Ueber *Aldrovanda vesiculosa*. Bot. Zeit. Jahrg. VIII. 1850, p. 600. Nachtrag zu *Aldrovandia*. Bot. Zeit. Jahrg. VIII. 1850, pp. 831, 832.

(These notes describe certain occurrences of this plant in the wild state, and give directions for its cultivation. The author shows that, in nature, it grows among reeds or protected by the leaves of Waterlilies, and that it can only be cultivated successfully if these shade conditions are reproduced.)

Hausleutner, (1850[2]) Ueber eine neue Nymphaea aus Schlesien. Bot. Zeit. Jahrg. VIII. 1850, pp. 905–908.

(This is the record of the occurrence of a new species of *Castalia*, called by the author *Nymphaea neglecta*.)

Hausleutner, (1851) [p. 111] Ueber die *Aldrovanda* in Schlesien. Bot. Zeit. Jahrg. IX. 1851, pp. 301–304.

(A discussion of the anomalous distribution of *Aldrovandia*. It has been found in Schlesia in two lakes alone. The author thinks that it is improbable that it is distributed by waterfowl, since it perishes so rapidly on being removed from the water.)

Hegelmaier, F. (1864) [pp. 169, 175, 216, 236, 311] Monographie der Gattung *Callitriche*. 64 pp., 4 pls. Stuttgart, 1864.

(In this memoir the anatomy and floral structure of the genus are fully treated and all the species are described. The geographical distribution and affinities are also discussed. The author returns to Robert Brown's opinion that this genus belongs to the Halorrhagideae, and he does not accept the newer view which relates it to the Euphorbiaceae.)

Hegelmaier, F. (1868) [pp. 73, 74, 75, 77, 80, 314 and Figs. 48, p. 76, 50, p. 79, 52, p. 81] Die Lemnaceen. Eine Monographische Untersuchung. 169 pp., 16 pls. Leipzig, 1868.

(This monograph deals with the family systematically and also discusses its affinities and distribution. The vegetative and floral morphology of the different genera and species, and their anatomy and biology, are also treated in detail.)

Hegelmaier, F. (1870) [p. 70] Ueber die Entwicklung der Blüthentheile von *Potamogeton*. Bot. Zeit. Jahrg. 28, 1870, pp. 281–289, 297–305, 313–319, 1 pl.

(An account of the morphology and development of the flowers and fruit of this genus, *P. crispus* being studied in the greatest detail.)

Hegelmaier, F. (1871) [pp. 73, 74, 80 and Fig. 47, p. 74] Ueber die Fructifikationstheile von *Spirodela*. Bot. Zeit. Jahrg. 29, 1871, pp. 621–629, 645–666, 1 pl.

(After writing his monograph of the Lemnaceae, the author obtained some of the very rare flowers of *Spirodela polyrrhiza* from N. America, on which the present illustrated account is based.)

Hegelmaier, F. (1885) [p. 73] *Wolffia microscopica*. Bot. Zeit. Jahrg. 43, 1885, pp. 241–249.

(A description of some material of this species from India.)

Henfrey, A. (1852) [p. 309] On the Anatomy of the Stem of *Victoria Regia*. Phil. Trans. Roy. Soc. Lond. 1852, pp. 289–294, 2 pls.

(An early account of the structure of this plant, which suffers from the fact that the only specimen available for study was partially decayed.)

Henslow, G. (1893) [pp. 142, 322] A Theoretical Origin of Endogens from Exogens, through Self-Adaptation to an Aquatic Habit. Journ. Linn. Soc. Bot. Vol. XXIX. 1893, pp. 485–528.

[An exposition of the author's theory of the aquatic origin of Monocotyledons. For a criticism of this paper see **Sargant, E.** (**1908**).]

Henslow, G. (1911) [pp. 322, 339] The Origin of Monocotyledons from Dicotyledons, through Self-Adaptation to a Moist or Aquatic Habit. Ann. Bot. Vol. XXV. 1911, pp. 717–744.

[A further development of the views put forward in **Henslow, G.** (**1893**) with a reply to the criticisms contained in **Sargant, E.** (**1908**).]

Hentze, W. (1848) Beschreibung einer neuen *Nymphaea*. Bot. Zeit. Jahrg. 6, 1848, pp. 601–603. Weitere Mittheilung über die Untersuchung deutscher Seerosen. Bot. Zeit. Jahrg. 6, pp. 697–702, 1848.

(These papers deal with several distinct forms of *Castalia alba*. The author leaves open the question whether these are, or are not, true species.)

Hiern, W. P. (1872) [p. 30] A Theory of the Floating Leaves in certain Plants. Proc. Camb. Phil. Soc. Vol. II. 1876 (for 1864–1876), Part XIII, read March 13, 1871, pp. 227–236.

(A mathematical paper, demonstrating the advantages conferred on a floating leaf by a circular form.)

Hildebrand, F. (1861) [p. 67] Einige Beobachtungen aus dem Gebiete der Pflanzen-Anatomie. Herrn Professor L. C. Treviranus zur Feier seines...Doctor-Jubiläums...dargebracht. 30 pp., 2 pls. Bonn, 1861.

(These miscellaneous notes include an account of the winter-buds of *Potamogeton crispus*, pp 24–26, with 1 figure.)

Hildebrand, F. (1870) Ueber die Schwimmblätter von *Marsilia* und einigen anderen amphibischen Pflanzen. Bot. Zeit. Jahrg. 28, 1870, pp. 1–8, 17–23, 1 pl.

(A record of the occurrence of floating leaves in *Marsilia quadrifolia, M. elata* and *M. pubescens*, and a comparison of the anatomy of the floating and aerial leaves in these cases, and also in *Sagittaria sagittifolia* and *Polygonum amphibium*.)

Hildebrand, F. (1885) [pp. 207, 228] Über *Heteranthera zosterifolia*. Engler's Bot. Jahrbüch. Bd. VI. 1885, pp. 137–145, 1 pl.

[Observations on living plants of this species grown from Brazilian seed. The author draws attention to the floating leaves, which were not noticed in **Solms-Laubach, H. Graf zu** (1883).]

Hiltner, L. (1886) [p. 233] Untersuchungen über die Gattung *Subularia*. Engler's Bot. Jahrbüch. Bd. 7, 1886, pp. 264–272, 1 pl., 1 text-fig.

(The author concludes, from a study of their morphology and anatomy, that *Subularia monticola* and the forms of *S. aquatica* are not true species, but owe their differences to their varying environments.)

Hochreutiner, G. (1896) [pp. 174, 204, 205, 245, 261, 281, 282 and Fig. 137, p. 206] Études sur les Phanérogames aquatiques du Rhône et du Port de Genève. Rev. gén. de Bot. T. VIII. 1896, pp. 90–110, 158–167, 188–200, 249–265, 1 pl., 15 text-figs.

[The first part of these studies consists of a detailed account of the morphology, anatomy and development of *Zannichellia palustris* (pp. 90–110). The remaining instalments deal with the following branches of the physiology of submerged plants:—the ascent of water (pp. 158–167), geotropism (pp. 188–200, 249–258), hydrotropism (pp. 258–263), rheotropism (pp. 263–264) and heliotropism (pp. 264–265).]

Hochreutiner, G. (1897) Notice sur la Répartition des Phanérogames dans le Rhône et dans le Port de Genève. Bull. de l'Herbier Boissier, Année V. No. 1, 1897, pp. 1–14, 1 pl.

(A study of the distribution and ecology of the water plants of this region.)

Hoffmann, J. F. (1840) [p. 78] Beiträge zur näheren Kenntniss von *Lemna arrhiza* nebst einigen Bemerkungen über *L. polyrrhiza, gibba, minor* und *trisulca*. Wiegmann's Archiv für Naturgeschichte, Berlin, Jahrg. 6, 1840, pp. 138–163, 2 pls.

(A translation of this memoir by Buchinger appeared in Ann. d. Sci. nat. Sér. II. T. XIV. Bot. pp. 223–242.)

Hofmeister, W. (1852) [p. 135] Zur Entwickelungsgeschichte der *Zostera*. Bot. Zeit. Jahrg. X. 1852, pp. 121–131, 137–149, 1 pl.

[An account of the development of the pollen, ovule and embryo of *Zostera*, which supplements and corrects **Grönland, J.** (1851). Some account of *Ruppia* is also given. For criticism see **Engler, A.** (1879).]

Hofmeister, W. (1858) [p. 82] Neuere Beobachtungen über Embryobildung der Phanerogamen. Pringsheim's Jahrbüch. f. wissenschaft. Bot. Bd. I. 1858, pp. 82–188, 4 pls.

[This memoir contains some account of the ovule and embryo of the following water plants:—*Alisma*, p. 147, *Lemna*, p. 152, *Naias*, p. 145, *Nelumbium*, p. 85, *Nuphar* (*Nymphaea*), p. 83, *Pistia*, p. 152, *Pontederia*, p. 166, *Trapa*, p. 105, *Zannichellia*, p. 147.]

Holm, T. (1885) [p. 129] Recherches anatomiques et morphologiques sur deux monocotylédones submergées (*Halophila Baillonii* Asch. et *Elodea densa* Casp.). Bihang till k. Svenska Vet.-Akad. Handlingar, Bd. 9, No. 13, 1885, 24 pp., 4 pls.

[These two species are described in some detail. In the case of *Halophila*, this paper may be regarded as supplementary to **Balfour, I. B.** (1879).]

Hooker, J. D. (1847) [pp. 233, 311] The Botany of the Antarctic Voyage of H.M. Discovery Ships *Erebus* and *Terror*. I. Flora Antarctica. Part II. 364 pp., 198 pls. London, 1847.

(On p. 334 there is an account of cleistogamy in *Limosella*.)

Hooker, J. D. (1887) On Hydrothrix, a new genus of Pontederiaceae. Ann. Bot. Vol. I. 1887–1888, pp. 89–94, 1 pl.

[This paper is chiefly of systematic interest. *Hydrothrix* is a reduced and aberrant member of the family. See also **Goebel, K.** (1913).]

Hope, C. W. (1902) [pp. 192, 214] The 'Sadd' of the Upper Nile: its Botany compared with that of similar Obstructions in Bengal and American waters. Ann. Bot. Vol. XVI. 1902, pp. 495–516.

(An account of the plants which play a part in the great vegetable accumulations that form barriers on the Nile, the floating vegetation of Bengal, etc.)

Horen, F. van (1869) [pp. 74, 75, 76] Observations sur la physiologie des Lemnacées. Bull. de la Soc. Roy. de Bot. de Belgique, T. VIII. 1869, pp. 15–88, 1 pl.

[These observations, which deal largely with the hibernation of the Lemnaceae, are intended by the author to supplement **Hegelmaier, F.** (1868). For an English version of this paper see **Horen, F. van** (1870).]

Horen, F. van (1870) On the Hibernation of Lemnaceae. Journ. Bot. Vol. VIII. 1870, pp. 36–40.

[This is an abridged translation by A. W. Bennett of **Horen, F. van** (1869).]

Hovelacque, M. (1888) [pp. 104, 107] Recherches sur l'appareil végétatif des Bignoniacées, Rhinanthacées, Orobanchées et Utriculariées. Paris, 1888.

(The fourth part of this book—pp. 635–745, 126 text-figs.—deals with the anatomy of the Utriculariaceae.)

Humboldt, A. de, and Bonpland, A. (1808) [p. 189] Plantae Aequinoctiales. T. I. vii + 232 pp., 65 pls. Paris, 1808.

[The aerenchyma of *Mimosa lacustris* (*Neptunia oleracea*, Lour.) is noticed on p. 56, but it is mistakenly described as a foreign body.]

Hutchinson, J. (1916) [pp. 313, 321, 324] Aquatic Compositae. Gard. Chron. Vol. 59, 1916, p. 305, 4 text-figs.

(An account of *Bidens Beckii*, *Cotula myriophylloides*, *Pectis aquatica* and *Erigeron heteromorphus*.)

Im Thurn, E. F. (1883) [pp. 118, 120, 300] Among the Indians of Guiana. xvi + 445 pp., 10 pls., 43 text-figs., 1 map. London, 1883.

(This book contains references to the Podostemaceae and *Victoria regia* living in Guiana waters.)

Im Thurn, E. F. and Oliver, D. (1887) [p. 109] The Botany of the Roraima Expedition of 1884: being Notes on the Plants observed, by Everard F. im Thurn; with a list of the Species collected, and Determinations of those that are new, by Prof. Oliver, F.R.S., F.L.S., and others. Trans. Linn. Soc. Lond. Ser. II. Vol. II. Bot. 1881–1887, Part XIII. 1887, pp. 249–300, 10 pls.

(This memoir contains an account of the curious mode of life of *Utricularia Humboldtii*, which lives in the water collected in the leaf axils of a Bromeliad.)

Irmisch, T. (1853) [pp. 26, 87] Kurze botanische Mittheilungen. 6. *Nymphaea alba* und *Nuphar luteum*. 7. *Potamogeton densus*. 8. Dauer der *Ceratophyllum*-Arten. Flora, N.R. Jahrg. XI. (G.R. Jahrg. XXXVI.) 1853, pp. 527–528, 1 pl.

[In these notes attention is called to the *stipula axillaris* of *Nymphaea* (*Castalia*) *alba*—the fact that the apparently opposite leaves of *Potamogeton densus* are really alternate—and the fact that the shoots of *Ceratophyllum* may vegetate through the winter.]

Irmisch, T. (1854) [Fig. 112, p. 173] Bemerkung über *Hippuris vulgaris* L. Bot. Zeit. Jahrg. 12, 1854, pp. 281–287, 1 pl.

(A detailed account of the mode of branching of the sympodial stems.)

Irmisch, T. (1858[1]) Botanische Mittheilungen. 1. Ueber *Utricularia minor*. Flora, Neue Reihe, Jahrg. XVI. (Ganz. Reihe, Jahrg. XLI.) 1858, pp. 33–37, 1 pl.

(An account of the morphology of this species; the author interprets the branching in connexion with the inflorescence axis as sympodial.)

Irmisch, T. (1858[2]) [pp. 52, 271] Ueber das Vorkommen von schuppen- oder haarförmigen Gebilden innerhalb der Blattscheiden bei monokotylischen Gewächsen. Bot. Zeit. Jahrg. 16, 1858, pp. 177–179.

(This paper records the occurrence of "squamulae intravaginales" in a number of Helobieae.)

Irmisch, T. (1858[3]) [p. 59] Ueber einige Arten aus der natürlichen Pflanzenfamilie der Potameen. Abhandl. d. naturwiss. Vereines f. Sachsen und Thüringen in Halle, Bd. II. Berlin, 1858 (Vol. published 1861), pp. 1–56, 3 pls.

(The morphology and life-history of *Potamogeton natans* L., *P. lucens* L., *P. crispus* L., *P. obtusifolius* M. et K., *P. pectinatus* L., and also of *Zannichellia* and *Ruppia*, are described and illustrated with the thoroughness characteristic of the author.)

Irmisch, T. (1859[1]) [pp. 169, 245] Bemerkungen über einige Wassergewächse. Bot. Zeit. Jahrg. 17, 1859, pp. 353–356.

(Notes on the morphology of *Myriophyllum, Callitriche, Potamogeton trichoïdes, Hydrocharis* and *Stratiotes*.)

Irmisch, T. (1859[2]) Zur Naturgeschichte des *Potamogeton densus* L. Flora, N.R. Jahrg. XVII. (G.R. Jahrg. XLII.) 1859, pp. 129–139, 1 pl.

[This paper is supplementary to **Irmisch, T. (1858[3])**.]

Irmisch, T. (1861) [p. 205] Ueber *Polygonum amphibium, Lysimachia vulgaris, Comarum palustre* und *Menyanthes trifoliata*. Bot. Zeit. Jahrg. 19, 1861, pp. 105–109, 113–115, 121–123, 1 pl.

(A description of the seedlings and of the development and morphology of the mature plant in the four species named.)

Irmisch, T. (1865) [p. 244] Beitrag zur Naturgeschichte des *Stratiotes Aloides*. Flora, N.R. Jahrg. XXIII. (G.R. Jahrg. XLVIII.) 1865, pp. 81–91, 1 pl.

(The fruit, seed and seedling of *Stratiotes aloides*, the seedling of *Naias major* and the axillary shoots of *Hydrocharis Morsus-ranae*, and *Vallisneria spiralis* are described and figured in this paper.)

Ito, T. (1899) [p. 83] Floating-apparatus of the Leaves of *Pistia Stratiotes*, L. Ann. Bot. Vol. XIII. 1899, p. 466.

(Notes on the structure and mode of flotation of the leaves of this plant which was studied in its native habitat.)

Jaensch, T. (1884[1]) [p. 192] Nachtrag zur Kenntniss von *Herminiera Elaphroxylon* G.P.R. Ber. d. deutsch. bot. Gesellsch. Bd. II. 1884, pp. 233–234.

(A note on the occurrence of *Aedemone mirabilis*, Kotschy, in Senegambia.)

Jaensch, T. (1884[2]) [p. 192] Zur Anatomie einiger Leguminosenhölzer. Ber. d. deutsch. bot. Gesellsch. Bd. II. 1884, pp. 268–292, 1 pl.

[This memoir includes an account of the structure of the wood in *Herminiera Elaphroxylon* (*Aedemone*), *Aeschynomene* and *Sesbania*. For a criticism see **Klebahn, H. (1891)**.]

Jäggi, J. (1883) [p. 302] Die Wassernuss, *Trapa natans* L., und der Tribulus der Alten. Zurich, 34 pp., 1 pl., 1883.

(This paper deals mainly with the history of *Trapa*, its distribution and uses, and the causes which are leading to its extinction in Switzerland and elsewhere. The author regards it as a plant of Southern Europe introduced into other localities at a very early period as a food plant.)

Jahn, E. (1897) [pp. 30, 31] Über Schwimmblätter. Fünfstück's Beiträge zur Wissens. Botanik, Bd. I. 1897, pp. 281–294, 1 pl.

(A general consideration of floating leaves, especially of the manner in which they are supposed to be adapted to their mechanical needs.)

Jeffrey, E. C. (1899) [p. 180] The Morphology of the Central Cylinder in the Angiosperms. Trans. Canad. Inst. Vol. VI. 1899, pp. 599–636, 5 pls.

[The section of this paper relating to polystely should be read in connexion with **Scott, D. H.** (1891).]

Johnston, G. (1853) [p. 210] The Botany of the Eastern Borders, London, 1853. xii + 336 pp., 13 pls.

(Pp. 191–192 give an early account of the spread of *Elodea canadensis* in this country.)

Jönsson, B. (1883) [p. 282] Der richtende Einfluss strömenden Wassers auf wachsende Pflanzen und Pflanzentheile (Rheotropismus). Ber. d. deutsch. Bot. Gesellsch. Bd. I. 1883, pp. 512–521.

(The author proposes the term "Rheotropismus" for the directive influence exerted upon plants by a water current.)

Jönsson, B. (1883–1884) [p. 236] Om befruktningen hos slägtet *Najas* samt hos *Callitriche autumnalis*. Lunds Univ. Års-skrift, Tom. xx. Part IV. 1883–1884, 26 pages, 1 pl.

(A Swedish paper with a résumé in German dealing with the pollination of *Naias* and *Callitriche*.)

Juel, O. (1910) [p. 312] *Cynomorium* und *Hippuris*. Svensk. Bot. Tidskrift, Bd. 4, 1910, pp. 151–159, 6 text-figs.

(A comparison of these two genera leads the author to the conclusion that there is little ground for assuming a relationship between them. He shows that it is not even certain that *Hippuris* belongs to the Choripetalae, but if placed in this group it is best treated as representing a distinct family allied to *Halorrhagideae*.)

Juel, O. (1911) [p. 312] Studien über die Entwicklungsgeschichte von *Hippuris vulgaris*. Nova Acta Regiae Societatis Scientiarum Upsaliensis, Ser. IV. Vol. 2, N. 11, 1907–1911, 26 pp., 3 pls.

(The author's study of the development of the ovule and embryo leads him to the conclusion that the systematic position of the genus is still uncertain.)

Jussieu, A.L.de (1789) [p. 315] Genera Plantarum. lxxii + 499 pp. Paris, 1789.

(In this work *Zostera* is included among the Aroids, see p. 24.)

Kalberlah, A. (1895) [p. 76] Das Blühen der Wasserlinsen. Zeitschrift. f. Naturwissenschaften, Bd. 68 (Folge v. Bd. 6), 1895, pp. 136–138.

(The author's observations suggest certain minor corrections in the accounts hitherto published of the flowering of *Lemna minor*.)

Kamienski, F. (1877) [pp. 100, 103 and Fig. 67, p. 100] Vergleichende Untersuchungen über die Entwickelungsgeschichte der Utricularien. Bot. Zeit. Jahrg. 35, 1877, pp. 761–776, 1 pl.

(This paper is concerned with the embryology, germination and anatomy of *Utricularia vulgaris*.)

Karsten, G. (1888) [pp. 146, 284] Ueber die Entwickelung der Schwimmblätter bei einigen Wasserpflanzen. Bot. Zeit. Jahrg. 46, 1888, pp. 565–578, 581–589.

[The author has repeated and extended the experiments on the regulation of growth of the petiole in floating leaves recorded by **Frank, A. B. (1872)** and he comes to conclusions differing from those of the latter author. He employed *Hydrocharis*, *Marsilea* and *Ranunculus sceleratus*.]

Keller, I. A. (1893) [p. 272] The Glandular Hairs of *Brasenia peltata* Pursch. Proc. Acad. Nat. Sci. Philadelphia, 1894 (for 1893), pp. 188–193, 1 pl.

(The author shows that the mucilaginous coat covering the younger parts of this member of the Nymphaeaceae is due to the secretory activity of ephemeral hairs.)

Kerner, A. and Oliver, F. W. (1894–1895) [p. 301] The Natural History of Plants. 2 vols., 1760 pp., 1000 figs., 16 pls.

(This well-known book includes a good deal of information about water plants.)

Kingsley, M. H. (1897) [p. 213] Travels in West Africa. xvi + 743 pp., 47 illustrations. London, 1897.

(Pp. 378–380 contain an account of the rapid multiplication of *Pistia Stratiotes* in the river Ogowé.)

Kirchner, O. von, Loew, E. and Schröter, C. (1908, etc.) [pp. 50, 59, 74, 81, 123, 205, 206, 276 and Figs. 49, p. 79, 136, p. 205] Lebensgeschichte der Blütenpflanzen Mitteleuropas. Bd. I. Abth. I. and III., and Bd. II. Abth. III. Stuttgart, 1908, 1909 and 1917.

(The life-history of the following aquatic groups is dealt with: Helobieae, Abth. I. pp. 394–714, 195 text-figs.; Lemnaceae, Abth. III. pp. 57–80, 23 text-figs.; Ceratophyllaceae, Bd. II. Abth. III. pp. 51–73, 16 text-figs.)

Kirchner, O. See **Schröter, C. and Kirchner, O. (1902)**.

Kirschleger, F. (1856) [p. 12] Etwas über fluthende Pflanzen (Plantae fluitantes) und sonstige Notizen in Bezug auf die rheinische Flora. Flora, N.R. Jahrg. XIV. (G.R. Jahrg. XXXIX.) 1856, pp. 529–536.

(Observations on the forms of *Sagittaria sagittifolia*, *Scirpus lacustris* and *Sparganium simplex* with floating leaves.)

Kirschleger, F. (1857) [p. 287] Nachtrag zu der Notiz über fluthende Pflanzen. Flora, N.R. Jahrg. XV. (G.R. Jahrg. XL.) 1857, pp. 193–194.

[A continuation of **Kirschleger, F. (1856)** giving further references, and an account of the dependence of *Scirpus lacustris* upon the nature of the soil.]

Klebahn, H. (1891) [p. 192] Ueber Wurzelanlagen unter Lenticellen bei *Herminiera Elaphroxylon* und *Solanum Dulcamara*. Nebst einem Anhang über die Wurzelknöllchen der ersteren. Flora, N.R. Jahrg. 49 (G.R. Jahrg. 74), 1891, pp. 125–139, 1 pl.

[The author shows, in opposition to **Jaensch, T.** (1884[2]) that the lenticels of *Herminiera Elaphroxylon*, G.P.R. (*Aedemone mirabilis*, Kotschy) are not "Markstrahlrindenporen," but are lenticels of normal structure. He also describes, both in this plant and in *Solanum Dulcamara*, the occurrence beneath the lenticels of rudimentary adventitious roots, which may develop under favourable circumstances.]

Klebs, G. (1884) [p. 245 and Fig. 158, p. 245] Beiträge zur Morphologie und Biologie der Keimung. Unters. bot. Inst. Tübingen, Bd. I. Heft 4, 1884 pp. 536–635, 24 text-figs.

(In this paper the seedlings of certain water plants come under consideration.)

Klinge, J. (1881) [pp. 15, 18] Ueber *Sagittaria sagittaefolia* L. Sitzungsber. d. Naturforscher-Gesellsch. bei d. Univ. Dorpat, Bd. v. Heft III. 1881 (for 1880), pp. 379–408.

(A description of the morphology and anatomy of this species and of the different forms in which it occurs.)

Klinsmann, F. (1860) [p. 54] Ein Beitrag zur Entwickelungsgeschichte von *Stratiotes aloides*. Bot. Zeit. Jahrg. 18, 1860, pp. 81–82, 1 pl.

(The author succeeded in germinating the seeds of this plant and describes and figures the seedling.)

Knoch, E. (1899) [p. 34] Untersuchungen über die Morphologie, Biologie und Physiologie der Blüte von *Victoria Regia*. Bibliotheca Botanica, Bd. IX. 1899, Heft 47, 60 pp., 6 pls.

(In this memoir the anatomy and morphology of the flower of *Victoria regia* are dealt with, and special attention is paid to the development of heat at the flowering period.)

Knupp, N. D. (1911) [p. 232] The Flowers of *Myriophyllum spicatum* L. Proc. Iowa Acad. Sci. (Des Moines), Vol. XVIII. 1911, pp. 61–73, 4 pls.

(A study of the development and general structure of the flowers of this species.)

Koch, K. (1852) [pp. 82, 316] Ueber *Pistia* in Allgemeinen und *Pistia Turpini* Blume insbesondere. Bot. Zeit. Jahrg. 10, 1852, pp. 577–585, 1 pl.

(The author describes the germination of *Pistia*, which he regards as differing from that of *Lemna*. He also emphasizes the dissimilarity of *Pistia* and the Aroids. He describes the flowers of *Pistia Turpini* which he observed in the living state.)

Koehne, E. (1884) Ueber Zellhautfalten in der Epidermis von Blumenblättern und deren mechanische Function. Ber. d. deutsch. bot. Gesells. Bd. II. 1884, pp. 24–29, 1 pl.

(The author shows that the folds in the lateral walls of epidermal cells of petals serve a mechanical purpose in strengthening the organ. This conclusion may have some bearing on the loss of folding in the epidermal cells of water leaves.)

Korzchinsky, S. (1886) [p. 244] Ueber die Samen der *Aldrovandia vesiculosa* L. Bot. Centralbl. Bd. XXVII. 1886, pp. 302–304, 334–335, 1 pl.
(An account of the structure of the ripe seed and the germination of this plant.)

Kotschy, T. (1858) [p. 192] Eine neue Leguminose vom weissen Nil. Oesterreichische Bot. Zeitschrift, Jahrg. VIII. 1858, No. 4, pp. 113–116, 1 pl.
(The first scientific description of *Aedemone mirabilis*, Kotschy = *Herminiera Elaphroxylon*, G.P.R.)

Krause, K. and Engler, A. (1906) [pp. 62, 142, 143, 154, 239, 305, 314] Aponogetonaceae, in Das Pflanzenreich, IV. 13 (herausgegeben von A. Engler), 24 pp., 71 text-figs. Leipzig, 1906.
(A monograph of this family of hydrophytes.)

Kurz, S. (1867) [pp. 73, 291] Enumeration of Indian Lemnaceae. Journ. Linn. Soc. Bot. Vol. IX. 1867, pp. 264–268, 1 pl.
(A systematic paper with some general notes on distribution of the group.)

Lamarck, J. B. P. A. (1809) [p. 155] Philosophie Zoologique. 2 vols. Paris, 1809.
(The heterophylly of the Water Crowfoot is discussed in Vol. I. Chapter VII. p. 230.)

Land, W. J. G. See **Coulter, J. M. and Land, W. J. G.** (1914).

Lebel, E. (1863) [pp. 134, 195, 216, 311] *Callitriche*. Esquisse Monographique. Mém. de la Soc. Imp. des Sci. Nat. de Cherbourg, T. XI. 1863, pp. 129–176.
(A systematic monograph of the genus, dealing also with its anatomy, relationships, etc.—a most vividly written and interesting memoir.)

Lewakoffski, N. (1873[1]) [p. 188] Ueber den Einfluss des Wassers auf das Wachsthum der Stengel und Wurzeln einiger Pflanzen. (Gelehrte Schriften der k. Universität in Kasan, 1873.) Abstracted in Just's Bot. Jahresbericht, Jahrg. I. 1873, p. 594.
(According to the abstract, this Russian paper deals with the development of aerenchyma in the stems and roots of *Epilobium hirsutum*, *Lycopus europaeus* and two species of *Lythrum*, when grown in water.)

Lewakoffski, N. (1873[2]) [p. 200] Zur Frage über den Einfluss des Mediums auf die Form der Pflanzen. (Gelehrte Schriften der k. Universität in Kasan, 1873.) Abstracted in Just's Bot. Jahresbericht, Jahrg. I. 1873, pp. 594, 595.
(According to the abstract this Russian paper deals with the effect of an aquatic life on *Rubus fruticosus*.)

Lewakoffski, N. (1877) [p. 200] Ueber den Einfluss des Wassers auf die Entwickelung einiger Arten von Salix (Beilage zu dem Protocolle der 91. Sitzung der Naturforsch. an der Universität zu Kazan). Abstracted in Just's Bot. Jahresber. Jahrg. V. 1879 (for 1877), pp. 575, 576.
(According to the abstract, this Russian paper deals chiefly with the effect of submergence on *Salix* shoots and demonstrates that very little effect is produced on their anatomy by water life.)

Lindberg, S. O. (1873) [p. 46] Is *Hydrocharis* really Dioecious? Trans. and Proc. Bot. Society, Edinburgh, Vol. XI. 1873, p. 389.
(The author suggests that *Hydrocharis*, instead of being dioecious as is commonly supposed, is really monoecious or monoico-female.)

Lister, G. (1903) [p. 113] On the Occurrence of *Tristicha alternifolia*, Tul., in Egypt. New Phyt. Vol. II. 1903, pp. 15–18, 1 pl.
(The author discovered *Tristicha alternifolia* Tul. var. *pulchella* Warmg. in rushing water below the first cataract on the Nile; this is the first record of the family Podostemaceae from Egypt.)

Loeselius, J. (1703) [pp. 11, 20] Flora Prussia, sive Plantae in Regno Prussiae sponte nascentes...Curante Johanne Gottsched...Regiomonti ...Sumptibus Typographiae Georgianae, 1703.
[*Plantago aquatica* (=*Alisma graminifolium*, Mich.), Pl. 62 and p. 199, and *Sagittaria aquatica foliis variis* (=*Sagittaria sagittifolia*, L.), Pl. 74 and p. 234, represented in both cases with the ribbon type of leaf and bearing an inflorescence.]

Loew, E. See **Kirchner, O. von, Loew, E. and Schröter, C.** (**1908, etc.**).

Loew, O. (1893) Worauf beruht die alkalische Reaction, welche bei Assimilationsthätigkeit von Wasserpflanzen beobachtet wird? Flora, Bd. 77, 1893, pp. 419–422.
(The red coloration obtained, when phenolphthalein is added to the water in which aquatic plants are living and assimilating, is attributed by the author to calcium carbonate held in a colloidal state by the presence of organic matter.)

Ludwig, F. (1881) [pp. 80, 84] Ueber die Bestäubungsverhältnisse einiger Süsswasserpflanzen und ihre Anpassungen an das Wasser und gewisse wasserbewohnende Insekten. Kosmos (Stuttgart), Jahrg. V. Bd. X. 1881, pp. 7–12, 17 text-figs.
(Observations on the pollination of *Lemna, Callitriche, Myriophyllum* and *Ceratophyllum*.)

Ludwig, F. (1886) [p. 243] Ueber durch Austrocknen bedingte Keimfähigkeit der Samen einiger Wasserpflanzen. Biol. Centralbl. Bd. VI. No. 10, 1887 (for 1886), pp. 299, 300.
(A note on the effect of drying on the seeds of *Mayaca fluviatilis*.)

Luetzelburg, P. von (1910) [pp. 91, 96, 102, 105 and Fig. 71, p. 105] Beiträge zur Kenntnis der Utricularien. Flora, Bd. 100, 1910, pp. 145–212, 48 text-figs.
(This paper is concerned, in part, with the aquatic Utricularias. The secretions of the bladders are investigated, and a number of cultural experiments are described.)

Lundström, A. N. (1888) [p. 62] Ueber farblose Oelplastiden und die biologische Bedeutung der Oeltropfen gewisser Potamogeton-Arten. Bot. Centralbl. Bd. XXXV. 1888, pp. 177–181.
(A discussion of the cause and significance of the oily surface possessed by the leaves of some submerged Potamogetons.)

Lyte, H. See **Dodoens, R.** (**1578**).

McCallum, W. B. (1902) [p. 160]
On the nature of the stimulus causing the change of form and structure in *Proserpinaca palustris*. Bot. Gaz. Vol. XXXIV. 1902, pp. 93–108, 10 text-figs.
[Experimental work on the "land type" and "water type" of leaf in *Proserpinaca palustris*. This paper should be read in conjunction with **Burns, G. P.** (1904).]

MacCaughey, V. (1917) [p. 182]
Gunnera petaloidea Gaud., a remarkable plant of the Hawaiian Islands. American Journ. Bot. Vol. IV. 1917, pp. 33–39.
(A "titanic herbaceous-perennial" belonging to a genus whose anatomy is of interest in relation to that of certain water plants.)

MacDougal, D. T. (1914) [p. 162]
The Determinative Action of Environic Factors upon *Neobeckia acquatica* Greene. Flora, N.F. Bd. VI. (G.R. Bd. 106), 1914, pp. 264–280, 14 text-figs.
(A study of the heterophylly of this plant under a variety of conditions.)

MacLeod, J. (1893 and 1894) [pp. 9, 230]
Over de bevruchting der bloemen in het kempisch gedeelte van Vlaanderen. Bot. Jaarboek, Vol. V. 1893, pp. 156–452, 58 text-figs.; Vol. VI. 1894, pp. 119–511, 65 text-figs.
(The second part of this elaborate memoir on the pollination of the plants of Flanders, concludes with an index and a summary in French.)

Magnin, A. (1893) [pp. 274, 279, 287, 290, 323]
Recherches sur la végétation des lacs du Jura. Rev. gén. de Bot. T. V. 1893, pp. 241–257, 303–316, 8 text-figs.
(An ecological survey of 62 out of the 66 lakes which occur in the Jura region.)

Magnus, P. (1870[1])
Beiträge zur Kenntniss der Gattung *Najas* L. viii + 64 pages, 8 pls. Berlin, 1870.
(This monograph of the genus contains an historical account of the literature, a description of the germination, general morphology, apical development and anatomy, and a discussion of the interpretation of the peculiar floral structure.)

Magnus, P. (1870[2]) [p. 135]
Ueber die Anatomie der Meeresphanerogamen. Sitzungs-Berichte d. Gesellsch. naturforsch. Freunde zu Berlin, Dec. 20, 1870, pp. 85–90.
[An anatomical account of some marine Phanerogams, which should be read in connexion with **Ascherson, P.** (1870).]

Magnus, P. (1871) [p. 169]
Einige Bemerkungen zu dem Aufsatze des Herrn J. Borodin "Ueber den Bau der Blattspitze einiger Wasserpflanzen." Bot. Zeit. Jahrg. 29, 1871, pp. 479–484.
[A criticism of **Borodin, J.** (1870). The author points out the analogy of the ephemeral stomates described by Borodin at the leaf apex of *Callitriche* with the stomates found at the nerve endings of the leaves of such land plants as *Crassula*. He shows, on the other hand, that the analogy, suggested by Borodin, with the outgrowths at the leaf apices of *Myriophyllum* and *Ceratophyllum* does not hold.]

Magnus, P. (**1872**) [p. 135] Untersuchungen über die Anatomie der Cymodoceen. Sitzungs-Ber. d. Gesellschaft naturforsch. Freunde zu Berlin, 1872, pp. 30–33.
(These notes are chiefly devoted to the occurrence of "Schlauchgefässe" in *Cymodoçea*.)

Magnus, P. (**1883**) [p. 332] Ueber eine besondere geographische Varietät der *Najas graminea* Del. und deren Auftreten in England. Ber. d. deutsch. bot. Gesellsch. Bd. I. 1883, pp. 521–524.
[This paper on a form of *Naias graminea* which grows in the Egyptian rice fields should be read in connexion with **Ascherson, P.** (**1874**) and **Bailey, C.** (**1884**).]

Magnus, P. (**1894**) Ueber die Gattung *Najas*. Ber. d. deutsch. bot. Gesellsch. Bd. XII. 1894, pp. 214–224, 1 pl., 3 text-figs.
[A reply to the criticisms on **Magnus, P.** (**1870**[1]) contained in **Schumann, K.** (**1892**).]

Magnus, W. and Werner, E. (**1913**) [p. 121] Die atypische Embryonalentwicklung der Podostemaceen. Flora, N.F. Bd. 5 (G.R. Bd. 105), 1913, pp. 275–336, 4 pls., 41 text-figs.
(A detailed comparative account of the embryo-sac and embryo in the Podostemaceae, with a general discussion of the ecological and morphological significance of the peculiarities observed.)

Maisonneuve, D. de (**1859**) [p. 110] Aldrovandia. Bull. de la Soc. bot. de France, T. VI. 1859, pp. 399–401.
(The author of this note shows that many plants of *Aldrovandia* may remain at the bottom of the water even in June, weighted down by the remnant of the turion.)

Marloth, R. (**1883**) [p. 241] Über mechanische Schutzmittel der Samen gegen schädliche Einflüsse von aussen. Engler's Bot. Jahrb. Bd. IV. 1883, pp. 225–265, 1 pl.
(A detailed account of the protective layers in seed coats, including references to certain water plants. The paper concludes with an index of the species studied.)

Marshall, W. (**1852**) [pp. 55, 210] Excessive and noxious Increase of *Udora Canadensis* (*Anacharis Alsinastrum*). Phytologist, Vol. IV. 1852, pp. 705–715.
(An historical account of the introduction of this plant.)

Marshall, W. (**1857**) [pp. 55, 210] The American Water-weed. *Anacharis Alsinastrum*. Phytologist, Vol. II. N.S. 1857–8, pp. 194–197.
[An additional note on the nuisance caused by this weed; see **Marshall, W.** (**1852**).]

Martens, G. von (**1824**) [p. 135] Reise nach Venedig. Ulm, 1824.
[This book contains (p. 623) an early reference to the heterophylly of *Sagittaria sagittifolia*. There is also a mention (p. 550) of the part played by *Zostera marina* in the Venetian lagoons, and its use from time immemorial in packing Venetian glass.]

Martins, C. (1866) [pp. 189, 192]
(1) Sur les racines aérifères ou vessies natatoires des espèces aquatiques du genre Jussiaea L. (2) Sur la synonymie et la distribution géographique du *Jussiaea repens* de Linné. Mémoires de la section d. sci. Acad. des Sci. et Lettres de Montpellier, Vol. VI. 1866, pp. 353–381, 4 pls.
(An account of the air roots of *Jussiaea*. Habit drawings of three species are given. The same papers appeared without illustrations in Bull. Soc. bot. de France, T. XIII. pp. 169–189, 1866.)

Massart, J. (1910) [pp. 198, 283, 291 and Figs. 13, p. 28, 99 and 100, p. 152]
Esquisse de la Géographie botanique de la Belgique. Recueil de l'Inst. bot. Léo Errera, Tome supplémentaire VII. bis. xi + 332 pp., 101 text-figs. Brussels, 1910.
(This work, which deals exhaustively with the ecology of Belgium, contains a certain amount of information about aquatics—see especially pp. 115–123. There is also a separate "annexe" with numerous photographs of the vegetation, including a number of pictures of water plants.)

Matthews, J. R. (1914) [p. 289 and Fig. 165, p. 288]
The White Moss Loch: A Study in Biotic Succession. New Phyt. Vol. XIII. 1914, pp. 134–148, 2 text-figs.
[An ecological study in which the aquatic formation of the Loch is dealt with (pp. 137–140).]

Matthiesen, F. (1908) [pp. 112, 114, 117, 122, 255 and Fig. 81, p. 119]
Beiträge zur Kenntnis der Podostemaceen. Bibl. Bot. Bd. XV. Heft 68, 1908, 55 pp., 9 pls., 1 text-fig.
(This memoir is chiefly occupied with a description of certain species of Podostemaceae from Venezuela, but it also includes a general account of the morphology and anatomy of the group.)

Meierhofer, H. (1902) [p. 103 and Figs. 61, p. 93, 62, p. 95, 73, p. 107]
Beiträge zur Anatomie und Entwickelungsgeschichte der Utricularia-Blasen. Flora, Bd. 90, 1902, pp. 84–114, 9 pls.
(The author describes the structure and development of the bladders of the European aquatic Utricularias and comes to the conclusion that these organs are foliar in nature.)

Meister, F. (1900) [pp. 100, 101, 299]
Beiträge zur Kenntnis der europäischen Arten von Utricularia. Mémoires de l'Herbier Boissier, No. 12, 1900, 40 pp., 4 pls.
(A systematic account with biological notes.)

Mellink, J. F. A. (1886) [p. 258]
Zur Thyllenfrage. Bot. Zeit. Jahrg. 44, 1886, pp. 745–753, 1 pl.
[An account of a petiole of *Nymphaea* (*Castalia*) *alba* which had at some time been wounded at various points. It was found that, in the neighbourhood of the wounds, the air canals were choked by hairs, which had grown out from the surrounding parenchyma cells in a thylose-like manner into the canals.]

Mer, É. (1880[1]) [pp. 163, 165, 279]
Des modifications de forme et de structure que subissent les plantes, suivant qu'elles végètent à l'air ou sous l'eau. Bull. de la Soc. bot. de France, T. XXVII. (Sér. II. T. II.) 1880, pp. 50–55.
(An analysis of the differences in morphology and structure exhibited by the land and water forms of *Ranunculus aquatilis*, *R. Flammula*, *Littorella lacustris*, etc. The author suggests a comparison between etiolated and submerged plants.)

Mer, É. (1880[2]) Des causes qui modifient la structure de certaines plantes aquatiques végétant dans l'eau. Bull. de la Soc. bot. de France, T. XXVII. (Sér. II. T. II.) 1880, pp. 194–200.

(This paper is concerned with the differences between the forms of *Littorella* and *Isoetes* growing under different conditions.)

Mer, É. (1881) Observations sur les variations des plantes suivant les milieux. Bull. de la Soc. bot. de France, T. XXVIII. (Sér. II. T. III.) 1881, pp. 87–90.

(Brief notes on the submerged and aerial forms of *Callitriche Littorella*, etc.)

Mer, É. (1882[1]) [pp. 32, 42, 165, 195] De la végétation à l'air des plantes aquatiques. Comptes rendus de l'acad. des sciences, Paris, T. 94, 1882, pp. 175–178.

(An account of the experimental production of land forms in the case of certain aquatic plants.)

Mer, É. (1882[2]) De quelques nouveaux exemples relatifs à l'influence de l'hérédité et du milieu sur la forme et la structure des plantes. Bull. Soc. bot. de France, T. XXIX. 1882, pp. 81–87.

(A study of the leaf characters of *Potamogeton rufescens* growing in deep or shallow water.)

Merz, M. (1897) [p. 100] Untersuchungen über die Samenentwickelung der Utricularieen. Flora, Bd. 84 (Ergänzungsband zum Jahrg. 1897), 1897, pp. 69–87, 34 text-figs.

(A detailed account of the embryo-sac and seed in ten species of the genus, which is characterised by the early disappearance of the nucellus and the development of endospermic haustoria at both ends of the sac.)

Micheli, M. (1881) Alismaceae, Butomaceae, Juncagineae. A. and C. de Candolle's Monographiae Phanerogamarum, Vol. III. 1881, pp. 7–112.

(A systematic account of these families with a discussion of their affinities, etc.)

Micheli, P. A. (1729) [pp. 76, 235] Nova Plantarum Genera....Florentiae, 1729.

(The flowers of *Lemna gibba* are figured on Pl. 11 under the name of "Lenticula," and the sterile plants of *L. minor*, *L. trisulca*, *Wolffia arrhiza* and *Spirodela polyrrhiza* under the name of "Lenticularia." *Vallisneria* with its floating ♂ flowers and spiral peduncles is shown on Pl. 10.)

Milde, (1853) *Wolffia Michelii* Hork. (*Lemna arrhiza* L.). Bot. Zeit. Jahrg. XI. 1853, pp. 896, 897.

(A note on the occurrence of this plant in Germany.)

Miller, G. S. and Standley, P. C. (1912) The North American Species of *Nymphaea*. Contributions from the U.S. National Herbarium, Vol. 16, Pt 3, 1912. Smithsonian Institution. U.S. National Museum. viii + 109 pp., 39 text-figs., 13 pls.

(This systematic monograph is fully illustrated, especially with photographs of fruits and with maps showing the distribution of the various species in N. America.)

Minden, M. von (1899) [pp. 83, 266, 268, 269 and Fig. 53, p. 82] Beiträge zur anatomischen und physiologischen Kenntnis Wasser-secernierender Organe. Bibliotheca Botanica, Bd. IX. Heft 46, 1899, 76 pp., 7 pls.

(Pp. 1–30 deal with the secretion of watery solutions from water pores and apical openings in the leaves of water plants, and form an exhaustive account of the subject so far as it had been worked out by the end of the nineteenth century.)

Möbius, M. (1895) [p. 281] Ueber einige an Wasserpflanzen beobachtete Reizerscheinungen. Biol. Centralbl. Bd. 15, 1895, pp. 1–14, 33–44, 8 text-figs.

(Observations on the effect of light and darkness on water plants.)

Moeller, J. (1879) [p. 191] *Aeschynomene aspera* Willd. (Papilionaceen). Bot. Zeit. Jahrg. 37, 1879, pp. 720–724, 1 text-fig.

(An account of the anatomy of the floating wood of this plant.)

Mönkemeyer, W. (1897) [p. 291] Die Sumpf- und Wasserpflanzen. Ihre Beschreibung, Kultur und Verwendung. iv + 189 pp., 126 text-figs. Berlin, 1897.

(A useful compendium of water and marsh plants arranged on Engler and Prantl's system, with notes on their cultivation. It is by the Inspector of the Leipzig Botanic Garden, and is primarily intended to help those who wish to grow water plants in an aquarium or water garden.)

Montesantos, N. (1913) [pp. 50, 51, 52, 157, 239, 282] Morphologische und biologische Untersuchungen über einige Hydrocharideen. Flora, N.F. Bd. V. (Ganze Reihe, Bd. 105), 1913, pp. 1–32, 5 pls.

(This paper deals with the genera *Limnobium*, *Blyxa*, *Ottelia* and *Stratiotes*. Certain experiments are described showing that the heterophylly of *Limnobium* and *Stratiotes* is not due to the direct action of the medium, and that the sinking of *Stratiotes* in the autumn is due to the increase of weight brought about by a deposition of chalk on the leaves.)

Monti, Gaetano (1747) [p. 109] De Aldrovandia novo herbae palustris genere. De Bononiensi Scientiarum et Artium Instituto atque Academia Commentarii. Tomi secundi Pars tertia, 1747, pp. 404–411, 1 pl.

[This old and rare memoir on *Aldrovandia* is analysed in **Augé de Lassu (1861)**.]

Mori, A. (1876) [p. 110] Nota sull' irritabilità delle foglie dell' *Aldrovandia vesiculosa*. Nuovo Giornale Botanico Italiano, Vol. VIII. 1876, p. 62.

(The author shows that the irritability of the *Aldrovandia* leaf is confined to the central glandular region.)

Moss, C. E. (1913) [p. 291] Vegetation of the Peak District. x + 235 pp., 36 figs., 2 maps. Cambridge, 1913.

(Chapter VI. contains an account of the marsh and aquatic associations of the district.)

Müller, F. (1877) [p. 311] Untersuchungen über die Struktur einiger Arten von *Elatine*. Flora, N.R. Jahrg. XXXV. (G.R. Jahrg. LX.) 1877, pp. 481–496, 519–526, 1 pl.

(A description of the anatomy and flower structure of *Elatine* and a discussion of the affinities of the genus.)

Müller, F. (1883) [pp. 239, 282]
Einige Eigenthümlichkeiten der *Eichhornia crassipes*. Kosmos, Jahrg. VII. Heft IV. 1883, pp. 297–300.
(Notes on the floral biology of this species.)

Münter, J. (1845) [p. 15]
Beobachtungen über besondere Eigenthümlichkeiten in der Fortpflanzungsweise der Pflanzen durch Knospen. III. Ueber die Knospen der *Sagittaria sagittaefolia* L. Bot. Zeit. Jahrg. 3, 1845, pp. 689–697.
[An account of the tuber formation in this species, with a mention of certain early references to the subject. The author discusses the statement in **Nolte, E. F.** (1825), that tuber formation was once found by him in *Alisma Plantago*, and decides that this is undoubtedly an error.]

Murray, H.
See **Weiss, F. E. and Murray, H.** (1909).

Nakano, H. (1911) [p. 291]
The Vegetation of Lakes and Swamps in Japan. I. Teganuma (Tega-Swamp). Bot. Mag. Tōkyō, Vol. XXV. 1911, pp. 35–51, 6 text-figs.
(The first ecological survey of a Japanese lake and swamp.)

Nolte, E. F. (1825) [pp. 15, 50, 52, 54 and Fig. 32, p. 53]
Botanische Bemerkungen über *Stratiotes* und *Sagittaria*. Kopenhagen, 44 pp., 2 pls., 1825.
[An admirable account of the life-history of these two genera, in which special attention is paid to their methods of vegetative reproduction, and, in the case of *Stratiotes*, to the distribution of the sexes and the structure of the fruit. For a criticism see **Münter, J.** (1845).]

Ohno, N. (1910) [p. 258]
Ueber lebhafte Gasausscheidung aus den Blättern von *Nelumbo nucifera* Gaertn. Zeitschrift f. Bot. Jahrg. II. 1910, pp. 641–644, 4 text-figs.
(In this paper—from a Japanese laboratory—the author draws the conclusion that the pressure which produces the streaming of gas bubbles from the leaves of *Nelumbo* can be explained on purely physical grounds, but that it also has a physiological significance.)

Oliver, D.
See **Im Thurn, E. F. and Oliver, D.** (1887).

Oliver, F. W. (1888) [pp. 151, 234, 266]
On the Structure, Development, and Affinities of *Trapella*, Oliv., a new Genus of Pedalineae. Ann. Bot. Vol. II. 1888–1889, pp. 75–115, 5 pls., 1 text-fig.
(A monograph of *Trapella sinensis*, Oliv., a Chinese water plant discovered by Dr Henry.)

Oliver, F. W. (1889)
On a new form of *Trapella sinensis*. Ann. Bot. Vol. III. 1889–1890, p. 134.
[A brief account of a land form of this species. This note is supplementary to **Oliver, F. W.** (1888).]

Oliver, F. W. (1894)
See **Kerner, A. and Oliver, F. W.** (1894).

Onslow, The Hon. Mrs Huia
See **Wheldale, M.** (1916).

Osbeck, P. (1771) [p. 17]
A Voyage to China....Translated...by John Reinhold Forster. Vol. I. xx + 367 pp., 4 pls. London, 1771.
(The author mentions on pp. 334, 335 that a species of *Sagittaria* is cultivated by the Chinese as a food plant.)

Osborn, T. G. B. (1914) [p. 127] Botany and Plant Pathology. Reprinted from Handbook of South Australia. British Association Visit. Adelaide, 1914, 27 pp., 7 figs.

(On p. 11 there is a brief reference to the marine Angiosperms of S. Australia, and a mention of the curious seedlings of *Cymodocea antarctica*.)

Ostenfeld, C. H. (1908) [pp. 123, 134] On the Ecology and Distribution of the Grass-Wrack (*Zostera marina*) in Danish Waters. Report of the Danish Biological Station to the Board of Agriculture. XVI. Translated from Fiskeri-Beretning for 1907, Copenhagen, 1908, 62 pp., 9 text-figs.

(The distribution of this plant is dealt with in detail, and special attention is paid to the algae and marine animals with which it is associated.)

Otis, C. H. (1914) [p. 261] The transpiration of emersed water plants: its measurement and its relationships. Bot. Gaz. Vol. 58, 1914, pp. 457–494, 3 text-figs. and 14 charts.

(An extremely elaborate study of the transpiration of water plants whose leaves are in contact with the atmosphere, e.g. *Castalia*, *Sagittaria* and *Pontederia*.)

Overton, E. (1899) [pp. 228, 280, 290] Notizen über die Wassergewächse des Oberengadins. Vierteljahrsschrift der Naturforschenden Gesellschaft in Zürich, Jahrg. 44, 1899, pp. 211–228.

(A record of certain experiments on the effect of light and temperature on the flowering of *Hydrocharis* and *Elodea*; and of the vertical distribution of the water plants of the Upper Engadine.)

Paillieux, A. and Bois, D. (1888) [pp. 17, 24] Les plantes aquatiques alimentaires. Bull. de la Soc. nat. d'acclimatation de France, Sér. IV. T. 5, Année 35, 1888, pp. 782–793, 924–929, 1028–1035, 1102–1108.

(An account of a number of aquatic plants which are used for food, including *Aponogeton*, *Trapa*, *Eleocharis*, *Sagittaria* and various Nymphaeaceae.)

Pallis, M. (1916) [pp. 207, 211] The Structure and History of Plav: the Floating Fen of the Delta of the Danube. Journ. Linn. Soc. Bot. Vol. 43, 1916, pp. 233–290, 15 pls., 1 text-fig.

(This paper deals with *Phragmites communis*, Trin., *β flavescens*, Gren. & Godr.)

Parkin, J. See **Arber, E. A. N. and Parkin, J. (1907)**.

Parmentier, P. (1897) [p. 312] Recherches anatomiques et taxinomiques sur les Onothéracées et les Haloragacées. Ann. des sci. nat. Sér. VIII. Bot. T. III. 1897, pp. 65–149, 6 pls.

(A discussion, based on the anatomical characters of stem and leaf, of the relation of the Onothereae, Ludwigieae, Halorageae and Gunnereae.)

Payne-Gallwey, R. See **Walsingham, Lord, and Payne-Gallwey, R. (1886)**.

Pearsall, W. H. (1917–1918) [p. 288] The Aquatic and Marsh Vegetation of Esthwaite Water. Journ. of Ecology, Vol. v. 1917, pp. 180–202 and Vol. vi. 1918, pp. 53–74, 12 text-figs.
(A detailed ecological survey of a Lancashire lake.)

Pearsall, W. H. (1918) [p. 288] On the classification of Aquatic Plant Communities. Journ. of Ecology, Vol. vi. 1918, pp. 75–83.
(The author regards aquatic, fen and moor successions as together forming a unit and leading up to the one *formation* moor.)

Perrot, É. (1900) [p. 169 and Fig. 110, p. 170] Sur les organes appendiculaires des feuilles de certains *Myriophyllum*. Journ. de Bot. T. xiv. 1900, pp. 198–202, 5 text-figs.
(An account of the peculiar processes borne by the leaf of *Myriophyllum verticillatum* and *M. spicatum*, which the author regards as pluricellular caducous trichomes.)

Pfeiffer, L. (1854) Ueber einige deutsche Nymphäen. Bot. Zeit. Jahrg. 12, 1854, pp. 172–175.
(A critical article in which special stress is laid on the importance for systematic purposes of ascertaining the characters of the ripe fruit.)

Pieters, A. J. (1894) [p. 288] The Plants of Lake St Clair. Bull. Michigan Fish Commission, No. 2, 1894, 12 pp., 1 map.
[An ecological study in which the zonation of the plants observed in Lake St Clair is compared with that recorded by **Magnin, A. (1893)** for the Jura lakes.]

Pieters, A. J. (1902) [p. 291] Contributions to the Biology of the Great Lakes. The Plants of Western Lake Erie, with Observations on their Distribution. Bull. United States Fish Commission, Vol. xxi. 1902 (for 1901), pp. 57–79, 10 pls., 9 text-figs.
(An ecological study.)

Planchon, J. E. (1844) [p. 314] Observations sur le genre *Aponogeton* et sur ses affinités naturelles. Ann. d. sci. nat. Sér. iii. Bot. T. i. 1844, pp. 107–120, 1 pl.
(The author describes *Aponogeton distachyus* and brings forward evidence for removing it from the neighbourhood of *Saururus* and placing it in a sub-order between the Alismaceae and Juncaginaceae.)

Planchon, J. E. (1853) Études sur les Nymphéacées. Ann. des sci. nat. Sér. iii. T. xix. 1853, pp. 17–63.
(A systematic account of this family.)

Pond, R. H. (1905) [p. 264] Contributions to the Biology of the Great Lakes. The Biological Relation of Aquatic Plants to the Substratum. University of Michigan. Inaug. Diss. Ann Arbor, 1905, 43 pp., 6 text-figs.
(This paper contains important experimental work relating to the function of the roots in aquatic plants.)

Porsch, O. (1903) [pp. 165, 166] Zur Kenntnis des Spaltöffnungsapparates submerser Pflanzenteile. Sitzungsber. d. Math.-naturwiss. Klasse d. k. Akad. d. Wissens. Wien, Bd. CXII. Abt. I. 1903, pp. 97–138, 3 pls.
(An account of the means by which the flooding of the intercellular spaces through the stomata of submerged organs is prevented.)

Porsch, O. (1905) [p. 166 and Fig. 107, p. 167] Der Spaltöffnungsapparat im Lichte der Phylogenie. xv + 196 pp., 4 pls. and 4 text-figs. Jena, 1905.
[Pp. 83–87 deal with the stomates of water plants and form a résumé of **Porsch, O. (1903)**.]

Praeger, R. L. (1913) [p. 297] On the Buoyancy of the Seeds of some Britannic Plants. Sci. Proc. Royal Dublin Soc., N.S., Vol. XIV. 1913–1915, pp. 13–62.
[This memoir is supplementary to the parts of **Guppy, H. B. (1906)** which relate to seed buoyancy.]

Prankerd, T. L. (1911) [pp. 181, 197, 216, 228, 233 and Fig. 127, p. 197] On the Structure and Biology of the Genus *Hottonia*. Annals of Bot. Vol. XXV. 1911, pp. 253–267, 2 pls. and 7 text-figs.
(A general account of *H. palustris* and *H. inflata*. The author shows that the mature plant of *H. palustris* is not rootless as generally supposed. Traces of polystely occur at the base of the inflorescence axis in both species.)

Preston, T. A. (1895) [pp. 232, 291] The Flora of the Cropstone Reservoir. Trans. Leic. Lit. and Phil. Soc. Vol. III. 1895, pp. 430–442.
(An account of the flora of a reservoir which had been in existence twenty-four years.)

Prillieux, E. (1864) [pp. 63, 173] Recherches sur la végétation et la structure de l'*Althenia filiformis* Petit. Ann. d. sci. nat. Sér. v. Bot. T. II. 1864, pp. 169–190, 2 pls.
(A general account of this Mediterranean member of the Zannichellieae.)

Pringsheim, N. (1869) [pp. 97, 106 and Fig. 72, p. 106] Über die Bildungsvorgänge am Vegetationskegel von *Utricularia vulgaris*. Monatsber. d. k. preuss. Akad. d. Wissens. Berlin, 1869, pp. 92–116, 1 pl.
(The classic account of the apical development of *Utricularia* and its bearing on the morphological perplexities presented by the genus.)

Pringsheim, N. (1888) [p. 51] Ueber die Entstehung der Kalkincrustationen an Süsswasserpflanzen. Pringsheim's Jahrb. f. wiss. Bot. Bd. XIX. 1888, pp. 138–154.
(The author shows experimentally that the chalk incrustation on the surface of so many fresh-water plants is due to the abstraction, during the process of assimilation, of CO_2 which has held the calcium carbonate in solution.)

Queva, C. (1910) [p. 244] Observations anatomiques sur le "*Trapa natans* L." Association Française pour l'avancement des sciences. Compte rendu de la 38e session, Lille, 1909 (1910), pp. 512–517, 2 text-figs.
(The author's anatomical study of the seedling leads him to the conclusion that the primary root is entirely unrepresented. The anatomy of the hypocotyl is modified by the insertion of numerous adventitious roots which are localised on the same side of the axis as the large cotyledon.)

Raciborski, M. (1893) Ueber die Inhaltskörper der Myriophyllumtrichome. Ber. d. deutsch. bot. Gesellsch. Bd. XI. 1893, pp. 348–351.

(The highly refractive bodies present in the trichomes of *Myriophyllum* are considered to be of the nature of a glucoside, and to be related to substances found in the trichomes of the leaves of *Ceratophyllum, Elatine,* etc.)

Raciborski, M. (1894¹) Die Morphologie der Cabombeen und Nymphaeaceen. Flora, Bd. 78, 1894, pp. 244–279, 9 text-figs.

(In this memoir special attention is paid to the ontogeny of the flower and the vegetative shoots.)

Raciborski, M. (1894²) [pp. 29, 35, 272, 338] Beiträge zur Kenntniss der Cabombeen und Nymphaeaceen. Flora, Bd. 79 (Ergänzungsband), 1894, pp. 92–108, 1 pl.

[This paper is supplementary to **Raciborski, M.** (1894¹) and includes a reply to the criticisms on the latter contained in **Schumann, K.** (1894).]

Raciborski, M. (1895) Die Schutzvorrichtungen der Bluthenknospen. Flora, Bd. 81 (Ergänzungsband), 1895, pp. 151–194, 30 text-figs.

(This paper contains a section, pp. 190–192, dealing with the protection of the flower-bud among water plants.)

Raffeneau-Delile, A. (1841) [pp. 38, 258] Évidence du mode respiratoire des feuilles de Nelumbium. Ann. d. sci. nat. Sér. II. T. XVI. Bot. 1841, pp. 328–332.

(This paper on the emission of air from the leaves of the Lotus is followed by nine pages of controversy on the subject with Dutrochet.)

Raunkiaer, C. (1896) [Figs. 34, p. 55, 159, p. 246, 161, p. 248, 166, p. 319, 167, p. 339, 168, p. 339] De Danske Blomsterplanters Naturhistorie. Bd. I. Enkimbladede I. Helobieae, 1896, 138 pp., 240 text-figs.

(This fully illustrated account of the biology of the Helobieae native to Denmark is in Danish.)

Raunkiaer, C. (1903) [pp. 62, 65, 331 and Fig. 38, p. 61] Anatomical Potamogeton-Studies and *Potamogeton fluitans*. Botanisk Tidsskrift, Vol. 25, 1903, pp. 253–280, 9 text-figs.

(In this paper, which is written in English, the author shows the value of anatomical characters of the leaf and stem in classifying the genus *Potagometon*.)

Ravn, F. K. (1894) Om Flydeevnen hos Frøene af vore Vandog Sumpplanter. Botanisk Tidsskrift, Vol. 19, 1894, pp. 143–188, 26 text-figs.

(This Danish paper on the floating power of the seeds of aquatic and marsh plants concludes with a French résumé.)

Regnard, P. (1891) [pp. 253, 255, 278] Recherches expérimentales sur les conditions physiques de la vie dans les eaux. vii + 500 pp., 236 text-figs., 4 pls. Paris, 1891.

(This book consists of a series of lectures on aquatic biology. The physical aspect is fully treated; the applications relate chiefly to animals, but plants are not excluded.)

Reid, C. (1892) [pp. 296, 298] On the Natural History of Isolated Ponds. Trans. Norfolk and Norwich Nat. Soc. Vol. v. 1894 (for 1889–1894), Part 3, 1892, pp. 272–286.

(This paper is chiefly based on a study of the dew ponds of the South Downs. It forms an important contribution to the subject of the methods of dispersal of water plants.)

Reid, C. (1893) [p. 54] On *Paradoxocarpus carinatus*, Nehring, an extinct fossil plant from the Cromer Forest-bed. Trans. Norfolk and Norwich Nat. Soc. Vol. v. 1894 (for 1889–1894), Part 4, 1893, pp. 382–386, 1 text-fig.

[An account of a fossil fruit which was eventually discovered to belong to *Stratiotes aloides* L. (*vide* note by same author in Trans. Norfolk and Norwich Nat. Soc. Vol. vi. Pt 3, 1897, p. 328).]

Reid, C. (1899) [p. 303] The Origin of the British Flora. vi + 191 pp. London, 1899.

(This classical study, based on the flora of the Newer Tertiary beds, contains many references to water plants.)

Reinsch, P. (1860) Morphologische Mittheilungen. 5. Ueber die dreierlei Arten der Blätter der *Sagittaria sagittaefolia* L. Flora, N.R. Jahrg. xviii. (G.R. Jahrg. xliii.) 1860, pp. 740–742, 1 fig.

(An account of the heterophylly of *Sagittaria* including a mention of the distribution of the stomates in the different types of leaf. The arrow-head leaves are distinguished as "Blüthezeitblätter.")

Rendle, A. B. (1899) [p. 315] A Systematic Revision of the Genus *Najas*. Trans. Linn. Soc. Lond. Ser. ii. Vol. v. 1895–1901, Part xii. 1899, pp. 379–436, 4 pls.

(This monograph includes a general introduction dealing with the morphology, structure and distribution of the genus.)

Rendle, A. B. (1900) Supplementary Notes on the Genus *Najas*. Trans. Linn. Soc. Lond. Ser. ii. Vol. v. 1895–1901, Part xiii. 1900, pp. 437–444.

[This paper supplements **Rendle, A. B.** (1899).]

Rendle, A. B. (1901) [p. 304] Naiadaceae, in Das Pflanzenreich, iv. 12 (herausgegeben von A. Engler). 21 pp., 71 text-figs. Leipzig, 1901.

(An authoritative account of all the species of the genus *Naias*; the general description of the group is in English.)

Rendle, A. B. (1904) [p. 314] The Classification of Flowering Plants. Vol. i. Gymnosperms and Monocotyledons. xiv + 403 pp., 187 text-figs. Cambridge, 1904.

(This instalment of a text book of systematic botany gives a very useful account of the Helobieae and other Monocotyledonous aquatics.)

Richard, L. C. (1808) [p. 311] Démonstrations Botaniques ou Analyse du Fruit. Paris, 1808. xii + 111 pp.

(On p. 33 the author makes the suggestion that *Callitriche* is related to the Euphorbiaceae by its seed structure.)

Rodier, É. (1877[1]) [p. 90] Sur les mouvements spontanés et réguliers d'une plante aquatique submergée, le *Ceratophyllum demersum*. Comptes rendus de l'acad. des sciences, Paris, T. 84, 1877, pp. 961–963.

[For an English account of this work see **Rodier, É.** (1877[2]).]

Rodier, É. (1877[2]) [p. 90] The Movements of a Submerged Aquatic Plant. Nature, Vol. XVI. 1877, pp. 554–555, 1 text-fig.

(This brief paper is translated from an article by the author in "La Nature" and contains substantially the facts recorded in **Rodier, É.** (1877[1]) with the addition of a text-figure showing the successive positions assumed in the course of two days by a branch of *Ceratophyllum demersum*.)

Rohrbach, P. (1873) Beiträge zur Kenntniss einiger Hydrocharideen. Abhandl. d. naturforsch. Gesellschaft zu Halle, Bd. XII. 1873, pp. 53–114, 3 pls.

(This memoir deals chiefly with the morphology and anatomy of *Hydrocharis Morsus-ranae*, *Stratiotes aloides* and *Vallisneria spiralis*. Special attention is paid to the shoot and inflorescence systems and to the development of the flower.)

Roper, F. C. S. (1885) [p. 146] Note on *Ranunculus Lingua*, Linn. Journ. Linn. Soc. Bot. Vol. XXI. 1886, pp. 380–384, 2 pls.

(An account of the submerged leaves of this species. The two types of leaf are clearly figured, and there is an historical account from the literature of the records of their occurrence.)

Rosanoff, S. (1871) [p. 189 and Fig. 123, p. 191] Ueber den Bau der Schwimmorgane von *Desmanthus natans* Willd. Bot. Zeit. Jahrg. 29, 1871, pp. 829–838, 1 pl.

[A study of the aerenchyma of *Desmanthus natans*, Willd. (*Neptunia oleracea*, Lour.).]

Rossmann, J. (1854) [p. 144] Beiträge zur Kenntniss der Wasserhahnenfüsse, *Ranunculus* sect. *Batrachium*. vi + 62 pp. Giessen, 1854.

(This memoir is divided into two parts; the first deals generally with the Water Buttercups, and discusses their heterophylly, while the second consists of descriptions of the species recognised at this date.)

Roux, M. le (1907) [p. 291] Recherches biologiques sur le lac d'Annecy. Annales de Biologie Lacustre, T. II. Fasc. 1 and 2, 1907, pp. 220–387, 6 pls., 14 text-figs.

(This memoir includes an ecological study of the flora of the lake.)

Roxburgh, W. (1832) [p. 110] Flora Indica. Vol. II. vi + 691 pp. Serampore, 1832.

(On p. 112 the author mentions that *Aldrovandia verticillata* is "Found swimming on ponds of water over Bengal during the cold and hot season.")

Royer, C. (1881–1883) [pp. 24, 27, 87, 216, 234, 236] Flore de la Côte-d'Or avec déterminations par les parties souterraines. 2 vols., 693 pp. (2 vols. paged as one). Paris, 1881–1883.

(This flora, of that Département of France which includes Dijon, is unusual in paying special attention to the biology and life-history of the plants enumerated. It contains a good many useful notes on water plants.)

Roze, E. (1887) [p. 71] Le mode de fécondation du *Zannichellia palustris* L. Journ. de Bot. T. I. 1887, pp. 296–299, 1 text-fig.
(Observations on the submerged pollination of this species.)

Roze, E. (1892) [p. 85] Sur le mode de fécondation du *Najas major* Roth et du *Ceratophyllum demersum* L. Bull. de la Soc. bot. de France, T. XXXIX. (Sér. II. T. XIV.) 1892, pp. 361–364.
[The pollination of *Naias* is described, and, in the case of *Ceratophyllum*, the observations of **Dutailly, G.** (1892) are confirmed.]

Russow, E. (1875) [pp. 107, 180] Betrachtungen über das Leitbündel- und Grundgewebe (Jubiläumschrift Dr Alexander von Bunge). 78 pp. Dorpat, 1875.
(The anatomy of water plants is dealt with in this memoir in some detail.)

Sanio, C. (1865) [pp. 65, 86, 175, 176, 179] Einige Bemerkungen in Betreff meiner über Gefassbündelbildung geäusserten Ansichten. Bot. Zeit. Jahrg. 23, 1865, pp. 165–172, 174–180, 184–187, 191–193, 197–200.
[This paper forms a reply to the criticism of the author's anatomical views by **R. Caspary** in Prings. Jahrb. Bd. 4, 1865–6, pp. 101–124. It includes an account of the anatomy of certain water plants—*Hippuris* (pp. 184–186), *Myriophyllum* (p. 186), *Elodea* (pp. 186, 187 and 191–192), *Ceratophyllum* (pp. 192, 193), *Trapa* (p. 193), and *Potamogeton* (p. 193).]

Sargant, E. (1903) [p. 320] A Theory of the Origin of Monocotyledons, founded on the Structure of their Seedlings. Ann. Bot. Vol. 17, 1903, pp. 1–92, 7 pls., 10 text-figs.
[This paper does not deal with water plants, but should be read in connexion with the theory of the aquatic origin of Monocotyledons proposed in **Henslow, G.** (1893).]

Sargant, E. (1908) [pp. 308, 320, 323] The Reconstruction of a Race of Primitive Angiosperms. Ann. Bot. Vol. XXII. 1908, pp. 121–186, 21 text-figs.
(This memoir contains a criticism (pp. 175–178) of Henslow's theory of the aquatic origin of Monocotyledons.)

Sauvageau, C. (1889[1]) [p. 208 and Fig. 140, p. 209] Sur la racine du *Najas*. Journ. de Bot. Vol. III. 1889, pp. 3–11, 7 text-figs.
(A detailed account of the extremely reduced anatomy of the roots of *Naias*.)

Sauvageau, C. (1889[2]) [p. 66] Contribution à l'étude du système mécanique dans la racine des plantes aquatiques; les *Potamogeton*. Journ. de Bot. Vol. III. 1889, pp. 61–72, 9 text-figs.
(A full comparative study of the root anatomy of the genus, bringing out the interesting point that lignin is as abundant in the roots of *Potamogeton* as in those of many land plants.)

Sauvageau, C. (1889[3]) [p. 135] Contribution à l'étude du système mécanique dans la racine des plantes aquatiques; les *Zostera*, *Cymodocea* et *Posidonia*. Journ. de Bot. Vol. III. 1889, pp. 169–181, 5 text-figs.

(A continuation of the author's detailed study of the roots of submerged plants. In *Zostera* and *Cymodocea* the mechanical tissue is of the nature of collenchyma, while in *Posidonia* it is sclerised.)

Sauvageau, C. (1890[1]) [pp. 124, 131, 164] Observations sur la structure des feuilles des plantes aquatiques; *Zostera*, *Cymodocea* et *Posidonia*. Journ. de Bot. T. IV. 1890, pp. 41–50, 68–76, 117–126, 128–135, 173–178, 181–192, 221–229, 237–245, 38 text-figs.

(The author's elaborate study of these three genera leads to the conclusion that anatomical data serve here to distinguish species.)

Sauvageau, C. (1890[2]) [p. 131] Sur la feuille des Hydrocharidées marines. Journ. de Bot. T. IV. 1890, pp. 269–275, 289–295, 3 text-figs.

(This memoir deals with the leaf structure of *Enhalus*, *Thalassia* and *Halophila*.)

Sauvageau, C. (1890[3]) [pp. 124, 131] Sur la structure de la feuille des genres *Halodule* et *Phyllospadix*. Journ. de Bot. Vol. IV. 1890, pp. 321–332, 7 text-figs.

(This paper forms the conclusion of the author's study of the leaves of marine Angiosperms, and includes a summary of his results.)

Sauvageau, C. (1891[1]) [pp. 123, 131, 254, 261, 264, 266, 331 and Figs. 84, p. 125, 85 and 86, p. 128, 88 and 89, p. 132, 108, p. 167, 162, p. 262] Sur les feuilles de quelques monocotylédones aquatiques. Ann. d. sci. nat. Sér. VII. Bot. T. XIII. 1891, pp. 103–296, 64 text-figs. (Also published as Thèses présentées à la faculté des sciences de Paris, Sér. A, No. 158, No. d'ordre 720, 1891.)

[An exhaustive account of the leaf structure of forty-eight species of the Potamogetonaceae (as defined by Ascherson), incorporating the results published in **Sauvageau, C. (1890[1])** and (**1890[3]**). The memoir contains some experimental work on the transpiration current in submerged plants.]

Sauvageau, C. (1891[2]) Sur la tige des *Zostera*. Journ. de Bot. T. V. 1891, pp. 33–45, 59–68, 9 text-figs.

(A description of the anatomy and morphology of the stems of the five species of *Zostera*, showing that the stem anatomy gives even better criteria for distinguishing the species than those deduced from the author's study of the leaves.)

Sauvageau, C. (1891[3]) [pp. 135, 331] Sur la tige des Cymodocées Aschs. Journ. de Bot. T. V. 1891, pp. 205–211, 235–243, 6 text-figs.

(The author shows that the different species of *Cymodocea* and *Halodule* can be distinguished by the anatomy of their stems just as they can by that of their leaves.)

Sauvageau, C. (1893) [p. 269 and Fig. 164, p. 270] Sur la feuille des Butomées. Ann. des sci. nat. Sér. 7, Bot. T. 17, 1893, pp. 295–326, 9 text-figs.

(Certain of the plants dealt with in this paper are aquatic, e.g. *Hydrocleis nymphoides*, whose leaf anatomy is fully described.)

Sauvageau, C. (1894) [pp. 59, 63, 71, 243 and Figs. 37, p. 60, 43, p. 68] Notes biologiques sur les Potamogeton. Journ. de Bot. T. VIII. 1894, pp. 1–9, 21–43, 45–58, 98–106, 112–123, 140–148, 165–172, 31 text-figs.
[An account of the anatomy and life-history of *Potamogeton crispus* L., *P. trichoides* Ch. et Schl., *P. pusillus* L., *P. gemmiparus* (Robbins) Morong, *P. acutifolius* Link, *P. perfoliatus* L., *P. polygonifolius* Pourr., *P. lucens* L., *P. pectinatus* L., *P. natans* L., *P. densus* L.]

Schaffner, J. H. (1896) [p. 19] The embryo-sac of *Alisma Plantago*. Bot. Gaz. Vol. XXI. pp. 123–132, 2 pls., 1896.
[An account of fertilisation and embryo development in this species.)

Schaffner, J. H. (1897) [p. 9] Contribution to the Life History of *Sagittaria variabilis*. Bot. Gaz. Vol. XXIII. 1897, pp. 252–273, 7 pls.
(This paper is confined to an account of the gametophytes, fertilisation and embryology of this species.)

Schaffner, J. H. (1904) [pp. 309, 314] Some Morphological Peculiarities of the Nymphaeaceae and Helobiae. The Ohio Naturalist, Vol. IV. 1904, pp. 83–92, 3 pls., 2 text-figs.
(The author attempts to show that the Nymphaeaceae are Monocotyledons.)

Schenck, H. (1884) [p. 202 and Fig. 133, p. 202] Ueber Structuränderung submers vegetirender Landpflanzen. Ber. d. deutsch. bot. Gesellsch. Bd. II. 1884, pp. 481–486, 1 pl.
(An account of the differences observed between the structure of the normal terrestrial form of *Cardamine pratensis*, and of the same species when growing submerged.)

Schenck, H. (1885) [Passim] Die Biologie der Wassergewächse. Verhandl. des naturhist. Vereines d. preuss. Rheinlande, Westfalens und des Reg.-Bezirks Osnabrück, Jahrg. 42 (Folge v. Jahrg. 2), 1885, pp. 217–380, 2 pls.
[This memoir, in conjunction with **Schenck, H.** (1886), forms one of the most important general contributions ever made to the study of water plants; it summarises the state of knowledge of a generation ago. Many of the more recent accounts of this biological group are based on Schenck's work.]

Schenck, H. (1886) [Passim and Figs. 40, p. 64, 41, p. 65, 51, p. 79, 56, p. 87, 74, p. 108, 106, p. 165, 109, p. 168, 111, p. 170, 114, p. 176, 138 and 139, p. 209] Vergleichende Anatomie der submersen Gewächse. Bibliotheca Botanica, Bd. I. Heft I. 1886, 67 pp., 10 pls.
(A detailed and fully illustrated account of the anatomy of those water plants which are most completely specialised for an aquatic life.)

Schenck, H. (1887) Beiträge zur Kenntniss der Utricularien. *Utricularia montana* Jacq. und *Utr. Schimperi* nov. spec. Pringsheim's Jahrb. Bd. XVIII. 1887, pp. 218–235, 3 pls.
(The two species described in this paper are terrestrial, but they are compared with the aquatic members of the genus.)

Schenck, H. (1889) [pp. 188, 189, 192 and Fig. 122, p. 190] Ueber das Aërenchym, ein dem Kork homologes Gewebe bei Sumpfpflanzen. Pringsheim's Jahrb. f. wissen. Bot. Bd. xx. 1889, pp. 526–574, 6 pls.

(A detailed account of the occurrence of aerenchyma in Onagraceae, Lythraceae, Melastomaceae, Hypericaceae, Capparidaceae, Euphorbiaceae, Labiatae and Leguminosae. The aerenchyma is regarded as primarily a breathing tissue.)

Scheuchzerus, J. (1719) [p. 154] Agrostographia sive Graminum, Juncorum, Cyperorum, Cyperoidum, iisque affinium Historia....Tiguri, Typis et Sumptibus Bodmerianis, 1719.

(This book contains an early reference to the floating leaves of *Scirpus lacustris*; see "*Scirpus paniculatus*," p. 354.)

Schiller, K. See **Schorler, B., Thallwitz, J. and Schiller, K. (1906)**.

Schilling, A. J. (1894) [p. 271] Anatomisch-biologische Untersuchungen über die Schleimbildung der Wasserpflanzen. Flora, Bd. 78, 1894, pp. 280–360, 17 text-figs.

(A full account of the mucilage organs of water plants; the author shows that they are all of the morphological nature of hairs. He believes that the function of the mucilage, which is formed in all cases at the expense of the cell wall, is to prevent excess of water from passing into the young tissues.)

Schindler, A. K. (1904) [pp. 181, 312] Die Abtrennung der Hippuridaceen von den Halorrhagaceen. Beiblatt zu den Bot. Jahrb. (Engler) Bd. xxxiv. Heft 3, 1904, pp. 1–77.

[A detailed study of the anatomy and morphology of these families from which the author concludes that Halorrhagaceae (Halorrhagideae + Gunnereae) and Hippuridaceae are entirely unrelated.]

Schlechtendal, D. F. L. von (1852) Einige Worte über *Nymphaea neglecta* und *biradiata*. Bot. Zeit. Jahrg. 10, 1852, pp. 557–559.

(The author shows that these two species of *Castalia* tend to approach one another, but he leaves open the question as to whether transitional forms exist.)

Schlechtendal, D. F. L. von (1854) Betrachtungen über die *Limosella*-Arten. Bot. Zeit. Jahrg. 12, 1854, pp. 909–918.

(A critical account of the species and varieties.)

Schleiden, M. J. (1837) [pp. 63, 84, 86] Beiträge zur Kenntniss der Ceratophylleen. Linnaea, Bd. 11, 1837, pp. 513–542, 1 pl.

(A very thorough account of the family; the author includes all the known forms under the single species *Ceratophyllum vulgare*, Schl.)

Schleiden, M. J. (1838[1]) [p. 316] Bemerkungen über die Species von *Pistia*. Allgemeine Gartenzeitung, Jahrg. 6, No. 3, 1838, pp. 17–20.

(A systematic account of the genus, with a discussion of its affinities.)

Schleiden, M. J. (1838[2]) Berichtigungen und Nachträge zur Kenntniss der Ceratophylleen. Linnaea, Bd. 12, 1838, pp. 344–346, 1 pl.

[Supplementary to **Schleiden, M. J. (1837)**. The germination of the seeds is described, and attention is drawn to the suppression of the primary root and the absence of adventitious roots.]

Schleiden, M. J. (1839) [p. 73] Prodromus Monographiae Lemnacearum oder Conspectus Generum atque Specierum. Linnaea, Bd. XIII. 1839, pp. 385–392.

(A systematic account of the Lemnaceae which are treated as a tribe of the Aroideae.)

Schoenefeld, W. de (1860) [p. 110] Sur le mode de végétation de l'*Aldrovanda vesiculosa* en hiver et au printemps. Bull. de la soc. bot. de France, T. VII. 1860, pp. 389–392.

(The author shows that when kept indoors the turions of this plant may float all the winter and then germinate in the spring.)

Schorler, B., Thallwitz, J. and Schiller, K. (1906) [p. 291] Pflanzen- und Tierwelt des Moritzburger Grossteiches bei Dresden. Annales de Biologie Lacustre, T. I. Fasc. 2, 1906, pp. 193–310.

(This work includes an ecological study of the vegetation of this lake by B. Schorler.)

Schrenk, J. (1888) [pp. 30, 205, 258, 266, 267, 272] On the Histology of the Vegetative Organs of *Brasenia peltata*, Pursch. Bull. Torr. Bot. Club, Vol. XV. 1888, pp. 29–47, 2 pls.

(The points to which special attention is paid in this paper are the nature and origin of the surface layer of mucilage, the internal hairs, and the submerged leaves.)

Schrenk, J. (1889) [p. 193 and Fig. 124, p. 193] On the Floating-tissue of *Nesaea verticillata* (L.), H.B.K. Bull. Torr. Bot. Club, Vol. XVI. 1889, pp. 315–323, 3 pls.

(An account of the biology and anatomy of the "aerenchyma" in a member of the Lythraceae. The author regards it primarily as a floating tissue which serves only secondarily, if at all, for purposes of aeration.)

Schröter, C. and Kirchner, O. (1902) [pp. 291, 322] Die Vegetation des Bodensees. T. II. Der "Bodensee-Forschungen" neunter Abschnitt. 86 pp., 3 pls., 1 map, 15 text-figs. Lindau i. B. 1902.

[In this book the water and marsh vegetation (higher plants) of the Bodensee is discussed from an ecological standpoint.]

Schröter, C. See **Kirchner, O. von, Loew, E. and Schröter, C.** (1908, etc.).

Schuchardt, T. (1853) Beiträge zur Kenntniss der deutschen Nymphaeen. Bot. Zeit. Jahrg. XI. 1853, pp. 497–510.

[A critical account of the species and varieties of *Nymphaea* (*Castalia*) native to Germany.]

Schultz, F. (1873) [p. 101] Beiträge zur Flora der Pfalz (Schluss). Flora, Neue Reihe, Jahrg. XXXI. (Ganz. Reihe, Jahrg. LVI.) 1873, pp. 247–251.

(The author mentions that *Utricularia intermedia* had at that time existed in Pfalz for forty years without flowering.)

Schumann, K. (1892) [p. 70] Morphologische Studien. Heft I. x + 206 pp., 6 pls. Leipzig, 1892.

[Pp. 119–186 contain an account of the ontogeny of the flowers of the Potamogetonaceae, Zannichelliaceae and Naiadaceae. The author criticises the views on the flower of *Naias* expressed by **Magnus, P.** (1870[1]).]

Schumann, K. (1894) Die Untersuchungen des Herrn Raciborski über die Nymphaeaceae und meine Beobachtungen über diese Familie. Ber. d. deutsch. bot. Gesellsch. Bd. XII. 1894, pp. 173–178.
[A criticism of **Raciborski, M.** (**1894**[1]).]

Scott, D. H. (1891) [p. 180] Origin of Polystely in Dicotyledons. Annals of Bot. Vol. V. 1890–1891, pp. 514–517.
(In this paper the hypothesis is put forward that the cases of polystely known to occur among Angiosperms may be associated with an aquatic ancestry.)

Scott, D. H. and Wager, H. (1888) [p. 191] On the Floating-Roots of *Sesbania aculeata*, Pers. Ann. Bot. Vol. I. 1887–1888, pp. 307–314, 1 pl.
(An account of the aerenchyma developed on the roots of this Leguminous plant. The spongy tissue is produced by a cortical phellogen.)

Scott, J. (1869) [p. 235] Note on the *Isoëtes capsularis*, Roxb. Journ. Linn. Soc. Bot. Vol. X. 1869, pp. 206–209, 1 pl.
(In this note, the curator of the Calcutta Botanic Garden shows that Roxburgh's so-called "*Isoetes capsularis*" is the detached male flower of *Vallisneria spiralis*, L.)

Seehaus, C. (1860) [p. 286] *Hydrilla verticillata* (L. fil.) Çasp. var. *pomeranica* (Rchb.) Casp. Verhandlung. d. bot. Vereins f. d. Provinz Brandenburg, Heft II. 1860, pp. 95–102.
(Observations on the life-history of this species.)

Seidel, C. F. (1869) [pp. 34, 309] Zur Entwickelungsgeschichte der *Victoria regia* Lindl. Nov. Act. Acad. Caes. Leopoldino-Carolinae Germanicae Naturae Curiosorum (Verhandl. d. Kais. Leop.-Car. deutschen Akad. d. Naturforscher), T. 35, 1870 (for 1869), No. 6, 26 pp., 1 table, 2 pls.
(A general account of *Victoria regia* with a discussion of its affinities. The author regards the Nymphaeaceae as Monocotyledons related to the Hydrocharitaceae.)

Serguéeff, M. (1907) [pp. 142, 244, 281, 314 and Fig. 91, p. 142] Contribution à la morphologie et la biologie des Aponogétonacées. Université de Genève. Thèse... docteur ès sciences, Institut de Botanique. Prof. Dr Chodat, 7me série, VIIIme fasc. 1907, 132 pp., 5 pls., 78 text-figs.
[A detailed study of *Aponogeton* (*Ouvirandra*) *fenestralis*, and *A. distachyus*. A general account of the family and a discussion of its affinities are included.]

Shull, G. H. (1905) [p. 162] Stages in the Development of *Sium cicutaefolium*. Carnegie Institution of Washington, Publication No. 30. Papers of Station for Experimental Evolution at Cold Spring Harbor, New York, No. 3, 1905, 28 pp., 7 pls., 11 text-figs.
(A study of heterophylly in this species.

Siddall, J. D. (1885) [pp. 55, 210, 211] The American Water Weed, *Anacharis Alsinastrum*, Bab.: Its Structure and Habit; with some Notes on its introduction into Great Britain, and the causes affecting its rapid spread at first, and apparent present diminution. Proc. Chester Soc. Nat. Sci. No. 3, 1885, pp. 125–134, 1 pl.

(This paper gives the early history of *Elodea canadensis*, Michx. in this country.)

Snell, K. (1908) [pp. 208, 260, 265] Untersuchungen über die Nahrungsaufnahme der Wasserpflanzen. Flora, Bd. 98, 1908, pp. 213–249, 2 text-figs.

(The author's main conclusion is that, in the case of rooted submerged plants, the greater part of the water supply is taken in by the roots, but that the leaves may also absorb water.)

Snell, K. (1912) [p. 266] Der Transpirationsstrom der Wasserpflanzen. Ber. d. deutschen bot. Gesellsch. Jahrg. XXX. 1912, pp. 361, 362.

[A note which should be read in connexion with **Snell, K. (1908)** and **Hannig, E. (1912)**.]

Snow, L. M. (1914) [p. 183] Contributions to the knowledge of the diaphragms of water plants. *I. Scirpus validus*. Bot. Gaz. Vol. 58, 1914, pp. 495–517, 16 text-figs.

(This paper contains a comprehensive review of the records in the literature relating to the occurrence of diaphragms in various groups of the higher plants.)

Solereder, H. (1913) [pp. 42, 46, 52, 135, 165, 169, 340] Systematisch-anatomische Untersuchung des Blattes der Hydrocharitaceen. Beihefte zum Bot. Centralbl. Bd. XXX. Abth. I. 1913, pp. 24–104, 53 text-figs.

(A highly detailed comparative study of the leaves of the Hydrocharitaceae. The author has examined all the genera belonging to this family.)

Solereder, H. (1914) [p. 42] Zur Anatomie und Biologie der neuen *Hydrocharis*-Arten aus Neuguinea. Mededeelingen van's Rijks Herbarium Leiden, No. 21, 1914, 2 pp.

(A description of the leaf structure of *H. parnassifolia* and *H. parvula*—the former has typical air leaves like *H. asiatica*, and the latter, swimming leaves like *H. Morsus-ranae*.)

Solms-Laubach, H. Graf zu (1883) [p. 317] Pontederiaceae. A. and C. de Candolle's Monographiae Phanerogamarum, Vol. IV. 1883, pp. 501–535.

(A systematic account of this group with a discussion of the geographical distribution, etc.)

Spenner, F. C. L. (1827) [p. 28] Ueber *Nuphar minima* Smith. Flora, Jahrg. X. Bd. I. 1827, pp. 113–119, 2 pls.

[In his account of this plant, the author describes the submerged leaves and figures them (Pl. I). He suggests that leaves of this type probably occur in other Nymphaeaceae, but that they have been overlooked.]

Spruce, R. (1908) [pp. 31, 99, 154, 190, 229, 290, 291, 311] Notes of a Botanist on the Amazon and Andes... during the years 1849–1864, edited by A. R. Wallace. 2 vols. London, 1908.

(These volumes contain a number of notes on the aquatic plants observed by Spruce in S. America.)

Stahl, E. (1900) [p. 164] Der Sinn der Mycorhizenbildung. Pringsheim's Jahrbüch. Bd. 34, 1900, pp. 539–668, 2 text-figs.

(The author does not deal with aquatics, but his classification of plants into "starch leaved" strong transpirers and "sugar leaved" weak transpirers seems to have a bearing upon the nature of the epidermis of submerged leaves.)

Standley, P. C. See **Miller, G. S. and Standley, P. C.** (1912).

Stein, B. (1874) [p. 111] Über Reizbarkeit der Blätter von *Aldrovanda vesiculosa*. Zweiundfünfzigster Jahres-Ber. d. Schlesischen Gesellsch. 1875 (1874), pp. 83–84.

(The author records the sensitiveness of the *Aldrovandia* leaf to contact.)

Stöhr, A. (1879) [pp. 165, 171, 279] Über Vorkommen von Chlorophyll in der Epidermis der Phanerogamen-Laubblätter. Sitzungsberichte der math.-naturwissens. Classe d. k. Akad. d. Wissens. Wien, Bd. LXXIX. Abth. I. 1879, pp. 87–118, 1 pl.

(In opposition to the current opinion, the author shows that chlorophyll is frequently present in the epidermis of the lower side of the leaf in terrestrial Dicotyledons, while it is absent in the case of terrestrial Monocotyledons.)

Strasburger, E. (1884) [p. 37] Das Botanische Practicum. xxxvi + 664 pp., 182 figs. Jena, 1884.

[This well-known text-book contains many references to the anatomy of water plants, e.g. *Vallisneria*, p. 54; *Nymphaea* (*Castalia*), p. 171; *Potamogeton*, p. 182; *Hippuris*, pp. 185 and 249; *Elodea*, p. 187.]

Strasburger, E. (1902) [pp. 85, 86, 272, 309] Ein Beitrag zur Kenntniss von *Ceratophyllum submersum* und phylogenetische Erörterungen. Prings. Jahrb. f. wiss. Bot. Bd. 37, pp. 477–526, 3 pls., 1902.

(This investigation supports the view that *Ceratophyllum* is allied to Nymphaeaceae. The development of embryo-sac and pollen-grain are described in detail. There is a discussion of the use of mucilage in water plants.)

Sykes, M. G. See **Thoday, D. and Sykes, M. G.** (1909).

Sylvén, N. (1903) Studier öfver organisationen och lefnadssättet hos *Lobelia Dortmanna*. Arkiv för Botanik utgifvet af K. Svenska Vetenskaps-Akad. Bd. I. 1903–1904, 1 pl., pp. 377–388.

(This Swedish paper is reviewed in the Bot. Centralbl. Bd. 93, 1903, pp. 613–614).

Täckholm, G. (1914) [p. 311] Zur Kenntnis der Embryosackentwicklung von *Lopezia coronata* Andr. Svensk Bot. Tidsk. Vol. 8, 1914, pp. 223–234, 5 text-figs.

(The author is in favour of removing *Trapa* from the Onagraceae on account of its embryo-sac structure.)

Täckholm, G. (**1915**) [p. 311] Beobachtungen über die Samenentwicklung einiger Onagraceen. Svensk Bot. Tidsk. Vol. 9, 1915, pp. 294–361, 16 text-figs.

(It is pointed out on p. 354 that in the true Onagraceae and in *Trapa* we have two highly differentiated and widely separated embryo-sac types.)

Tansley, A. G. (**1911**) [pp. 286, 287, 288, 290] Types of British Vegetation, by Members of the Central Committee for the Survey and Study of British Vegetation, edited by A. G. Tansley. Cambridge, 1911.

(The following sections of this book deal with the ecology of British water plants: pp. 187–203, Aquatic Vegetation; pp. 223–229, The Aquatic Formation of the River Valleys of East Norfolk.)

Tepper, J. G. O. (**1882**) [pp. 127, 205] Some Observations on the Propagation of *Cymodocea antarctica* (Endl.). Trans. and Proc. and Rep. of the Royal Society of South Australia, Vol. IV. 1882 (for 1880–81), pp. 1–4, 1 pl. Further Observations on the Propagation of *Cymodocea antarctica*. Ibid. pp. 47–49, 1 pl.

(The first account of the viviparous growth of this plant.)

Terras, J. A. (**1900**) [pp. 48, 280] Notes on the Germination of the Winter Buds of *Hydrocharis Morsus-Ranae*. Trans. and Proc. of the Bot. Soc. of Edinburgh, Vol. XXI. 1900 (for 1896–1900), Part IV. 1900, pp. 318–329.

(An account of experiments upon the germination of winter buds of this species, with special reference to conditions of illumination.)

Thallwitz, J. See **Schorler, B., Thallwitz, J. and Schiller, K.** (**1906**).

Theophrastus (**Hort**) (**1916**) [p. 208] Enquiry into Plants, with an English translation by Sir Arthur Hort. 2 vols. London, 1916.

(Book IV. Chapter 9 contains an exceedingly clear description of *Trapa natans*.)

Thiébaud, M. (**1908**) [p. 291] Contribution à la Biologie du Lac de Saint-Blaise. Annales de Biologie Lacustre, T. III. Fasc. 1, 1908, pp. 54–140, 5 pls.

(This work contains a short section dealing with the plants of the Lake.)

Thoday, D. and Sykes, M. G. (**1909**) [pp. 262, 266] Preliminary Observations on the Transpiration Current in Submerged Water-plants. Ann. Bot. Vol. XXIII. 1909, pp. 635–637.

(The authors have demonstrated by experiments on plants of *Potamogeton lucens*, *in situ* in the River Cam, that an unexpectedly rapid water current occurs in detached, rootless stems, and that this current is to a great extent dependent on the leaves.)

Thurn, E. F. Im See **Im Thurn, E. F.**

Tieghem, P. van (1866) [p. 256] Recherches sur la respiration des plantes submergées. Bull. de la Soc. bot. de France, T. XIII. 1866, pp. 411–421.

[An account of some experiments dealing with assimilation (not respiration in the modern sense). The author claims to show that if, in the case of certain submerged plants, the decomposition of CO_2 is initiated in direct sunlight, it may continue actively for some hours after the plant has been placed in darkness; see, however, **Tieghem, P. van (1869²)**.]

Tieghem, P. van (1867) Note sur la respiration des plantes aquatiques. Comptes rendus de l'acad. des sciences, Paris, T. 65, 1867, pp. 867–871.

[A further communication dealing with the same results as **Tieghem, P. van (1866)**.]

Tieghem, P. van (1868) [pp. 107, 108, 227] Anatomie de l'Utriculaire commune. Bull. de la Soc. bot. de France, T. XV. 1868, pp. 158–162.

(An account of the anatomy of the submerged and aerial parts of *Utricularia vulgaris*.)

Tieghem, P. van (1869¹) [p. 107] Anatomie de l'Utriculaire commune. Ann. d. sci. nat. Sér. V. Bot. T. X. 1869, pp. 54–58.

[See **Tieghem, P. van (1868)**.]

Tieghem, P. van (1869²) Sur la respiration des plantes submergées. Comptes rendus de l'acad. des sciences, Paris, T. 69, 1869, pp. 531–535.

[In this paper the author withdraws his previously expressed opinion (**Tieghem, P. van (1866)** and **(1867)** that assimilation in submerged plants continues after the removal of the light.]

Tittmann, J. A. (1821) [p. 34] Die Keimung der Pflanzen. viii + 200 pp., 27 pls. Dresden, 1821.

[The seedlings of a number of water plants are described and figured: *Alisma Plantago*, *Nymphaea* (*Castalia*) *alba* and *Nuphar luteum* (*Nymphaea lutea*), *Potamogeton natans*, *Trapa natans*.]

Trécul, A. (1845) [pp. 37, 309] Recherches sur la structure et le développement du *Nuphar luteum*. Ann. des sci. nat. Sér. III. Bot. T. IV. 1845, pp. 286–345, 4 pls.

(The anatomy of the stem, roots and leaves, and the structure of the reproductive organs, are described in detail. Attention is drawn to the points in the anatomy and mode of germination which recall the Monocotyledons.)

Trécul, A. (1854) [pp. 33, 37, 38, 309] Études anatomiques et organogéniques sur la *Victoria regia*, et anatomie comparée du *Nelumbium*, du *Nuphar* et de la *Victoria*. Ann. d. sci. nat. Sér. IV. Bot. T. I. 1854, pp. 145–172, 3 pls.

[From his study of *Victoria regia* and other Waterlilies the author concludes that the Nelumbiaceae differ widely from the Nymphaeaceae. Among the points to which he draws attention are the operculum of the seed of *Victoria* and *Nuphar* (*Nymphaea*) and the succession of leaf types in the seedling of *Victoria*. For a criticism see **Blake, J. H. (1887)**.]

Treviranus, L. C. (1821) [p. 164] Vermischte Schriften. Bd. 4. ii + 242 pp., 6 pls. Bremen, 1821.

(The "absence of an epidermis" on the lower side of the iea of *Potamogeton crispus* is alluded to on p. 76.)

Treviranus, L. C. (1848[1]) [pp. 93, 99, 154] Noch etwas über die Schläuche der Utricularien. Bot. Zeit. Jahrg. 6, 1848, pp. 444–448.

(Notes on the bladders of *Utricularia* which the author regards as of foliar nature.)

Treviranus, L. C. (1848[2]) Observationes circa germinationem in Nymphaea et Euryale. Abhandl. d. Math.-Phys. Classe d. königＧＧ. bay. Akad. d. Wiss. Bd. v. Abt. II. 1848, pp. 397–403, 1 pl.

[A description in Latin of the germination of *Nymphaea* (*Castalia*) *caerulea* and *Euryale ferox*. In the latter case the author figures the four outgrowths which were later described by Goebel as breathing organs.]

Treviranus, L. C. (1853) De germinatione seminum Euryales. Bot. Zeit. Jahrg. XI. pp. 372–374, 1853.

[This short paper should be read in connexion with **Treviranus, L. C.** (**1848**[2]) since it consists of corrections of the latter, based on a better supply of material of *Euryale ferox*.]

Treviranus, L. C. (1857) [p. 67] Vermischte Bemerkungen. 1. Hybernacula des *Potamogeton crispus*. 2. Hybernacula der *Hydrocharis Morsus Ranae* L. Bot. Zeit. Jahrg. 15, 1857, pp. 697–702, 1 pl.

[As regards *Potamogeton crispus* Treviranus confirms the observations recorded by **Clos, D.** (**1856**). He also gives a short description of the winter buds of *Hydrocharis*.]

Tulasne, L. R. (1852) [p. 112] Podostemacearum Monographia. Archives du Muséum d'hist. nat. T. VI. 1852, 208 pp., 13 pls.

(This highly important Latin monograph is illustrated with a series of exquisite plates, giving a clear idea of the peculiarities of this anomalous family.)

Unger, F. (1849) Die Entwickelung des Embryo's von *Hippuris vulgaris*. Bot. Zeit. Jahrg. 7, 1849, pp. 329–339, 2 pls.

(A description of the embryology of *Hippuris*. The spherical multicellular embryo becomes sunk in the endosperm by means of a long suspensor.)

Unger, F. (1854[1]) Einiges über die Organisation der Blätter der *Victoria regia* Lindl. Sitzungsber. d. k. Akad. d. Wissenschaften, Math.-Naturwissens. Classe, Bd. XI. Wien, 1854 (for 1853), pp. 1006–1014, 1 pl.

(The author describes the minute perforations which are characteristic of the leaves of *Victoria regia*.)

Unger, F. (1854[2]) [p. 256] Beiträge zur Physiologie der Pflanzen. I. Bestimmung der in den Intercellulargängen der Pflanzen enthaltenen Luftmenge. Sitzungsber. d. k. Akad. d. Wissenschaften, Math.-Naturwissens. Classe, Bd. XII. Wien, 1854, pp. 367–378.

(The author shows experimentally how much air is contained in various plant tissues. One of the organs investigated was the leaf of *Pistia*.)

Unger, F. (1862) [p. 260]
Beiträge zur Anatomie und Physiologie der Pflanzen. XII. Neue Untersuchungen über die Transspiration der Gewächse. Sitzungsberichte der math.-naturwiss. Classe der k. Akad. der Wissens. Wien, Bd. XLIV. Abth. II. 1862, pp. 327–368, 1 text-fig.
(Pp. 364–367 contain an account of experiments on water plants demonstrating the existence of a definite transpiration stream even in submerged plants.)

Ursprung, A. (1912) [p. 258]
Zur Kenntnis der Gasdiffusion in Pflanzen. Flora, N.F. Bd. 4 (G.R. Bd. 104), 1912, pp. 129–156.
(The greater part of this memoir is occupied with a critical account of the literature dealing with the bubbling of gas which takes place, under certain conditions, from the leaves of the Nymphaeaceae. The writer also brings forward some fresh experimental evidence.)

Uspenskij, E. E. (1913) [pp. 139, 195]
Zur Phylogenie und Ekologie der Gattung Potamogeton. I. Luft-, Schwimm- und Wasserblätter von *Potamogeton perfoliatus* L. Bull. des Naturalistes de Moscou, N.S., Vol. 27, 1913, pp. 253–262, 3 text-figs.
(An account of the land form of this species, followed by a general comparison between dissected and thin flat laminae, regarded as adaptations to aquatic life.)

Vaucher, J. P. (1841) [pp. 216, 219]
Histoire physiologique des plantes d'Europe. T. II. 743 pp. Paris, 1841.
(On p. 358 the winter buds of *Myriophyllum* are described. Observations on other water plants are also included.)

Vöchting, H. (1872) [p. 178 and Figs. 116 and 117, p. 179]
Zur Histologie und Entwickelungsgeschichte von *Myriophyllum*. Nova Acta Acad. Caesareae Leopoldino-Carolinae Germanicae Naturae Curiosorum. T. XXXVI. 1873, 18 pp., 4 pls.
(An account of the anatomy and apical development of this genus.)

Volkens, G. (1883) [p. 267]
Ueber Wasserausscheidung in liquider Form an den Blättern höherer Pflanzen. Jahrb. d. k. bot. Gartens und d. bot. Museums zu Berlin, Bd. II. 1883, pp. 166–209, 3 pls.
(This paper is the earliest general account of the excretion of water in liquid form from the leaves of the higher plants. The structure and development of the apical opening in the leaf of *Alisma Plantago* are described and figured, p. 206 and Pl. VI, figs. 5 and 6.)

Vries, H. de (1873) [p. 284]
Die vitalistische Theorie und der Transversal-Geotropismus. Flora, N.R. Jahrg. XXXI. (G.R. Jahrg. LVI.) 1873, pp. 305–315.
[A criticism of **Frank, A. B. (1872)**.]

Wächter, W. (1897[1]) [pp. 12, 117, 156, 266]
Beiträge zur Kenntniss einiger Wasserpflanzen. I. and II. Flora, Bd. 83, 1897, pp. 367–397, 21 text-figs.
(The first part of this paper deals with the results of experimental work on the production of the different forms of leaves in *Sagittaria natans*, Michx., *S. chinensis*, Sims, *Eichhornia azurea*, Kth., *Heteranthera reniformis*, R. et P., *Hydrocleis nymphoides*, Buchenau. The second part deals with the morphology and anatomy of *Weddellina squamulosa*, Tul., one of the Podostemaceae.)

Wächter, W. (1897²) [p. 159] Beiträge zur Kenntniss einiger Wasserpflanzen. III. Flora, Bd. 84, Ergänzungsband zum Jahrgang 1897, pp. 343–348.

[This paper is a continuation of the first part of **Wächter, W.** (1897¹). It contains an account of experiments upon the heterophylly of *Castalia*, showing that the production of the different forms of leaf in this genus is dependent upon external conditions, just as in the case of the Monocotyledons previously investigated.]

Wager, H. See **Scott, D. H. and Wager, H.** (1888).

Wagner, R. (1895) [p. 39 and Fig. 23, p. 41] Die Morphologie des *Limnanthemum nymphaeoides* (L.) Lk. Bot. Zeit. Jahrg. 53, Abt. I. 1895, pp. 189–205, 1 pl., 1895.

(A general descriptive paper dealing with the development, branching, etc. of this species.)

Walker, A. O. (1912) [p. 212] The Distribution of *Elodea canadensis*, Michaux, in the British Isles in 1909. Proc. Linn. Soc. London, 124th session, 1912, pp. 71–77.

(This paper gives the result of enquiries made in 1909 among local natural history societies as to the degree of success attained by *Elodea* in establishing itself in this country.)

Walsingham, Lord, and Payne-Gallwey, R. (1886) [pp. 135, 302] Shooting (Moor and Marsh). Badminton Library. xiii + 348 pp. London, 1886.

(The authors mention, pp. 158 and 165, that Brent Geese feed on *Zostera*, and that these birds are almost confined to those parts of the coast where *Zostera* occurs.)

Walter, F. (1842) [pp. 15, 17] Bemerkungen über die Lebensweise einiger deutschen Pflanzen. Flora, Jahrg. XXV. Bd. II. 1842, pp. 737–748, 1 pl.

(A picturesque account of Walter's discovery of tuber-formation and heterophylly in *Sagittaria sagittifolia*.)

Warming, E. (1871) [p. 135] Forgreningen hos *Pontederiaceae* og *Zostera*. Videnskab. Meddel. fra den naturhist. Forening i Kjöbenhavn for Aaret 1871, pp. 342–346, 1 text-fig.

(This Danish paper deals with the nature of the shoot system in the plants mentioned.)

Warming, E. (1874) [p. 100] Bidrag til Kundskaben om Lentibulariaceae. Videnskab. Meddel. fra den naturhist. Forening i Kjöbenhavn for Aaret 1874 (1874–5), pp. 33–58, 3 pls.

(This paper, which is in Danish, deals with *Genlisea* and *Utricularia*. The germination of *Utricularia* is described.)

Warming, E. (1881, 1882, 1888, 1891) [pp. 112, 118, 310] Familien Podostemaceae. Kongel. Dansk. Videnskab. Selskabs Skrifter. Sjette Raekke. 1. Vol. II. 1881, pp. 1–34, 6 pls. 2. Vol. II. 1882, pp. 77–130, 9 pls. 3. Vol. IV. 1888, pp. 443–514, 12 pls. 4. Vol. VII. 1891, pp. 133–179, 185 text-figs.

(This important monograph is in Danish, but each part is followed by a French résumé.)

Warming, E. (1883[1]) [p. 245] Botanische Notizen. Bot. Zeit. Jahrg. 41, 1883, pp. 193–204.

(In section 2, "Zur Biologie der Keimpflanzen," pp. 200–203, the author refers to the development of long root-hairs, at the junction of root and hypocotyl, which attach the seedlings of certain water plants to the substratum.)

Warming, E. (1883[2]) [Figs. 76 and 77, p. 115, 79, p. 116] Studien über die Familie der Podostemaceae. Engler's Bot. Jahrbüch. Bd. IV. 1883, pp. 217–223, 5 figs.

(A German version of part of the author's work on this family.)

Warming, E. (1909) [p. 291] Œcology of Plants. xi + 422 pp. Oxford, 1909.

(This English version of the author's well-known book contains sections dealing with aquatic and marsh plants; see especially pp. 97–100 and 149–190.)

Webber, H. J. (1897) [p. 213] The Water Hyacinth, and its relation to navigation in Florida. U.S. Depart. of Agriculture. Division of Botany. Bulletin, No. 18, 1897, 20 pp., 1 pl., 4 text-figs.

[An account of the excessive luxuriance of *Piaropus crassipes*, (Mart.) Britton, = *Eichhornia speciosa*, Kunth, = *Eichhornia crassipes*, (Mart.) Solms.]

Weddell, H. A. (1849) [pp. 80, 300] Observations sur une espèce nouvelle du genre *Wolffia* (Lemnacées). Ann. des sci. nat. Sér. III. Bot. T. 12, 1849, pp. 155–173, 1 pl.

(The author discovered in Brazil a minute species of *Wolffia*, which he named *W. brasiliensis*. Twelve of the flowering plants could be accommodated on one frond of *Lemna minor*.)

Weddell, H. A. (1872) [pp. 113, 295] Sur les Podostémacées en général, et leur distribution géographique en particulier. Bull. de la Soc. bot. de France, T. XIX. 1872, pp. 50–57.

(This paper is based upon the author's own observations in Brazil. Stress is laid upon the very local distribution of many of the Podostemaceae.)

Weinrowsky, P. (1899) [pp. 261, 266, 269] Untersuchungen über die Scheitelöffnung bei Wasserpflanzen. Fünfstück's Beiträge zur Wissensch. Bot. Bd. III. 1899, pp. 205–247, 10 text-figs.

(An extremely important account of the apical openings of the leaves of water plants.)

Weiss, F. E. and Murray, H. (1909) [p. 303] On the Occurrence and Distribution of some Alien Aquatic Plants in the Reddish Canal. Mem. and Proc. of the Manchester Lit. and Phil. Soc. Vol. 53, 1909, No. 14, 8 pp., 1 map.

[The authors show that *Naias graminea* (Del.) var. *Delilei* (Magnus), recorded in **Bailey, C. (1884)** as occurring in the warm water of this canal, has now disappeared. Certain alien Algae are also discussed, and the distribution of *Vallisneria spiralis*, which was planted here forty years ago.]

Went, F. A. F. C. (1910) [pp. 114, 122] Untersuchungen ueber Podostemaceen. Verhandelingen d. Konin. Akad. van Wetenschappen te Amsterdam, Tweede Sectie, Dl. XVI. No. 1, 1910, 88 pp., 15 pls.

[In this memoir, based upon the results of the author's travels in Surinam, the following members of the Podostemaceae are described: 6 sp. of *Oenone* of which 3 are new, 3 new species of *Apinagia*, *Lophogyne* (1 sp.), *Mourera* (1 sp.) and *Tristicha* (1 sp.). The anatomy, and the development of the ovules, are treated, as well as the general morphology.]

Werner, E. See **Magnus, W. and Werner, E.** (1913).

West, G. (1905) [p. 287] A Comparative Study of the dominant Phanerogamic and Higher Cryptogamic Flora of Aquatic Habit, in Three Lake Areas of Scotland. Proc. Roy. Soc. Edinb. Vol. XXV. Part II. 1906 (for 1905), pp. 967–1023, 55 pls.

(A general ecological survey of certain Scottish Lakes.)

West, G. (1908) [pp. 287, 290] Notes on the Aquatic Flora of the Ness Area. Bathymetrical Survey of the Fresh-water Lochs of Scotland. VIII. The Geogr. Journal, Vol. XXXI. 1908, pp. 67–72.

[This brief paper, which is of a general nature, should be read in conjunction with the author's detailed studies of the Scottish lakes—**West, G.** (1905) and (1910).]

West, G. (1910) [pp. 20, 87, 145, 200, 234, 287, 299, 325] A Further Contribution to a Comparative Study of the dominant Phanerogamic and Higher Cryptogamic Flora of Aquatic Habit in Scottish Lakes. Proc. Roy. Soc. Edinb. Vol. XXX. 1910 (Session 1909–10), pp. 65–181, 62 pls.

[A continuation of **West, G.** (1905).]

Wettstein, R. von (1888) [p. 38] Beobachtungen über den Bau und die Keimung des Samens von *Nelumbo nucifera* Gärtn. Verhandl. d. k. k. zool.-bot. Gesellsch. in Wien, Bd. 38, 1888, pp. 41–48, 1 pl.

(The structure and germination of the seed of this plant, which has no endosperm or perisperm, is figured with great clearness.)

Wheldale, M. (The Hon. Mrs Huia Onslow) (1916) [p. 277] The Anthocyanin Pigments of Plants. x + 318 pp. Cambridge, 1916.

[Chapter VI. (Physiological Conditions and Factors Influencing the Formation of Anthocyanins) and Chapter VIII. (The Significance of Anthocyanins), may be consulted in connexion with the red coloration so prevalent in water plants.]

Wheldon, J. A. and Wilson, A. (1907) [p. 299] The Flora of West Lancashire. 511 pp., 15 pls., 1 map. Eastbourne, 1907.

(On p. 339 a reference is made to a pond which was dug experimentally in order to see what water plants would colonise it.)

Wigand, A. (1871) [p. 37] *Nelumbium speciosum*, W. Bot. Zeit. Jahrg. 29, 1871, pp. 813–826, 1 text-fig.
(An account of the development, morphology, anatomy and starch distribution in this member of the Nymphaeaceae.)

Wight, R. (1849) [p. 99] Conspectus of Indian Utriculariae. Hooker's Journal of Botany and Kew Garden Miscellany, Vol. I. 1849, pp. 372–374.
(The author records the occurrence of a whorl of floats below the flower in *U. stellaris*.)

Willdenow, C. L. (1806) [p. 85] Determination of a new aquatic vegetable Genus, called *Caulinia*, with general Observations on Water-plants. Annals of Botany (edited by C. Konig and J. Sims), Vol. II. 1806, pp. 39–51.
(A translation of a paper by this author who was the first to suggest that the pollination of *Ceratophyllum* was hydrophilous.)

Willis, J. C. (1902) [Passim and Figs. 78, p. 115, 80, p. 118, 82, p. 121] Studies in the Morphology and Ecology of the Podostemaceae of Ceylon and India. Ann. Roy. Bot. Gard. Peradeniya, Vol. I. 1902, pp. 267–465, 34 pls.
(An important general work dealing with the structure and biology of this group.)

Willis, J. C. (1914[1]) [pp. 112, 286, 327, 329] On the Lack of Adaptation in the Tristichaceae and Podostemaceae. Proc. Roy. Soc. Vol. 87, B. 1914, pp. 532–550.
(The detailed development of a thesis, to which the author has been led in the course of seventeen years' study of these families in India, Ceylon and Brazil—namely, that the natural selection of infinitesimal variations is quite incompetent to explain their evolution.)

Willis, J. C. (1914[2]) [p. 305] The Endemic Flora of Ceylon, with Reference to Geographical Distribution and Evolution in General. Phil. Trans. Roy. Soc. London, Ser. B, Vol. 206, 1914, pp. 307–342.
(This paper does not deal with water plants, but is quoted here because it is the first of the series of contributions in which the author has developed his "Age and Area" hypothesis, which has an important bearing on the study of aquatics.)

Willis, J. C. (1915[1]) [p. 112] A New Natural Family of Flowering Plants—Tristichaceae. Linn. Soc. Journ. Bot. Vol. 43, 1915, pp. 49–54.
(A proposal to separate the Podostemaceae into two families—Tristichaceae = Chlamydatae, and Podostemaceae = Achlamydatae.)

Willis, J. C. (1915[2]) [pp. 112, 327] The Origin of the Tristichaceae and Podostemaceae. Ann. Bot. Vol. XXIX. 1915, pp. 299–306.
(A reconstruction of the type of ancestor from which these groups are probably derived.)

Willis, J. C. (1917) [p. 306] The Relative Age of Endemic Species and other Controversial Points. Ann. Bot. Vol. XXXI. 1917, pp. 189–208.
(See pp. 201, 202 for a consideration of the Podostemaceae and Tristichaceae from the point of view of the author's "Age and Area" Law of plant distribution.)

Willis, J. C. and Burkill, I. H. (1895) [p. 230] Flowers and Insects in Great Britain. Ann. Bot. Vol. IX. 1895, pp. 227–273.
(This paper includes observations on the pollination of *Peplis* and *Mentha aquatica*.)

Wilson, A. See **Wheldon, J. A. and Wilson, A.** (1907)

Wilson, W. (1830) [p. 76] *Lemna gibba*. Remarks on the Structure and Germination. Hooker's Botanical Miscellany, Vol. I. 1830, pp. 145–149, 1 pl.
(A description, with clear figures, of the seedlings of this species.)

Wydler, H. (1863) Morphologische Mittheilungen. *Alisma Plantago*, L. Flora, N.R. Jahrg. XXI. (G.R. Jahrg. XLVI.) 1863, pp. 87–90, 97–100, 2 pls.
(A detailed study of the shoot relations and the inflorescence of *Alisma Plantago*, L.)

Wylie, R. B. (1904) [pp. 55, 57] The Morphology of *Elodea canadensis*. Bot. Gaz. Vol. XXXVII. 1904, pp. 1–22, 4 pls.
(An account of the gametophytes, pollination, etc. in this species.)

Wylie, R. B. (1912) [pp. 55, 86 and Fig. 35, p. 56] A long-stalked *Elodea* flower. Bull. from the Labs. of Nat. Hist. State University Iowa, Vol. VI. 1912, pp. 43–52, 2 pls.
(A description of a new species of *Elodea*, *E. ioensis*, in which the male flower reaches the surface through great elongation of its stalk.)

Wylie, R. B. (1917[1]) [p. 234 and Fig. 153, p. 234] Cleistogamy in *Heteranthera dubia*. Bull. from the Labs. of Nat. Hist. State University Iowa, Vol. VII. No. 3, 1917, pp. 48–58, 1 pl.
(The cleistogamy of this species, which is very thoroughly described, is considered by the author to be 'largely accidental.')

Wylie, R. B. (1917[2]) [p. 235] The Pollination of *Vallisneria spiralis*. Bot. Gaz. Vol. 63, 1917, pp. 135–145, 1 pl. and 6 text-figs.
(The author corrects a number of errors in earlier accounts of this plant, and lays great stress upon the part played in pollination by the surface film.)

Zacharias, O. (1891) Die Tier- und Pflanzenwelt des Süsswassers, Vol. I. x + 380 pp., 79 text-figs. Leipzig, 1891.
(F. Ludwig contributes a section, pp. 65–134, dealing with the Phanerogams of fresh waters.)

INDEX TO BIBLIOGRAPHY

GENERA AND FAMILIES NAMED IN THE BIBLIOGRAPHY, EITHER IN TITLES OR ABSTRACTS

INDEX

[The names of authors which occur in the bibliography are not included in the following index, since page references are given in connexion with the titles in the bibliography, which thus also serves as an index of authors' names]

CAMBRIDGE: PRINTED BY J. B. PEACE, M.A., AT THE UNIVERSITY PRESS

Printed in the United States
By Bookmasters